THE ASSURANCE SCIENCES

An Introduction
To Quality Control
And Reliability

THE ASSURANCE SCIENCES

An Introduction

To Quality Control

And Reliability

Siegmund Halpern

Technical Assurance Department, RCA Corporation

PRENTICE-HALL, INC., *Englewood Cliffs, New Jersey   07632*

Library of Congress Cataloging in Publication Data

Halpern, Siegmund, 1918–
    The assurance sciences.

    Bibliography : p.
    Includes index.
    1. Quality control. 2. Reliability (Engineer-
ing) 3. Maintainability (Engineering) I. Title.
TS156.H3   620'.0045   77-2967
ISBN 0-13-049601-4

Printed in the United States of America

10 9 8 7 6

PRENTICE-HALL INTERNATIONAL, INC., *London*
PRENTICE-HALL OF AUSTRALIA PTY. LIMITED, *Sydney*
PRENTICE-HALL OF CANADA, LTD., *Toronto*
PRENTICE-HALL OF INDIA PRIVATE LIMITED, *New Delhi*
PRENTICE-HALL OF JAPAN, INC., *Tokyo*
PRENTICE-HALL OF SOUTHEAST ASIA PTE. LTD., *Singapore*
WHITEHALL BOOKS LIMITED, *Wellington, New Zealand*

To
Marianne

CONTENTS

RELIABILITY

RELATED SUBJECTS

APPENDICES

PREFACE

This book is an introductory text to the assurance sciences. It covers the fundamental aspects of the subjects of quality control and reliability, with many calculated examples to illustrate the application of the principles to realistic equipment problems. It also includes a chapter on each of the allied disciplines of maintainability and integrated logistic support engineering. The book will not only be found useful in a college curriculum, but will prove equally valuable for technical institutes, industrial applications, seminars, and self-study.

It has been the author's experience that engineering graduates have little, if any, background in the assurance sciences. This constitutes a decided disadvantage upon joining an up-to-date company utilizing these disciplines. More emphasis on teaching the assurance sciences as part of a regular engineering or technical curriculum in colleges and technical institutes, as well as in company-sponsored training courses, would appear to be most desirable.

The first chapter gives a general overview of each of the assurance disciplines cited above. The second and third chapters deal with probability and statistical inference, which are the basic mathematical tools for the assurance sciences. Chapters four through seven are devoted to quality control, and chapters eight through eleven to reliability. Chapter twelve is devoted to maintainability, and chapter thirteen covers integrated logistic support (ILS).

A special effort has been made to incorporate reprints or pertinent portions of Government specifications and standards, including tables and graphs, in order to familiarize the reader with the various existing documents and to show how they are applied in the solution of practical problems. Many of these problems are as pertinent to modern spacecraft equipments as they are to military and commercial applications.

The book covers a broad area of topics which have proven to be pertinent in the practice of the assurance sciences. Rigorous mathematical derivations generally are not included. Appropriate texts are referenced where considered useful for a fuller mathematical development. The author hopes that this book will help in developing both an understanding and appreciation for these disciplines.

The author wishes to express his appreciation to E. S. Shecter, G. Branin and L. Caplan of RCA Corporation for their review of the manuscript and for their helpful comment; and to F. A. Eble for his most valuable suggestions and contribution. Gratitude is also expressed to the literary executor of the late Sir Ronald A. Fisher, F.R.S., to Dr. Frank Yates, F.R.S., and to Longman Group, Ltd., London, for permission to adapt Table IV from their book *Statistical Tables for Biological, Agricultural and Medical Research*, 6th Ed., 1974. Special thanks go to Marie Chapman for her painstaking typing of the manuscript.

SIEGMUND HALPERN

Cherry Hill, New Jersey

THE ASSURANCE SCIENCES

An Introduction
To Quality Control
And Reliability

INTRODUCTION

The Assurance Sciences are engineering disciplines which govern the quality, safety, economy, serviceability and long-term dependability of products. These disciplines perform vital roles in the development of today's complex equipment by influencing the design and by predicting and demonstrating the results.

This book introduces the Assurance Sciences of quality control and reliability, with supplementary material on maintainability and integrated logistic support (ILS) engineering.[1] Other members of the assurance family, which are beyond the scope of this book, include safety, value engineering, human factors and standardization.

There is no universal definition for the quality-related assurance discipline. The responsibilities delegated to quality departments may be limited to the traditional quality-control inspection in the manufacturing area, but they may also be extended to form a more comprehensive quality function (quality assurance). In the latter case the basic quality-control inspection activity is augmented by quality engineering and quality program auditing.

The more inclusive term *product assurance* also may be used by some companies. Since the title of the book categorizes quality control as an *Assurance Science* its treatment is consistent with the broader term *Quality Assurance*.

Quality control is not a new discipline. Attempts to improve and maintain the quality of manufactured goods date back to the very beginning of organized industry, but the methods were highly subjective, they lacked uniformity within industry, and they were strictly nonscientific. *Statistical quality control*, on the other hand, uses the

---

[1] ILS engineering also is known as *ILE*; the term *ILS* will be used in this book to denote the overall ILS activity and its engineering arm.

scientific principles of probability and statistics as a basis for decisions concerning the acceptability of a product.

Walter A. Shewart of the Bell Telephone Laboratories is generally credited with the introduction of statistical quality control methods in the United States. His initial interest was in the development of quality-control charts.[2] In 1929, H. F. Dodge and H. G. Romig applied statistical methods to sampling and prepared a variety of sampling inspection plans.[3] Slow at first in its acceptance by industry, statistical quality control, with the encouragement of technical societies and the armed services, soon gained in popularity. When the general mobilization effort for World War II began, the armed services, faced with giant procurement programs, intensively promoted the use of organized statistical-quality-control techniques, thereby contributing greatly to their general acceptance by both producers and consumers.

Quality control alone cannot guarantee the ability of the design to perform for long periods of time under the stresses of field operation. Occasionally the basic design is not adequate for trouble-free operation in field use. This is primarily due to debilitating operational or environmental stresses that may not have been considered by the designer. This is where the discipline of *reliability engineering* (also referred to as *product reliability*, or simply *reliability*) makes its contribution.

As in the case of quality control, reliability is based on probability and statistics and is introduced in the early stages of the design. The objective of reliability is to assure that the product will perform adequately with a predicted probability of success for a given time. This concept was promulgated by the Advisory Group on Reliability of Electronic Equipment (AGREE),[4] which formulated basic reliability concepts in 1957 under the sponsorship of the Assistant Secretary of Defense. Recognition of the need for a formal reliability discipline dates back to the early 1950s.

Concurrently with the introduction of statistical quality control and reliability, increasing attention was being paid to the need for more efficient diagnosis and replacement of failed parts and equipment maintenance in general. In the case of military electronic equipments, which need to be maintained in operational readiness, maintenance costs in the field skyrocketed as a result of the poor accessibility and repairability of earlier designs.

What was needed was a new approach to effective equipment maintenance and support. The introduction of the *maintainability* concept by the military services dates back to 1954, when it became apparent that, in addition to quality control and reliability, a separate formal discipline was needed to assure implementation of maintainability requirements. Within 5 years the U.S. Air Force was among the first to issue a formal maintainability specification.[5] The maintainability discipline places a

2 "Quality Control Charts," *Bell System Technical Journal*, October 1926, pp. 593–603.

3 H. F. Dodge and H. G. Romig, "A Method of Sampling Inspection," *Bell System Technical Journal*, October 1929, pp. 613–631.

4 "Reliability of Military Electronic Equipment," report by the Advisory Group on Reliability of Electronic Equipment, Office of the Assistant Secretary of Defense, June 4, 1957.

5 MIL-M-26512A Military Specification, "Maintainability Requirements for Weapon Systems and Subsystems," June 18, 1959.

limit on the mean time it takes to repair a given piece of equipment. Included in this time is the diagnosis, actual repair, and retest.

Finally, provisions were needed for the support of operational equipment in the form of maintenance instructions, tools, facilities, spares provisioning, and training. The discipline that is concerned with this support function is ILS. It was formally recognized by the U.S. Department of Defense in 1964.

Because of the particular organizational structure of various industrial concerns, the duties, objectives, and responsibilities of quality control and reliability, as well as those of the allied disciplines of maintainability and ILS, overlap. The literature contains differing opinions as to the relationships among these disciplines. A frequently used approach is to lump quality control, reliability, and maintainability under the organizational heading of *quality assurance* or *product assurance*, with ILS activities either not required or absorbed by related activities. In larger organizations product assurance may be at the top level, directing the quality assurance, ILS and reliability/maintainability efforts. There are also cases in which all assurance activities are directed by ILS. Figures 4-1 through 4-4, pages 75 to 77 give a few examples of typical organizational charts.

I

THE ASSURANCE SCIENCES:

AN OVERVIEW

This chapter gives a general overview of the purpose and the techniques used by the Assurance Sciences. The major topics are presented in greater detail later.

The military services pioneered in the development and use of the assurance disciplines; therefore, most of the specifications, standards and guidelines in the field are of military origin. However, these disciplines are becoming increasingly important in a broad spectrum of non-military applications.

Any reputable manufacturer will find it an economic necessity to institute an adequate assurance program. The principle of "caveat emptor" no longer applies; instead, today's rule is: "Let the seller beware!" There are many instances on record where consumers have instituted successful lawsuits against suppliers for faulty products. Increasing public demand for better, safer, longer-lasting and more serviceable products has stimulated the powerful consumer advocacy movement. Coupled with recent consumer legislation, this will increasingly encourage manufacturers to develop full-fledged assurance programs. Hence the call is for more comprehensive assurance programs.

1-1 QUALITY CONTROL[1]

Based on the principles of probability and statistics, quality control is a modern decision-making tool employed by management to assure a desired quality level of manu-

[1] The introductory material on quality control is based on Chapter 1, "The Scope of the Quality Function," by E. S. Shecter, in the booklet *Quality Control and Reliability Management*, copyright 1969 by the Education and Training Institute of the American Society for Quality Control, Inc., Milwaukee, Wisconsin.

factured goods. The objective of the quality-control department is not to eliminate all variability of the items produced—which would be an impossible task—but to constrain this variability to economically feasible limits. The basic quality-control plan should provide for the control of the product throughout its development and production cycle. It should also contain a flow chart identifying points of inspection and testing.

Manufacturing activity depends on adequate drawings and specifications. A review of specifications and drawings by the quality-control department is highly desirable and should concentrate on the following criteria:

1. Completeness and clarity.
2. Inspectability of product.
3. Establishment of tolerances.
4. Compliance with special customer requirements.
5. Incorporation of adequate evaluation criteria.
6. Identification of test requirements.
7. Previous knowledge and experience on processes to be used.

Consideration must be given to the method of developing quality reports that furnish inspection and test results on a current basis and in a format that is easily intelligible. The display technique used to depict pertinent aspects of performance can aid significantly in deciding what corrective action is necessary. Tabulated periodic reports are not nearly so effective as graphical presentations, which are far simpler and enable trends to be identified more rapidly.

Vital contributors to the success of any program are the vendors selected to support the program. Selecting the most suitable vendors is not an easy task. Each major vendor must be evaluated to determine his ability to fully comply with specification requirements. It is desirable to establish both vendor monitoring at the place of manufacture and the extent of evaluation of the product to be performed during incoming inspection. Meeting with the vendor to explain the quality requirements and exactly what is expected can avoid costly mistakes and unnecessary delays.

The inspection and testing plans that are developed should identify the points in the production line where specific quality-control operations are to occur and the criteria the product is to be evaluated against. The record-keeping and reporting procedures must also be properly developed so that information flow is accurate and prompt. It is necessary to reevaluate the scope of the test and evaluation plans as production proceeds. It is wasteful, for example, to continue inspection when virtually no defects are found. Particular trouble spots may, on the other hand, require the institution of special inspection points or criteria. The danger in introducing new inspection or test points is that they have a tendency to become permanent after they have outlived their usefulness. It is therefore desirable that periodic scheduled reviews of inspection and test performance be made to evaluate their effectiveness.

The traditional approach to product cost evaluation involves the use of material, labor, and overhead costs. The quality cost approach is one that subdivides the costs still further by considering failure costs as well as prevention costs. Failure costs generally fall into the category of rework, retest, and scrap. There are potential savings to the company in this area.

Cost of prevention includes efforts to provide adequate training for operators in order to impart the necessary skills, to provide operators with the proper equipment for the job, and to motivate them to do the job correctly every time. Also included in these costs are machine-capability studies, used to determine the best equipment utilization, the development of process controls, and the establishment of automatic measuring and corrective programs for the manufacturing cycle.

Another important task of quality control is participation in engineering design reviews to assure that adequate process controls are considered, that the item being manufactured is inspectable and repairable, and that adequate specifications are provided. Also reviewed for effectiveness and adequacy are the test equipment design and test methods.

The following tabulation lists those quality tasks which apply during the production cycle:

1. Process control.
2. Inspection and test performance.
3. Data reporting and analysis.
4. Vendor control.
5. Training.
6. Product evaluation.
7. Scrap and rework control.
8. Field performance.
9. Customer liaison.
10. Audit.
11. Corrective action.

Frequently, the imposition of government specifications automatically causes a vendor to raise prices. And yet, upon deeper reflection, the basic elements in any good quality-control system are essentially the same, regardless of whether the application is commercial, industrial, or military. Perhaps when dealing with the government, more documentation is required to substantiate that quality-control operations have actually been carried out.

Once a policy decision has been made to adopt a quality-control system, management must support and stand behind the rules, procedures, and decisions issued by the quality-control department, even though the quantity of product released for shipment could, at times, be adversely affected. The need to meet production schedules without a sacrifice in quality demands close cooperation among the various departments of the company and a conscientious effort to comply with the imposed quality controls. By controlling variability and increasing uniformity of the product, scrap and rework costs are reduced, thereby lowering the overall material and labor cost. The resulting savings may be passed on to the customer or may be used to enhance profit.

1-2 RELIABILITY

Recent history contains many instances of electronic equipment which failed outright, operated intermittently, or degraded prematurely in performance when exposed to such environmental stresses as heat, humidity, fungus, cold, salt atmosphere, vibration, and shock. While the repair or replacement of faulty equipment may involve unexpected costs, its unavailability when needed, may have even more serious consequences.

Past problems in this area were largely due to the fact that the concept of reliability, as we use it today, was not yet formally recognized as a performance characteristic of equipment.

Let us examine more closely the term *reliability*. Dictionaries generally define *reliable* in terms of something that is trusty, authentic, consistent, and honest. When we speak of a *reliable product*, we usually expect such adjectives to apply. The problem with this manner of expressing the performance capability of a product is that it is very subjective. Different users may have different expectations as to a product's performance or life. There may also be a diversity of opinion as to what exactly constitutes degraded performance in contrast to failure.

In the Assurance Sciences, the term *reliability* assumes a more definitive character: reliability may be defined, computed, tested, and verified. It may thus be specified in equipment procurement documents and contractually enforced. Most commonly, reliability is defined as *the probability that a device will perform its intended function for a specified period of time under stated conditions*. A probability of 99 %, for example, means that, on the average, a device will properly perform (as defined above) 99 of 100 times; or else that, on the average, 99 of 100 devices will perform properly. Another well-known reliability measurement parameter is MTBF (*mean time between failures*).

The term *intended function* used to describe equipment performance makes it possible to identify what constitutes nonperformance of the equipment (i.e., failure).

The specified period of time during which the equipment is to perform reliably may vary from the instantaneous operation of one-shot devices (e.g., explosive bolts) and operations lasting only a few hours to space missions lasting for years.

The performance *under stated conditions* refers to the operational and environmental conditions, or stresses, that the equipment may experience during its required lifetime. Operational conditions vary from one piece of equipment to another, so it is important that the conditions are identified fully. They may, for instance, include unusual voltage transients entering the equipment through the supply lines. Environmental stresses, such as heat and humidity, were mentioned earlier in the chapter. They must not cause or contribute to equipment failure.

There were, of course, earlier attempts to improve the life span of equipment.

Designers learned from previous mistakes and conducted a considerable amount of testing; but in trying to solve equipment problems and prolong equipment life, they frequently resorted to the practice of extensive overdesign, resulting in excessively bulky and costly equipments. When the demand for extensive miniaturization and light weight was added to the ever-increasing circuit and system complexity, the problems multiplied. The frequently used intuitive approach to design and the luxury of making samples of everything for tryout had to be replaced by a more effective and updated method. The new reliability discipline entered the field of equipment design and development in the early 1950s. It increased steadily in importance and refinement during the 1960s and 1970s.

Reliability is incorporated into a design by a series of techniques that are part of an organized plan. Management support is mandatory in this effort. First, it is necessary to analyze the customer's specifications and ascertain all the worst-case operational and environmental requirements. Based on this review, reliability concepts may be introduced in the design-formulation phase. This effort includes a preliminary reliability analysis based on initial design engineering data. This quick-look analysis will show whether the proposed design has a chance of meeting the desired probability of success. If the latter cannot be met, changes are necessary in the design, such as simplification in the configuration, reduced circuit complexity, or the use of more reliable parts. It may also be necessary to use redundancy techniques (a method that provides automatic substitution for a failed item). All these possible modifications require a cost and trade-off analysis before implementation.

Once the basic design approach is settled, the detailed parts-selection and parts-application effort can begin. The parts engineering and standardization specialist in the reliability group draws on his extensive knowledge of part types, military part specifications, failure histories, and vendor sources to render valuable assistance to the designers in this area. Parts application usually includes the use of a derating policy to assure longer life by limiting the electrical stress on parts. In the case of semiconductor devices, a burn-in may be specified to weed out weak performers or defective parts.

As the design progresses, the reliability engineer, in cooperation with the designer, will conduct a worst-case analysis of the circuits. This analysis is a method of determining whether a circuit will perform within limits when its parts are at the extreme of their tolerance limits and the ambient temperature varies from one possible extreme to the other. Periodically, the reliability group will update and refine the initial reliability estimate as more finalized circuit and subsystem information becomes available.

A technique used by the reliability engineer to ascertain the effect of the failure of a part, circuit or subsystem on the next higher assembly, and ultimately on the entire piece of equipment, is *failure-modes-and-effects analysis*. Overly sensitive areas in the equipment which appear to be unusually prone to failure by themselves, or through secondary effects, may thus be systematically identified and modified or removed.

The reliability group also participates in design reviews with the design engineering group. In these reviews the design is critically examined for its ability to meet the performance and reliability requirements. Also subject to review are the proposed performance and environmental tests planned to qualify the end item. As a result of the final design review the design is "frozen" and the drawings and specifications are finalized. The reliability tasks include a review of these documents to assure that they reflect the appropriate reliability requirements. In the case of procurement drawings for parts and materials, where certain reliability requirements must be passed on to the vendors, this review is quite essential.

After the parts and materials, procured from outside vendors, pass through incoming inspection, production commences. When failures occur in circuits or assemblies during factory testing, the reliability group performs a complete analysis to determine the cause of failure and appropriate corrective action. The latter may at times involve drastic changes in the design if a trend toward failure develops.

Finally, the reliability department may conduct reliability verification tests on the completed product to ascertain its compliance with the probability of survival requirement. This test is based on statistical theory and will confirm that the eqiupment meets the reliability specification. It must be conducted in a realistic operating environment.

It should be noted that the quality-control activities described earlier, as well as the maintainability tasks, are carried out concurrently with the reliability effort. Each discipline pursues its own plan of action but with a similar commitment—the assurance of product integrity.

The reliability program described is the type invoked frequently on military contracts. For commercial applications, the level of the reliability program used is a function of management attitudes, cost considerations, competition, and company reputation. Lately it has been stimulated by an environment of consumer awareness and the related legal climate.

1-3 MAINTAINABILITY

Since no equipment is composed of ideal (failure-free) parts, there will always be downtime due to repairs and maintenance. The more complex a piece of equipment, the longer it takes for the diagnosis and repair of failures. Downtime is of extreme concern to the military services, as it could mean that vital equipment will be unavailable for deployment when needed. The consequences in a case of national emergency could be disastrous. Military readiness, therefore, depends heavily on equipment readiness.

The problem of excessive downtime is not limited to military equipment but also applies to the commercial field. As examples, in the case of computer systems, where time is commonly rented on an hourly basis, or commercial aircraft fleets, considerable financial loss would be incurred if there were excessive downtime. The technique that is dedicated to limiting downtime and support requirements for equipment is *maintainability*. This discipline is relatively new; it began to appear in the late 1940s and has since assumed a position of major importance.

Design techniques that enhance maintainability include easy access, modular, replaceable assemblies and strategically placed test points. To be most effective, it is important that these features be introduced early in the design. As the design is finalized, continuous refining of maintainability features is necessary. This action requires frequent communication with design engineering and participation in design reviews.

Probably the best source for maintainability design principles is a widely used U.S. Navy handbook[2] of maintainability design criteria. A maintainability design checklist, contained in this report, is included in Chapter 12. It covers maintainability guidelines for accessibility, identification, standardization, component selection, safety, hardware mounting, and similar items.

Quantitative measures are required to define maintainability objectives. The most commonly used quantitative measure of maintainability is the *mean time to repair* (MTTR). Another frequently used quantitative measure, which merges the reliability and maintainability disciplines, is *inherent availability*. Inherent availability is the probability that a piece of equipment, when used under stated conditions in an ideal support environment, will operate satisfactorily at any given time. The mathematical expression for inherent availability is a function of MTTR and MTBF. More detailed information on inherent availability and the parameters on which it depends are presented in Chapter 12.

It is necessary that a rough maintainability analysis and prediction be made early in the conceptual phase of the design and that it be continually updated. The emphasis and timing of the efforts of the maintainability engineer must, by necessity, be coordinated with the design, reliability, and quality-control activities. None of these activities can afford to work in isolation. The practitioners of the assurance sciences must always work as members of a team.

Having designed maintainability into the equipment and mathematically predicted its MTTR, we must prove that the actual meantime to repair does not exceed the predicted value. Test verification is conducted in an environment that simulates the operational maintenance environment of the equipment in terms of working conditions, tools, test equipment, spares, facilities, and instruction manuals.

1-4 INTEGRATED LOGISTIC SUPPORT (ILS)

ILS blends together the ingredients needed to support a product or a system over its entire life span. ILS is concerned with maintenance procedures and personnel, repairshops and supply depots, test procedures and service manuals. It is also concerned with the maintenance features of the equipment and in this area interacts with the related discipline of maintainability.

Control of *life cycle cost* (LCC), the total price tag on a system over its expected life, is a prime function of ILS. LCC is minimized by mathematical (often computerized) optimization studies ranging from repair versus discard policy to allocation of spares.

Maintenance engineering analysis (MEA) is a central element of the ILS program. The MEA precisely defines the corrective and preventive maintenance tasks that are

---

[2] "Maintainability Design Criteria Handbook for Designers of Shipboard Electronic Equipment," Department of the Navy, NAVSHIP 0967-312-8010, July 1972.

necessary, and documents task frequencies, task times, and logistic resources required to accomplish each task.

Mathematical techniques used to allocate logistic resources in an optimum fashion include probability analysis, queueing theory, and linear programming.

After this general overview of the assurance sciences, we are now ready to explore them in more detail. Since the field of probability and statistics is basic to these disciplines, it is appropriate that we address this subject in the next two chapters.

PROBLEMS

1-1 What are the assurance sciences and what are their purposes?

1-2 Explain the difference between quality control and quality assurance.

1-3 What causes variability in manufactured products?

1-4 Why are the military services strong advocates of the assurance sciences?

1-5 What are some quality-control criteria used in the review of drawings and specifications?

1-6 What are some considerations used in the preparation of quality-control reports?

1-7 Why is it important to periodically review the scope of inspection plans?

1-8 What is included in failure costs and cost of prevention?

1-9 List six quality-control tasks performed during production.

1-10 How does consumer awareness influence product-quality programs?

1-11 Give examples of environmental and operational stresses and discuss why it is important that they be considered in the design.

1-12 Can the reliability of a piece of equipment be quantitatively specified? Give an example.

1-13 Define reliability.

1-14 State the purpose of the following reliability techniques:
(a) Worst-case analysis.
(b) Parts derating.
(c) Failure-modes-and-effects analysis.

1-15 Discuss the purpose of design reviews and the contribution made by representatives of the assurance sciences.

1-16 What is the purpose of failure analysis?

1-17 Why are reliability verification tests conducted?

1-18 Discuss briefly the problems related to military equipment that has inadequate maintainability and support features.

1-19 Give several design techniques that promote maintainability.

1-20 Give a quantitative measure for maintainability.

1-21 What is inherent availability?

1-22 What is ILS concerned with? What is life cycle cost?

1-23 Summarize briefly the main tasks of quality control, reliability, maintainability and ILS.

2

PROBABILITY

2-1 BASIC CONCEPTS

In order to understand the mathematical techniques used by the Assurance Sciences, it is helpful to review some basic probability concepts. This material has been selected for introduction to the quality control and reliability problems found in later chapters. Predictions made on the basis of a hunch or intuition lack a rational foundation and are therefore scientifically useless. Here, we are concerned with probabilities that have some statistical basis. For example, the probability is 100% that a ball thrown in the air will come down again; there is ample historical evidence to support this contention.

Or, if there is one winning ticket in a bucket containing 100 tickets, we know that there are also 99 nonwinning tickets among the 100 tickets. The probability of picking the winning ticket is obviously 1 in 100, or 1%. Picking the winning ticket represents the favorable outcome, or success, P_S. In our case,

$$P_S = \frac{1}{100} = 0.01$$

In the case of rolling a die, the possibilities are that we obtain a 1, a 2, a 3, a 4, a 5, or a 6. Each of the six cases is equally likely to occur, assuming a true die; hence, the probability of rolling any of the six numbers is

$$P_S = \frac{1}{6}$$

When tossing a coin we will obtain either a head or a tail. The probability is the same for each case, $\frac{1}{2}$.

$$P_{\text{head}} = P_{\text{tail}} = \frac{1}{2}$$

When a person's name appears on 6 of 10 tickets thrown in a hat, and one ticket is drawn, the person's chances of winning are 6 in 10.

The idea is to extract from a whole set of equally likely and possible outcomes, the subset of outcomes that are favorable. We can therefore state, in general terms, that

$$P_{\text{(favorable outcome or success)}} = \frac{\text{number of favorable outcomes (successes)}}{\text{number of possible outcomes}} \quad (1)$$

The probability P ranges between 0 and 1.0 (expressed as a percentage: 0–100%). The sum of the probabilities of a favorable outcome and an unfavorable outcome must always equal 1 (100%).

If $P_{\text{(favorable outcome)}}$ is represented as P_F and $P_{\text{(unfavorable outcome)}}$ as P_U, then

$$P_F + P_U = 1 \quad (2)$$

Given P_F, we can determine P_U by the relation

$$P_U = 1 - P_F \quad (3)$$

EXAMPLE 2-1 The probability that it will not rain in the evening is 85%. What is the probability that it will rain in the evening?

Solution

$$P_{\text{rain}} + P_{\text{no rain}} = 1$$
$$P_{\text{rain}} + 0.85 = 1$$
$$P_{\text{rain}} = 1 - 0.85 = 0.15 \quad (\text{or } \underline{15\%})$$

EXAMPLE 2-2 A game is played by two players using a die. If the rules are set up so that one player always wins when he rolls an odd number and the other player wins when he rolls an even number, what are each player's chances of winning?

Solution Each number between 1 and 6 has an equal chance of being rolled. Hence, the chance of rolling any one number is 1 in 6:

$$P_1, P_2, P_3, P_4, P_5, P_6 = \frac{1}{6}$$

The probabilities of rolling odd numbers are

$$P_1, \quad P_3, \quad \text{and} \quad P_5$$

The overall probability of rolling an odd number is

$$P_1 = \frac{1}{6}$$

$$P_3 = \frac{1}{6}$$

$$P_5 = \frac{1}{6}$$

$$\overline{\phantom{P_5 = \frac{1}{6}}}$$

$$P_{\text{odd}} = \frac{3}{6}$$

The sum of the probabilities of rolling an even number and an odd number must be 1.0. Hence, the probability of rolling an even number is

$$P_{\text{even}} = 1.0 - P_{\text{odd}}$$

$$= 1 - \frac{3}{6} = \frac{3}{6}$$

2-2 MUTUALLY EXCLUSIVE EVENTS

Events are said to be mutually exclusive if they cannot possible happen simultaneously. For example, if a coin is tossed, we cannot get a head and a tail at the same time. Or, a boy has a bike or he does not; the answer to a multiplication problem is either right or wrong. All of these are examples of mutually exclusive events.

If event E_1 and event E_2 are mutually exclusive, the probability that either event will occur is

$$P_{E_1 \text{ or } E_2} = P_{E_1} + P_{E_2} \tag{4}$$

> **EXAMPLE 2-3**   If the probability that it rains but does not snow tomorrow is 0.30 and the probability that it snows but does not rain is 0.40, what is the probability of either rain or snow, but not both, tomorrow?
>
> **Solution**
>
> $$P_{\text{rain or snow}} = P_{\text{rain}} + P_{\text{snow}}$$
>
> $$= 0.3 + 0.4 = \underline{0.7}$$

Hence, there is a 70% chance of either rain or snow, but not both.

> **EXAMPLE 2-4**   Rolling two dice, what is the probability of rolling a 6 or an 11?
>
> **Solution**   There are 36 different number combinations possible between the two dice. Hence, the chance of rolling any one combination is $\frac{1}{36}$. A 6 can be rolled in five different ways:

Die 1	Die 2
1	5
2	4
3	3
4	2
5	1

Therefore, the probability of drawing a 6 is

$$P_6 = \frac{5}{36}$$

Similarly, an 11 can be rolled in two different ways:

Die 1	Die 2
5	6
6	5

The probability of rolling an 11 is therefore

$$P_{11} = \frac{2}{36}$$

To determine the probability of rolling a 6 or 11, we use the formula for mutually exclusive events:

$$P_{6 \text{ or } 11} = \frac{5}{36} + \frac{2}{36} = \frac{7}{36} = \underline{0.194}$$

Therefore, on the average, 19.4% of the time, or 194 of 1000 rolls of two dice, either a 6 or an 11 will be obtained.

2-3 NOT MUTUALLY EXCLUSIVE EVENTS

When events E_1 or E_2 or both can occur simultaneously, we speak of not mutually exclusive events.

EXAMPLE 2-5 On Easter Sunday it may rain or snow, or it may rain and snow on the same day. The probability of rain on Easter Sunday is 70% ($P_R = 0.7$), and the probability of snow is 40% ($P_S = 0.4$). What is the probability that there will be either rain or snow, or rain and snow, on Easter Sunday, assuming that a combination of rain and snow occurs 20% of the time ($P_{R \text{ and } S} = 0.2$)?

Solution Figure 2-1 shows the situation graphically. The probability of rain and the probability of snow are depicted in the figure by circles. Unless the shaded area, which is common to both probabilities, is deducted from the overall probability, it would be incorrectly counted twice.

To solve our problem, we therefore need to consider the probability that it will rain and snow on Easter Sunday:

$$P_{R \text{ and } S} = 0.2$$

Then

$$P_{R \text{ or}^1 S} = P_R + P_S - P_{R \text{ and } S}$$
$$= 0.7 + 0.4 - 0.2 = \underline{0.9}$$

(5)

----
[1] The *or* in equation (5) stands for R alone, S alone and R and S together.

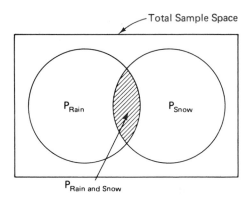

Figure 2-1 *Not mutually exclusive events.*

2-4 INDEPENDENT EVENTS

Event E_1 is said to be independent of event E_2 if the probability of the occurrence of event E_1 is not affected by the occurrence or nonoccurrence of event E_2. The probability that both events will occur in a trial is

$$P_{E_1 \text{ and } E_2} = P_{E_1} \times P_{E_2} \tag{6}$$

EXAMPLE 2-6 If during 2000 hours (h) of operation of an amplifier, the probability of transistor Q_1 failing is 5% and that of diode CR_1 is 9%, what is the probability of a failure of both Q_1 and CR_1 during 2000 h of operation? (The failure of either device is assumed to be independent of the failure of the other.)

Solution

$$P_{Q_1 \text{ and } CR_1} = P_{Q_1} \times P_{CR_1}$$
$$= 0.05 \times 0.09$$
$$= \underline{0.0045}$$

The probability of both the transistor and the diode failing during the 2000 h of operation is 0.45% (or 45 in 10,000 cases).

EXAMPLE 2-7 Three identical resistors are connected in series as shown in Figure 2-2. The probability of survival of one resistor is $P_S = 0.98$ for 10,000 h of operation. What is the probability that no failure will occur in the arrangement shown during 10,000 h of operation?

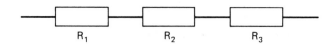

Figure 2-2 *Resistors connected in series.*

Solution   The failure-free operation of any of the three resistors is independent of the others. Hence,

$$P_{3 \text{ resistors surviving}} = P_{R_1} \times P_{R_2} \times P_{R_3}$$
$$= 0.98 \times 0.98 \times 0.98$$
$$= \underline{0.94}$$

EXAMPLE 2-8   A system consists of the four subassemblies shown in Figure 2-3. The probability of survival of each subsystem is given for a period of 1 year. The survival or failure of any subsystem is assumed to be independent of the state of the other subsystems. What is the probability of survival of the system during 1 year of operation?

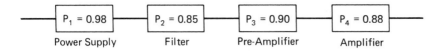

| $P_1 = 0.98$ | $P_2 = 0.85$ | $P_3 = 0.90$ | $P_4 = 0.88$ |
| Power Supply | Filter | Pre-Amplifier | Amplifier |

Figure 2-3 *Block diagram of four subassemblies in series.*

Solution

$$P_{\text{system}} = P_1 \times P_2 \times P_3 \times P_4$$
$$= 0.98 \times 0.85 \times 0.90 \times 0.88$$
$$= \underline{0.6597}$$

2-5 CONDITIONAL PROBABILITY

Event E_2 is said to be conditional on event E_1 when the probability of occurrence of event E_2 depends on the prior occurrence of event E_1. The probability that E_2 will occur provided E_1 has occurred is denoted by

$$P_{E_2 | E_1} \tag{7}$$

The probability that both events will occur is the probability that E_1 will occur multiplied by the probability that E_2 will occur, given that E_1 has occurred:

$$P_{E_1 \text{ and } E_2} = P_{E_1} \times P_{E_2 | E_1} \tag{8}$$

EXAMPLE 2-9   In a class of 40 students there are 25 boys and 15 girls. Of the 25 boys, 10 come from religious homes, and of the 15 girls, 8 come from religious homes. What is the probability $P_{B \text{ and } R}$ that a student elected randomly by lottery is both a boy and from a religious home?

Solution   Let us make a matrix that combines all the data (Table 2-1). Inspection of the table permits the postulation of a number of probability relations:

           The probability of drawing a boy is $P_B = \dfrac{25}{40}$

The probability of drawing a girl is $\qquad P_G = \dfrac{15}{40}$

The probability of drawing a religious person is $\qquad P_R = \dfrac{18}{40}$

The probability of drawing a nonreligious person is $\qquad P_{NR} = \dfrac{22}{40}$

The probability that the person drawn is religious, given that he is a boy, is $\qquad P_{R\mid B} = \dfrac{10}{25}$

The probability that the person drawn is religious, given that she is a girl, is $\qquad P_{R\mid G} = \dfrac{8}{15}$

The probability that the person drawn is a boy, given that he is religious, is $\qquad P_{B\mid R} = \dfrac{10}{18}$

The probability that the person drawn is a girl, given that she is religious, is $\qquad P_{G\mid R} = \dfrac{8}{18}$

Table 2-1 *Matrix of Sex and Religion of Students.*

	Boys	Girls	Total
Religious	10	8	18
Nonreligious	15	7	22
Total	25	15	40

Applying equation (8), the probability that the randomly selected student is both religious and a boy is

$$P_{R \text{ and } B} = P_R \times P_{B\mid R}$$

$$P_{R \text{ and } B} = \frac{18}{40} \times \frac{10}{18}$$

$$= \underline{0.25}$$

An alternative way to get the same result is to find the probability that the randomly selected student is found to be both a boy and religious, is

$$P_{B \text{ and } R} = P_B \times P_{R\mid B}$$

$$P_{B \text{ and } R} = \frac{25}{40} \times \frac{10}{25}$$

$$= \underline{0.25}$$

Both approaches give the same results.

EXAMPLE 2-10 Pursuing Example 2-9, what is the probability that a student selected at random is both a girl and from a nonreligious home?

Solution

$$P_{NR} = \frac{22}{40}$$

The conditional probability that the randomly selected individual is a girl given she is nonreligious is, from Table 2-1,

$$P_{G \mid NR} = \frac{7}{22}$$

$$P_{NR \text{ and } G} = P_{NR} \times P_{G \mid NR}$$

$$= \frac{22}{40} \times \frac{7}{22}$$

$$= \underline{0.175}$$

2-6 PROBABILITY DISTRIBUTIONS

(a) Discrete Probability Distributions

1. General Principles

A *probability distribution* gives the probabilities of all the possible events that can result from a given trial. For example, a bowl contains 50 green and 100 white marbles. In selecting one marble from among the 150 marbles in the bowl, only two events are possible relative to the green marbles:

Event 1: a green marble is chosen
Event 2: the marble chosen is not green

The chances of picking a green marble are 50 in 150, or 1 in 3. Therefore, the probability of picking a green marble is

$$P_{\text{green}} = \frac{1}{3}$$

Similarly, the probability of not picking a green marble, which is the same as the probability of picking a white marble, is 100 in 150, or 2 in 3. Thus,

$$P_{\text{not green}} = \frac{2}{3}$$

Observe that the sum of the probabilities of the possible events occurring is 1.0. One hundred percent of the possibilities are thereby encompassed.

$$P_{\text{green}} + P_{\text{not green}} = \frac{1}{3} + \frac{2}{3} = 1.0$$

Figure 2-4 is a plot of these probabilities.

In the example above we restricted the possible outcomes (events) of the trial to the simplest possible form: either one event or the other occurs; that is, either we pick a green marble or we do not pick a green marble. When the events are discrete, as in the example above, the probability distribution is referred to as a *discrete probability distribution*.

Distributions become a bit more complicated when the number of possible events is increased. For example, assume that we flip two coins, one in our left hand and one

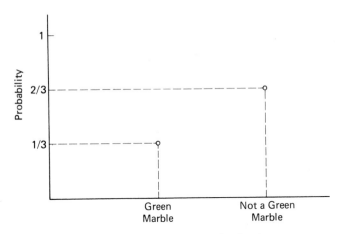

Figure 2-4 *Simple discrete probability distribution.*

in our right hand. There is an equal chance for a head or tail in either hand. The possible combinations are as follows:

Left Hand	Right Hand	
H	H	event 1
H	T	event 2
T	H	event 3
T	T	event 4

If the probability of flipping a head is p and the probability of flipping a tail is q, the probability of flipping two heads (event 1) is p^2. The other probabilities are found in the same manner:

H	H	p^2
H	T	pq
T	H	qp ⎱ $2pq$
T	T	q^2

Since $p^2 + 2pq + q^2$ contains all possible events in this trial, its sum equals 1.0:

$$p^2 + 2pq + q^2 = 1 \qquad (9)$$

The meaning of the individual terms of equation (9) is

p^2: the probability that both coins are heads
$2pq$: the probability that one coin is a head and the other coin a tail
q^2: the probability that both coins are tails

EXAMPLE 2-11   What is the probability of obtaining two tails when flipping two coins?

Solution   From equation (9), the probability of flipping two tails is q^2. Since the probability of flipping one tail is 0.50, the answer is

$$q^2 = 0.5^2 = \underline{0.25}$$

EXAMPLE 2-12   What is the probability of obtaining a head and a tail when flipping two coins?

Solution   From equation (9) the probability of flipping a head and a tail is $2pq$. Accordingly, the answer is

$$2pq = 2(0.5)(0.5) = \underline{0.50}$$

EXAMPLE 2-13   Determine the probability of obtaining at least one tail when flipping three coins simultaneously.

Solution   The possible outcomes, together with the corresponding probabilities, are as follows:

Possible Outcomes	Probability	
H H H	p^3	
H H T	p^2q	
H T H	p^2q	$3p^2q$
T H H	p^2q	
H T T	pq^2	
T H T	pq^2	$3pq^2$
T T H	pq^2	
T T T	q^3	

$$\text{Total}\quad p^3 + 3p^2q + 3pq^2 + q^3 = 1.0 \qquad (10)$$

where

$$p = \text{probability of flipping a head} = 0.5$$
$$q = \text{probability of flipping a tail} = 0.5$$

The outcomes that have at least one tail are added. Their sum in terms of probabilities is

$$p^2q + p^2q + p^2q + pq^2 + pq^2 + pq^2 + q^3 = 3p^2q + 3pq^2 + q^3$$

From equation (10) the sum of the probability terms above equals

$$1 - p^3 = 1 - 0.5^3 = \underline{0.875}$$

EXAMPLE 2-14   Plot a probability distribution expressing the outcomes in terms of obtaining or not obtaining tails when flipping three coins. The possibilities are:

3 tails	q^3
2 tails	$3pq^2$
1 tail	$3p^2q$
0 tail	p^3

Solution   The probability distribution is plotted in Figure 2-5.

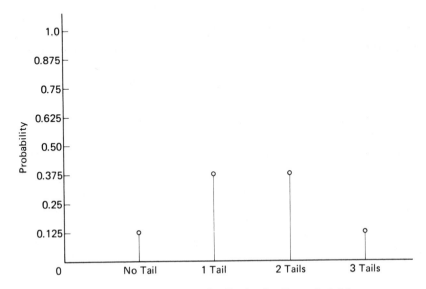

Figure 2-5 *Discrete probability distribution for Example 2-14.*

EXAMPLE 2-15   What is the probability of obtaining three equal coin faces when flipping three coins?

Solution   Three equal coin faces could be either three heads or three tails. The sum of the corresponding probabilities therefore constitutes the desired answer. From equation (10) we extract

$$p^3 + q^3$$
$$(0.5)^3 + (0.5)^3 = 0.125 + 0.125 = \underline{0.25}$$

Hence, when flipping three coins, the probability of obtaining either three heads or three tails is 25%.

2. Binomial Expansion

Examination of equations (9) and (10) reveals that they are expansions of the familiar binomial term

$$(p + q)^n \tag{11}$$

$p^2 + 2pq + q^2$ is immediately recognized as an expansion of $(p + q)^n$, when $n = 2$.

$$(p + q)^2 = p^2 + 2pq + q^2 \tag{12}$$

Similarly, $p^3 + 3p^2q + 3pq^2 + q^3$ is an expansion of $(p + q)^n$, when $n = 3$.

$$(p + q)^3 = p^3 + 3p^2q + 3pq^2 + q^3 \tag{13}$$

The general binomial expansion is

$$(p + q)^n = p^n + np^{n-1}q + n_1 p^{n-2}q^2 + n_2 p^{n-3}q^3 + \cdots + q^n \tag{14}$$

In using the binomial expansion, care must be taken to apply it only to problems where p and q are essentially unchanged by the drawing of the sample.

Where a population contains a fixed proportion of articles in the p and q categories, the sample must be very small relative to the population to satisfy this constraint. However, where p and q are inherent properties of the article drawn (such as the 0.5 probability of obtaining a head in an honest coin toss), this restriction does not apply. Notice that the first and last terms of equation (14), p^n and q^n, have coefficients of 1.0. The second term has a coefficient that equals n. The coefficients for the second term to the term preceding the last term may be found from Table 2-2.

Table 2-2 *Binomial Coefficients Excluding First and Last Terms.*

	2				$(p + q)^2$
	3	3			$(p + q)^3$
4		6	4		$(p + q)^4$
5	10		10	5	$(p + q)^5$

EXAMPLE 2-16   Expand $(p + q)^5$ using equation (14) and Table 2-2.

Solution   Per equation (14), for $n = 5$,

$$(p + q)^5 = p^5 + np^{5-1}q + n_1 p^{5-2}q^2 + n_2 p^{5-3}q^3 + n_3 p^{5-4}q^4 + q^5$$

From Table 2-2 the binomial coefficients, excluding the first and last terms, are

$$5 \quad\quad 10 \quad\quad 10 \quad\quad 5$$
$$(n) \quad\quad (n_1) \quad\quad (n_2) \quad\quad (n_3)$$

Substituting in equation (14), we have

$$(p + q)^5 = p^5 + 5p^4q + 10p^3q^2 + 10p^2q^3 + 5pq^4 + q^5$$

EXAMPLE 2-17   A lot of 15,000 transistors is known to be 10% defective. Determine the probability that a random sample of four transistors drawn from this lot will contain

(a) 0 defective transistors.
(b) 1 defective transistor.
(c) 2 defective transistors.
(d) 3 defective transistors.
(e) 4 defective transistors.

Solution   Since the sample drawn consists of four pieces, a binomial expansion with $n = 4$ applies:

$$p = \text{probability of no defect} = 90\% = 0.9$$

$$q = \text{probability of a defect} = 10\% = 0.1$$

$$(p + q)^4 = p^4 + 4p^3q + 6p^2q^2 + 4pq^3 + q^4$$

probability of 0 defects $= p^4$	$= (0.9)^4$		$= 0.6561$
probability of 1 defect $= 4p^3q$	$= 4(0.9)^3(0.1)$		$= 0.2916$
probability of 2 defects $= 6p^2q^2$	$= 6(0.9)^2(0.1)^2$		$= 0.0486$
probability of 3 defects $= 4pq^3$	$= 4(0.9)(0.1)^3$		$= 0.0036$
probability of 4 defects $= q^4$	$= (0.1)^4$		$= 0.0001$
		Total	$= 1.0000$

Therefore, the probability that the sample of four contains no defects is 65.61%. The probability that the sample contains one defect is 29.16%, and so on. The probability that all four transistors in the sample are defective will occur only once in 10,000 trials, since the probability was found to be 0.0001.

The probability that the sample of four transistors will, for example, have less than 2 defectives is simply the sum of the probabilities of having 0 defectives and 1 defective:

0 defectives	0.6561
1 defective	0.2916
Total	0.9477 or 94.77%

3. Binomial Probability Distribution

The binomial expansion may be expressed by the following probability equation:

$$P_{(r, n, p)} = \frac{n!}{(n - r)!r!} p^r (1 - p)^{n-r} \tag{15}$$

where

$P_{(r, n, p)}$ = probability of r failures in a sample of n units when the probability of the occurrence is p

r = number of occurrences

n = sample size

p = probability of the occurrence

$!$ = factorial (*example:* $4! = 4 \times 3 \times 2 \times 1$;   *note:* $0! = 1$)

EXAMPLE 2-18   Apply the binomial formula per equation (15) to Example 2-17. Specifically, find the probability of finding two defective transistors in a sample consisting of four transistors drawn randomly from a lot that is known to be 10% defective.

Solution

$$n = 4$$

$$r = 2$$

$$p = 0.1$$

$$P(2, 4, 0.1) = \frac{4!}{(4-2)!2!} 0.1^2 (1 - 0.1)^{4-2}$$

$$= \frac{4 \times 3 \times 2 \times 1}{(2 \times 1)(2 \times 1)} 0.01(0.9)^2$$

$$= \underline{0.0486}$$

This is the result that was obtained previously for $6p^2q^2$.

EXAMPLE 2-19 Suppose that we have a jar with three white and seven black marbles. While blindfolded we draw three marbles, one at a time, replacing each marble in the jar[2] before drawing the next and win 1.0 dollar for each white marble contained in the group drawn. How many chances in 1000 tries are there of winning 0, 1, 2, or 3 dollars?

Solution   Tablulating the conditions under which the dollar values above apply:

0 white, 3 black	0 dollars
1 white, 2 black	1 dollar
2 white, 1 black	2 dollars
3 white, 0 black	3 dollars

We shall next solve this example by the use of each of three methods.

METHOD 1

Picking white marbles means success. The total number of marbles in the jar is 10. The probability of picking a white marble is 3 chances out of 10, or

$$p = 0.3$$

The probability of picking a black marble equals 7 chances in 10, or

$$q = 0.7$$

$$(p + q)^3 = p^3 + 3p^2q + 3pq^2 + q^3$$

p^3	(3 successes, 0 failures) $= 0.3^3$		$= 0.027$
	WH BLK		
$3p^2q$	(2 successes, 1 failure) $= 3(0.3)^2(0.7)$		$= 0.189$
	WH BLK		
$3pq^2$	(1 success, 2 failures) $= 3(0.3)(0.7)^2$		$= 0.441$
	WH BLK		
q^3	(0 successes, 3 failures) $= 0.7^3$		$= \underline{0.343}$
	WH BLK		

Total = 1.000

---

[2] This process is called *drawing items with replacement* to distinguish it from the *without replacement* case, in which items drawn are not returned to the lot. (See Section 6 of this Chapter).

The computation shows a probability of 0.027 of successively picking three white marbles from the jar with replacement. Expressed differently, the chances of successively picking three white marbles are 27 times in 1000 tries. Table 2-3 gives a full listing of the winning chances:

Table 2-3 *Tabulation of Winning Chances.*

Number of White Marbles in One Draw from the Jar	Chances in 1000 Tries
3	27 chances to win $3
2	189 chances to win $2
1	441 chances to win $1
0	343 chances to win nothing
	Total = 1000 chances

METHOD 2

$$P(r, n, p) = \frac{n!}{(n-r)!r!}p^r(1-p)^{n-r}$$

(a) For 3 successes (white marbles), 0 failures (black marbles):

$$r = 3$$
$$n = 3$$
$$p = 0.3 \quad \text{(probability of drawing a white marble)}$$

$$P(3, 3, 0.3) = \frac{3 \times 2 \times 1}{(3-3)!3 \times 2 \times 1}0.3^3(1-0.3)^{3-3}$$

$$= \frac{1}{(0)!1}(0.027)(0.7)^0$$

Since $(0)!$ is 1.0 and $(0.7)^0$ is 1.0, we have

$$P(3, 3, 0.3) = \frac{1}{1}(0.027)(1.0)$$

$$= \underline{0.027}$$

(b) For 2 successes (white marbles), 1 failure (black marble):

$$r = 2$$
$$n = 3$$
$$p = 0.3$$

$$P(2, 3, 0.3) = \frac{3 \times 2 \times 1}{(3-2)!2 \times 1}0.3^2(1-0.3)^{3-2}$$

$$= \frac{3}{1!}0.09(0.7)^1 = 3 \times 0.09 \times 0.7$$

$$= \underline{0.189}$$

(c) For 1 success (white marble), 2 failures (black marbles):

$$r = 1$$
$$n = 3$$
$$p = 0.3$$

$$P(1, 3, 0.3) = \frac{3 \times 2 \times 1}{(3 - 1)!1!} 0.3^1 (1 - 0.3)^{3-1}$$

$$= \frac{6}{2 \times 1 \times 1} (0.3)(0.7)^2$$

$$= 3 \times 0.3 \times 0.49 = \underline{0.441}$$

(d) For 0 successes (white marble), 3 failures (black marbles):

$$r = 0$$
$$n = 3$$
$$p = 0.3$$

$$P(0, 3, 0.3) = \frac{3 \times 2 \times 1}{(3 - 0)!0!} 0.3^0 (1 - 0.3)^{3-0}$$

$$= \frac{6}{3 \times 2 \times 1 \times 1} (1.0)(0.7)^3$$

$$= \frac{6}{6} (1.0)(0.343)$$

$$= \underline{0.343}$$

The answers are the same as in method 1.

METHOD 3

Use Table A9(a), Appendix A, for values of the individual terms of the binomial distribution.

(a) For 3 successes (white marbles), 0 failures (black marbles):

$$r = 3$$
$$n = 3$$
$$p = 0.3$$

Looking up $n = 3$ in first column, we find 027 under $p = 0.3$ for $r = 3$.

All numbers in the table have a decimal point in front of them; hence, 027 is actually 0.027, which checks with the results obtained by the other two methods.

(b) For 2 successes (white marbles), 1 failure (black marble):

$$r = 2$$
$$n = 3$$
$$p = 0.3$$

For $n = 3$ in the first column, find 189 under $p = 0.3$ for $r = 2$; 0.189 checks with the results obtained with the other two methods.

The other two cases are found in a similar manner.

4. Cumulative Binomial Terms

In Example 2-19, we determined the individual binomial terms, using method 1:

$$p^3 = 0.027 \text{ (3 successes, for 3 white marbles)}$$
$$3p^2q = 0.189 \text{ (2 successes, for 2 white marbles)}$$
$$3pq^2 = 0.441 \text{ (1 success, for 1 white marble)}$$
$$q^3 = 0.343 \text{ (0 success, for 0 white marbles)}$$
$$\text{Total} = \overline{1.000}$$

The sum of the terms is, of course, 1.000. We may be interested to determine the probability of drawing one or more white marbles with replacement. To solve this problem we recognize that the statement "one or more white marbles" implies three possibilities:

1 white marble,   or
2 white marbles,   or
3 white marbles

The probability of drawing one or more white marbles is the sum of the probabilities of the above three cases, or

$$0.441$$
$$0.189$$
$$0.027$$
$$\text{Total} = \overline{0.657}$$

Hence, the probability of drawing one or more white marbles is 65.7%.

Table A9(b), Appendix A, gives the summation of the binomial terms. For the example above,

$$r = 1 \text{ or more}\quad (\text{actually } r = 1 \text{ or 2 or 3})$$
$$n = 3$$
$$p = 0.3$$

Look up $n = 3$ in the left column and find 0.657 for $r = 1$ under the column $p = 0.3$. Note that r in this table means r or more, up to $r = n$.

EXAMPLE 2-20 Use the cumulative binomial table to determine the probability of finding at least three Democrats in a randomly chosen sample of eight persons from the voter registration rolls of a city. It is known that 50% are Democrats. (The sample must be small compared to the total population so that drawing of the sample will not significantly alter the proportion of Democrats in the remaining population).

Solution The probability of success, which is the probability of picking a Democrat, is $p = 0.5$.

$$n = 8$$
$$r = 3 \text{ or more}\quad (\text{actually } r = 3 \text{ or 4 or 5 or 6 or 7 or 8})$$

Look up $n = 8$ in the left column of Table A9(b) and find 0.855 for $r = 3$ under column $p = 0.5$. Hence, the probability is $85\frac{1}{2}\%$ of finding at least three Democrats in the sample of eight persons. From the same column we may also conveniently ascertain the probabilities of at least 1, 2, 3, 4, 5, 6, or 7 Democrats contained in the sample of eight persons, as well as the probability of having no Democrat in the sample, or having the sample consist of Democrats only. Tabulating these cumulative binomial probabilities from Table A9(b), we obtain Table 2-4.

Table 2-4 *Tabulation of Probabilities.*

Number of Democrats	Probability
0 or more	1.000
1 or more	0.996
2 or more	0.965
3 or more	0.855
4 or more	0.637
5 or more	0.363
6 or more	0.145
7 or more	0.035
8	0.004

Note the small chance of 0.004, or 4 in 1000, that all eight members of the group are Democrats. On the other hand, there is a 63.7% chance that at least half the group is composed of Democrats.

The saving in time when using the tables for individual or cumulative binomial terms is obvious.

5. *Practical Limitation of the Binomial Distribution*

Referring to Example 2-20, let us assume that a sample of 50 persons is to be chosen randomly from a population of 150,000 persons. Again, half of the population are Democrats. What is the probability that at least 30 persons among the 50 are Democrats?

$$n = 50$$

$$r = 30 \text{ or more} \quad (\text{actually } r = 30 \text{ or } 31 \text{ or } 32 \ldots \text{ or } 50)$$

$$p = 0.5$$

To use method 1 as in Example 2-19 requires

$$(p + q)^n$$

where $q = 1 - p$. In this case we have

$$(p + q)^{50}$$

The binomial expansion would not be practical without computer assistance since it would involve over 50 terms.

The use of method 2 as in Example 2-19 would also require enormously tedious calculations; and finally, many binomial tables do not include such high n values. In addition to large values of n, small values of p also tend to make the calculation burdensome. Fortunately, when n is large and p is small, we may use a very good approximation to the binomial distribution. This approximation is the Poisson distribution, which was developed by the French mathematician S. D. Poisson (1781–1840). This distribution is discussed in section 7 of this chapter.

6. Drawing Samples without Replacement

In Example 2-19 we drew three marbles successively from a jar, returning the marble drawn before drawing the next. When drawing marbles without replacement, the probability of drawing a white marble will vary with each successive draw. For example, the chance of drawing three successive white marbles from the jar containing a total of 10 marbles is now expressed as

$$\frac{3}{10} \times \frac{2}{9} \times \frac{1}{8} = 0.0083$$

This calculation is based on the fact that on the first draw three white marbles were available out of a total group of 10. Once the first white marble had been drawn only two more were available out of the remaining group of nine, and so on. The result of 0.0083 compares with a 0.0270 chance of successively drawing three white marbles with replacement.

Problems involving more than one item on a single draw, such as a handful of marbles, are treated as successive draws without replacement. Therefore, the probability of drawing a group of three white marbles is 0.0083.

The probability of drawing one white marble and one black marble in two tries is

$$\left(\frac{3}{10} \times \frac{7}{9}\right) + \left(\frac{7}{10} \times \frac{3}{9}\right) = 0.4667$$

In this case we had to consider the mutually exclusive probabilities of the first draw being either white or black. The comparable result for two draws with replacement would have been

$$\left(\frac{3}{10} \times \frac{7}{10}\right) + \left(\frac{7}{10} \times \frac{3}{10}\right) = 0.4200$$

Without-replacement models are frequently used to describe multiple draws of playing cards from a deck.

7. The Poisson Distribution

The examples on binomial distribution which were presented all had one thing in common: we were dealing with a sample of definite size and were determining the potential number of times a certain event (such as the number of defects) may be observed in the sample. For instance, we calculated examples involving the flipping of three coins and determined the chances of obtaining two tails, one head and one tail, at least one tail, and so on. In each of these instances we are dealing with the potential number of times the event will and will not occur. This approach encompasses all the possible outcomes within the given sample space. This is confirmed by the fact that the sum of the probabilities of the occurrence of the event and the nonoccurrence of the event equals 1.0.

Consider a problem in which the number of occurrences of an event can be ascertained but not the number of nonoccurrences? For example, a private nurse in a hospital, who spends a full eight-hour shift with a patient and who is concerned about the patient's severe coughing spells, reports at the end of her shift that the patient had 6 spells. It is obvious that she could not report how many times the patient did not have spells. We are dealing here with isolated events that occur in a continuum of time. The events do not necessarily have to relate to time. We could observe the number of flaws in a roll of cloth; here, the isolated events (flaws) are observed in a continuum of length. We could observe the number of bacteria in a drop of water coming from a swimming pool; here, the isolated events (bacteria) are observed in a continuum of volume.

In all these cases it is assumed that the events are equally likely to occur in any part of the continuum. For example, the bacteria found in the one drop are equally likely, or have an equal chance to be located in any other drop of the liquid in the pool. Furthermore, it is assumed that the presence of any bacteria is independent of the presence or absence of other bacteria in the drop.

In addition to the use of the Poisson distribution for the estimation of isolated events in a continuum of time, area, or volume, as described above, we may use this distribution for instances where the use of the binomial distribution would entail excessively lengthy mathematical computations. An example of the latter would be the previously cited instance where n, the sample size is large (i.e., $n = 50$).

8. Applying the Poisson Distribution

The knowledge of the expected (or average) number of occurrences of an event is needed before the Poisson distribution can be applied. We have shown in Examples 2-18 and 2-19 typical examples useful in the determination of the expected number of occurrences of an event. Let us designate the number of occurrences by the letter a:

$$\text{expected number of occurrences of an event} = a \qquad (16)$$

The Poisson distribution is analogous to the binomial distribution, in that it also consists of a number of probability terms which add up to 1.0. The manner in which the Poisson distribution is applied is as follows. If we know, for example, that the expected number of defects in a randomly drawn sample is three, we are able to determine the probability that the sample contains a greater or smaller number of defects.

EXAMPLE 2-21 Assume that a continuous production line of 100-ohm (Ω) resistors is sampled every day by checking the resistance of a random sample of 50 resistors picked from that day's production. Past experience has shown that, on the average, 10% of the produced resistors are defective. Determine the expected number of defectives.

Solution   The expected number of defectives in the sample of 50 resistors is

$$p = 0.1$$
$$n = 50$$
$$a = np = 0.1(50) = 5$$

Obviously, the actual number of defects in the daily samples will not always be exactly 5. In fact, the number of defects in the sample could theoretically vary from 0 to 50.

Since, on the average, only 1 in 10 units produced is defective, it is highly unlikely that a sample of 50 will contain many more than 5 defectives, or for that matter, much fewer than 5 defectives. The probability of finding 10, 20, 30, 40, or 50 defective units is progressively lower. Similarly, the probability of finding 4, 3, 2, 1, or 0 defective units is progressively lower.

Table 2-5 lists some of the probabilities for the occurrence of number of defectives in the range from 0 through 50 units in a sample of 50 randomly chosen units from a lot that is 10% defective. An inspection of the table reveals that the expected number of defects, 5, will only occur 1755 of 10,000 times, or roughly once in six times. On the other hand, 15 defectives in the sample of 50 should, on the average, occur 2 times in 10,000.

Table 2-5 *Tabulation of the Probability of Finding Defectives.*

Number of Defective Units	Probability
0	0.0067
1	0.0337
3	0.1404
5	0.1755
10	0.0181
15	0.0002
30	0.0000
40	0.0000
50	0.0000

The manner in which the use of the Poisson distribution may help in the detection of a production problem is illustrated in the following example.

EXAMPLE 2-22 A production line is normally 10% defective. A sample of 50 units is drawn randomly from production on a daily basis. Table 2-6 lists the defectives found on a daily basis. A review of the table shows that production is indeed as good as anticipated for the first 9 days, with an average defective rate of 5 per day. But something goes wrong beginning with the tenth day. The defect rate for the tenth and eleventh day is double and triple the average value. Table 2-5 shows that, on the basis of chance alone, 10 defectives, out of a sample of 50, would, on the average, occur 181 in 10,000 times and 15 defectives only 2 out of 10,000 times. The probability of 10 and 15 defectives in two successive days is only (0.0181 × 0.0002), or less than 4 chances in a million. It is concluded that this increase in defectives cannot be due to chance alone, but that some production problem has developed. A decision is made to stop production and analyze the problem.

Table 2-6 *Tabulation of Daily Defectives.*

Day	Number of Defectives
1	5
2	4
3	5
4	6
5	5
6	4
7	6
8	5
9	5
10	10
11	15
	Production stopped

9. The Poisson Formula

We have designated the expected number of occurrences of an event by the letter a. This will be required in the Poisson formula. Another important quantity needed in the formula is the familiar mathematical constant identified by the letter e. The constant e is the sum of an infinite series of the following terms:

$$e = \frac{1}{0!} + \frac{1}{1!} + \frac{1}{2!} + \frac{1}{3!} + \frac{1}{4!} + \cdots \tag{17}$$

We already know that

$$0! = 1$$

and

$$1! = 1$$

Hence,

$$e = \frac{1}{1} + \frac{1}{1} + \frac{1}{2 \times 1} + \frac{1}{3 \times 2 \times 1} + \frac{1}{4 \times 3 \times 2 \times 1} + \cdots \tag{18}$$

$$= 1 + 1 + 0.5 + 0.1667 + 0.0416 + \cdots$$

When enough terms of the series are summed, we obtain

$$e = 2.7183\ldots \tag{19}$$

The constant e is the base of the *natural logarithm* (also called the *Naperian logarithm*).

To derive the Poisson formula as an equivalent expression to the binomial expansion $(p + q)^n = 1$, we make use of the constant e. If we expand e^x into an infinite series, we get

$$e^x = 1 + x + \frac{x^2}{2!} + \frac{x^3}{3!} + \frac{x^4}{4!} + \cdots \tag{20}$$

We know, from working with the binomial formula, that summing the terms of the binomial yields 1.0. To achieve the same result with the infinite series of equation (20), we use a simple mathematical device which converts the series into a probability distribution whose terms add to 1.0. We multiply e^x by e^{-x}, giving the product a value of 1.0:

$$e^x \cdot e^{-x} = \frac{e^x}{e^x} = 1.0 \tag{21}$$

Multiplying equation (20) by e^{-x} gives

$$e^{-x}\left(1 + x + \frac{x^2}{2!} + \frac{x^3}{3!} + \frac{x^4}{4!} + \frac{x^5}{5!} + \cdots\right) = 1 \tag{22}$$

Multiplying out and expanding the factorials,

$$e^{-x} + xe^{-x} + \frac{x^2}{2}e^{-x} + \frac{x^3}{6}e^{-x} + \frac{x^4}{24}e^{-x} + \frac{x^5}{120}e^{-x} + \cdots = 1.0 \tag{23}$$

Let us now substitute the quantity a, which is the expected number of occurrences of an event, for the x in equation (23) and tabulate the terms together with their distributional meaning, as shown in Table 2-7.

Table 2-7 *Poisson Terms.*

Poisson Term	Probability of the Occurrence of Event
e^{-a}	Probability of 0 occurrences of an event
ae^{-a}	Probability of 1 occurrence of an event
$\frac{a^2}{2}e^{-a}$	Probability of 2 occurrences of an event
$\frac{a^3}{6}e^{-a}$	Probability of 3 occurrences of an event
$\frac{a^4}{24}e^{-a}$	Probability of 4 occurrences of an event
$\frac{a^5}{120}e^{-a}$	Probability of 5 occurrences of an event
.	.
.	.
.	.
$\frac{a^k e^{-a}}{k!}$	Probability of k occurrences of an event

The calculation of the terms e^{-a} is not necessary since convenient tabulations are available (see Table A10, Appendix A).

EXAMPLE 2-23 If the expected number of occurrences of an event is $a = 0.5$, what is the probability of zero occurrences of the event?

Solution

$$P = e^{-a} = e^{-0.5}$$

From Table A10, Appendix A,

$$e^{-0.5} = 0.6065$$

Hence,

$$P = 0.6065$$

What are the probabilities of 1, 2, and 3 occurrences of the event?

Solution

$$P = ae^{-a} = 0.5e^{-0.5}$$

$$= 0.5(0.6065)$$

$$= \underline{0.3033} \qquad \text{probability of 1 event occurring}$$

$$P = \frac{a^2}{2}e^{-a} = \frac{0.5^2}{2}e^{-0.5}$$

$$= 0.125 \times 0.6065$$

$$= \underline{0.0758} \qquad \text{probability of 2 events occurring}$$

$$P = \frac{a^3}{6}e^{-a} = \frac{0.5^3}{6}e^{-0.5}$$

$$= 0.0208 \times 0.6065$$

$$= \underline{0.0126} \qquad \text{probability of 3 events occurring}$$

What is the probability of up to 2 events occurring?

Solution To solve this problem, we add the probabilities for the Poisson terms for 0, 1, and 2 events occurring:

	Probability
0 event	0.6065
1 event	0.3033
2 events	0.0758
	Total = 0.9856, say 0.986

The cumulative Poisson probability is $P = 0.986$ that either 0, 1, or 2 events will occur when the expected number of events is $a = 0.5$.

EXAMPLE 2-24 What is the probability of no failure (same as probability of survival) for a system with an expected number of failures of 0.65 during its operational period?

Solution The appropriate Poisson term for zero occurrence of an event (same as no failure) is

$$P = e^{-a}$$
$$a = 0.65$$
$$P = e^{-0.65}$$

Using Table A10, we find that

$$P = \underline{0.522}$$

EXAMPLE 2-25 Refer to Example 2-24. What is the maximum expected number of failures if the probability of survival of the equipment is to be 90%?

Solution Look in Table A10 and find the exponent a for which $e^{-a} = 0.9$. The nearest value for e^{-a} is seen to be 0.90032 (note that λt in the table [3] is equivalent to a) and the value of $\lambda t = a = 0.105$. Hence, the maximum expected number of failures is 0.105.

10. Cumulative Poisson Tables

In Example 2-23, we calculated the cumulative Poisson probability of 0, 1, and 2 events occurring. Table A11, Appendix A, gives cumulative Poisson probability terms. Let us find the cumulative Poisson probability of 0, 1, and 2 events occurring if the expected number of events is $a = 0.5$. Look in the left column of Table A11 and find $a = 0.5$. On the $a = 0.5$ line under the column $c = 2$ (which means 2 or less events, equaling 0, 1, and 2 events), find the answer $P = 0.986$. This checks with the results obtained in Example 2-23.

EXAMPLE 2-26 Again referring to Example 2-23, determine the cumulative Poisson probability for 1 and 2 events occurring.

Solution Referring to Table A11, we find 0.986 on the $a = 0.5$ line under the column $c = 2$. However, this result is for the sum of the probabilities of 0, 1, or 2 events; since we are only interested in 1 or 2 events, we must subtract the probability for 0 events.

cumulative probability for 0, 1, 2 events	0.986
less probability for 0 events	0.607
probability for 1 or 2 events	0.379

(b) Continuous Probability Distributions

1. General Principles

The discrete probability distributions discussed earlier involved the binomial and the Poisson distributions. In both of these distributions, we were dealing with n

---

[3] The concept of expected number of failures a as the product of failure rate λ and time t will be introduced in Chapter 10.

discrete events. For example, we determined the probability of finding 0, 1, 2, 3, etc., defectives in a lot of 25 items drawn from a production quantity that runs 10% defective.

In a continuous probability distribution, there are an infinite number of events possible within a defined range. For example, the amplification factor, beta, of a certain transistor may vary from 50 to 150. Within these two limits, beta may assume an infinite number of values when we consider that we do not need to restrict ourselves to integers only. The most important continuous probability distribution is the *normal distribution*, also called the *Gaussian distribution*.

2. Normal Probability Distribution

Many events in nature may be described by the normal distribution: for example, the heights of military recruits, the IQ of students in a college, and so forth. The normal curve has the well-known bell shape and is symmetrical about the center (see Figure 2-6). The area under the normal curve is 1.0; this means that it encompasses 100% of the events described by the distribution. Theoretically, the curve extends to infinity in both directions—it approaches the horizontal axis but never really reaches it.

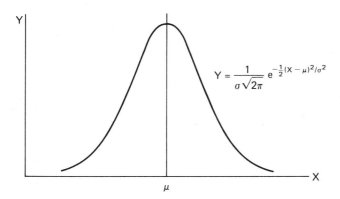

$$Y = \frac{1}{\sigma\sqrt{2\pi}}\, e^{-\frac{1}{2}(X-\mu)^2/\sigma^2}$$

Figure 2-6 *The normal curve.*

The midpoint μ (mu), the mean of the population, occurs under the highest point on the curve. Since the total area under the curve represents all possible events, portions of the area under the curve represent the relative frequency of events falling within that portion.

In determining significant portions of the area under the normal curve, the standard deviation is used. The standard deviation is designated by the letter σ (sigma) and is a measure of the deviation of the data from the mean.

The mean μ and the standard deviation σ are population parameters. The corresponding measures for a sample quantity drawn from the population are the statistics $\bar{X}$ and s.

Although it would be most desirable to be able to determine the true values for the mean and the standard deviation of a population, we are in most cases forced to estimate them on the basis of the mean and standard deviation of a sample drawn from the population. Thus, $\bar{X}$ and s are the best estimates of the true values of μ and σ of the population. The larger the size of the sample drawn, the closer will $\bar{X}$ and s approach the true parameters of the population.

When sampling plans are used, the size of the sample drawn from the population determines the sampling error. The latter represents the differences between $\bar{X}$ and s of the sample compared to the μ and σ of the population. When the size of the sample is small, the chances that the sample statistics will deviate from the true population parameters are considerably greater than for large sample sizes.

Sample sizes of 30 or more reduce the sampling error to tolerable values. When the sample size approaches the population size, the sampling error approaches zero. The reason we must be content with small sample sizes is that checking every unit in a population, or even checking large sample sizes, just is not practical from the standpoint of cost and time involved. Furthermore, some tests are destructive in nature, precluding large samples.

For sample sizes of 30 or more the normal distribution may be used with sufficient accuracy. When sample sizes are much lower than 30, the use of the normal curve is not advisable, since considerable bias would be introduced. A more suitable distribution for these instances is the *Student t distribution*. This distribution is similar to the normal distribution, but the areas under the curve depend on a factor referred to as the "number of degrees of freedom." The Student t distribution compensates for the sampling error when small sample sizes are used. Detailed application of this distribution is discussed in Section 3.3.

Specific percentages of the total area under the normal curve may be identified by multiples of the standard deviation. For example, the area under the normal curve bounded by $\mu \pm 1\sigma$ encompasses 68.26% of the items depicted by the distribution. Similarly, $\mu \pm 2\sigma$ encompasses 95.45% and $\mu \pm 3\sigma$ encompasses 99.73% of the items. Table 2-8 lists these σ values plus an additional two common values.

Table 2-8 *Tabulation of Popular Standard Deviation Values Versus Area Under the Normal Curve.*

σ Limits	Percent of Area Under Normal Curve
$\mu \pm 1\sigma$	68.26
$\mu \pm 1.645\sigma$	90.00
$\mu \pm 1.96\sigma$	95.00
$\mu \pm 2\sigma$	95.45
$\mu \pm 3\sigma$	99.73

Figure 2-7 locates the positive and negative standard deviation values $\mu \pm 1\sigma$, $\mu \pm 2\sigma$, and $\mu \pm 3\sigma$ on the horizontal axis of the normal curve. Figure 2-8 illustrates the specific areas under the curve for $\mu \pm 1\sigma$, $\mu \pm 2\sigma$, and $\mu \pm 3\sigma$.

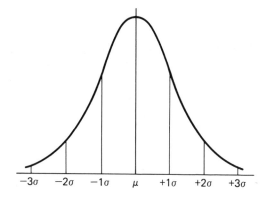

Figure 2-7 *Location of standard deviation values on the normal curve.*

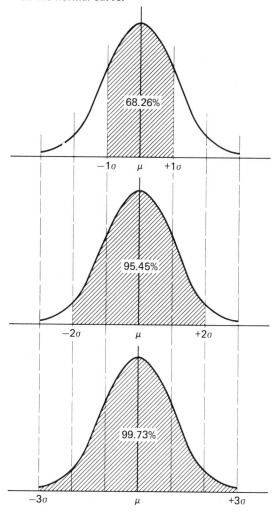

Figure 2-8 *Areas under the normal curve related to standard deviations.*

The application of the standard deviation σ and the corresponding areas under the curve will be illustrated in Example 2-27. The examples that follow deal with $\bar{X}$ and s rather than with μ and σ because samples drawn from the populations are used. Example 2-30 is an exception since the population parameters are known.

EXAMPLE 2-27 The IQ of 40 students from a large group was measured and a mean of $\bar{X} = 120$ was determined. The standard deviation was found to be $s = 14$. What percentage of the total group of students may be expected to have IQ's falling within the following limits?

(a) 134–106.
(b) 148–92.
(c) 162–78.

Solution We know that $\pm 1s$ encompasses 68.26%, $\pm 2s$ encompasses 95.45%, and $\pm 3s$ encompasses 99.73% of the population. Let us now determine how many multiples of s on either side of the $\bar{X} = 120$ are represented by the IQ ranges given above. Since s is given as 14, we will add the number 14 successively to $\bar{X}$ and subtract 14 successively from $\bar{X}$ and inspect the result:

$$120 + 1(14) = 134 \quad 120 - 1(14) = 106$$
$$120 + 2(14) = 148 \quad 120 - 2(14) = 92$$
$$120 + 3(14) = 162 \quad 120 - 3(14) = 78$$

Then

IQ Range

$$\bar{X} \pm 1s = 120 \pm 1(14) = 134\text{–}106$$
$$\bar{X} \pm 2s = 120 \pm 2(14) = 148\text{–}92$$
$$\bar{X} \pm 3s = 120 \pm 3(14) = 162\text{–}78$$

IQ Range	Standard Deviation Multiples	Percent of Area under Normal Curve (= % of total students)
134–106	$\bar{X} \pm 1s$	68.26
148–92	$\bar{X} \pm 2s$	95.45
162–78	$\bar{X} \pm 3s$	99.73

Hence, 68.26% of the students have an IQ between 134 and 106, 95.45% of the students have an IQ between 148 and 92, and 99.73% of the students, or nearly the total group, have an IQ between 162 and 78.

In Example 2-28, we will determine, step by step, the mean and standard deviation from grouped data.

EXAMPLE 2-28 Determine the mean of the distribution and the standard deviation for the following listing of the length of service of 40 office workers:

Years of Service	Number of Office Workers (Frequency)
15	1
14	5
13	7
12	13
11	11
10	1
9	1
8	1

Solution The mean and the standard deviation s is determined in the following manner: we prepare a tabulation as shown in Table 2-9. The first two columns are merely a tabulation of the data given. Column 3 requires that a 0 point (assumed origin) be assigned; this 0 point usually corresponds to the highest frequency in column 2 ($= 13$). The values entered in column 3 above and below the 0 point represent the respective numerical differences in the first column from the value associated with the 0-point (12 in this case).

For example, the value in column 3 directly above 0 is $+1$, since the corresponding difference in column 1 is $13 - 12 = +1$. The bottom value in column 3 is -4, since the corresponding difference in column 1 is $8 - 12 = -4$. Column 4

Table 2-9 *Calculation of Mean and Standard Deviation.*

Years	Frequency (f)	Deviation from 0 point (d)	d^2	fd	fd^2
15	1	$+3$	9	3	9
14	5	$+2$	4	10	20
13	7	$+1$	1	7	7
12	13	0	0	0	0
11	11	-1	1	-11	11
10	1	-2	4	-2	4
9	1	-3	9	-3	9
8	1	-4	16	-4	16
Total $N = 40$				$fd_{tot} = 0$	$fd^2_{tot} = 76$

$\bar{X} =$ assumed origin $+ i\left(\dfrac{fd_{tot}}{N}\right) = 12 + 1\left(\dfrac{0}{40}\right) = 12$ where i represents the cell interval (in this case $i = 1.0$, since the time interval in column 1 is 1 year)

$$s = i\sqrt{\frac{fd^2_{tot}}{N} - \left(\frac{fd_{tot}}{N}\right)^2}$$

$$= (1)\sqrt{\frac{76}{40} - 0} = (1)\sqrt{1.9} = 1.38$$

simply contains the squares of the values in column 3. For example, the top value in column 3 is $+3$; the square of $+3$ is 9. Similarly, the bottom value in column 3 is -4 is $+16$. Column 5 is the product of columns 2 and 3. Column 6 is the product of columns 2 and 4.

3. Z Scale

We have identified several specific σ values on the horizontal axis of the normal distribution as listed in Table 2-8. It is convenient to be able to read the horizontal scale in terms of standard deviation multiples. This is accomplished by converting to the Z scale. The conversion from the X scale to the Z scale is performed by using the following formula:

$$Z = \frac{X - \mu}{\sigma} \quad \text{④ footnote}$$ (24)

EXAMPLE 2-29   Determine the Z value for the IQ range used in Example 2-27.

Solution

$$\bar{X} \text{ is } 120$$
$$s = 14$$

(a) $X_1 = 134$ $Z_1 = \dfrac{134 - 120}{14} = \dfrac{14}{14} = 1$

       $X_2 = 106$ $Z_2 = \dfrac{106 - 120}{14} = \dfrac{-14}{14} = -1$

(b) $X_3 = 148$ $Z_3 = \dfrac{148 - 120}{14} = \dfrac{28}{14} = 2$

       $X_4 = 92$ $Z_4 = \dfrac{92 - 120}{14} = \dfrac{-28}{14} = -2$

(c) $X_5 = 162$ $Z_5 = \dfrac{162 - 120}{14} = \dfrac{42}{14} = 3$

       $X_6 = 78$ $Z_6 = \dfrac{78 - 120}{14} = \dfrac{-42}{14} = -3$

The results obtained above (positive and negative) use one standard deviation as the unit of measure. Frequently, calculated Z values are not whole-number multiples of s but decimal numbers.

Table A12, Appendix A, is a tabulation of Z values with the corresponding area under the normal curve. For simplicity the table only contains the Z values corresponding to half the normal curve (50% of the area). The use of the table of areas under the normal curve versus Z values is demonstrated in the following example.

EXAMPLE 2-30   A resistor company mass-produces a carbon film resistor. The population parameters are $\mu = 20,000\ \Omega$ and $\sigma = 100\ \Omega$. The acceptable resistance

----

[4] In the case of sample lots drawn from the population, we use

$$Z = \frac{X - \bar{X}}{s}$$

range is from 19,750 to 20,125 Ω. On the average, what is the likely number of rejected resistors for each 10,000 resistors produced?

Solution Population parameters μ and σ are used.

$$\mu \doteq 20,000 \ \Omega$$

$$X_1 = 19,750 \ \Omega$$

$$X_2 = 20,125 \ \Omega$$

$$\sigma = 100 \ \Omega$$

$$Z_1 = \frac{X_1 - \mu}{\sigma} = \frac{19,750 - 20,000}{100} = -2.5$$

$$Z_2 = \frac{X_2 - \mu}{\sigma} = \frac{20,125 - 19,750}{100} = +1.25$$

Look in Table A-12 and find the percent area contained between the mean μ and a Z value of -2.5, and between the mean and a Z value of $+1.25$ (for $Z = -2.5$ a value 0.4938 is found in the column adjacent to this Z value. In the case of $Z = -2.5$, the minus sign is ignored. A value of 0.3944 is read off at the intersection of the $Z = 1.2$ line and the 0.05 column for $Z = +1.25$.

$$\text{area between } \mu \text{ and } -2.5Z = 0.4938$$
$$\text{area between } \mu \text{ and } +1.25Z = 0.3944$$
$$\overline{\text{Total Area} \quad 0.8882}$$

Hence, 88.82% of the 10,000 transistors (or 8882 resistors) may be expected to fall within the acceptable resistance limits of 19,750 to 20,125 Ω. Therefore, since $10,000 - 8882 = 1118$, we may on the average expect 1118 rejected resistors.

The method of calculating the standard deviation as used in Example 2-28 is especially useful in instances where the raw data can be grouped. There were 13 workers with 12 years of service, 7 workers with 13 years of service, and so on.

When grouping is not feasible, the following method is used to determine the mean and the standard deviation:

$$\text{mean } \bar{X} = \frac{\Sigma X}{n} = \frac{\text{sum of the values of all items}}{\text{number of items}} \tag{25}$$

The standard deviation s is determined by tabulating the following data and performing the indicated computations:

Column 1: list all given X values of the distribution.
Column 2: calculate the value $X - \bar{X}$ for each value in column 1.
Column 3: square each $(X - \bar{X})$ value in column 2.
Then: sum all $(X - \bar{X})^2$ values, divide by $(n - 1)$, and take the square root.[5]

$$s = \sqrt{\frac{\Sigma (X - \bar{X})^2}{n - 1}} \tag{26}$$

[5] By using $n - 1$ rather than n in the denominator, the bias introduced by sampling without replacement from a finite population is removed.

The method above is applied in Example 2-31. In this example we shall attempt to determine whether a certain statistical acceptance sampling plan has a good chance of rejecting a product that is outside the specification limits.

Statistical sampling, in lieu of 100% testing, is a well-established technique and is based on the willingness to assume a defined risk of accepting a small percentage of a nonconforming product in exchange for reduced inspection cost and time.

> **EXAMPLE 2-31**   The Ajax Amplifier Company purchases transformers for a medium-power amplifier from S&H Transformer Company. Let us assume that the amplifier requires transformers whose excitation current must fall between 150 and 225 mA. Ajax uses a statistical sampling scheme which assures that a lot which is 3% or more defective will in 9 out of 10 cases be rejected.
>
> Before shipping a lot of 280 transformers the inspection department at S&H selects 32 transformers at random from production and records the following excitation currents in milliamperes.

200	160	210	175	195
155	184	202	180	200
170	172	164	178	180
180	178	176	165	175
205	154	157	190	190
190	198	211	205	192
222	212			

Based on these readings is the Ajax sampling scheme likely to accept or reject the lot of 280 transformers?

Solution

X	$X - \bar{X}$	$(X - \bar{X})^2$
200	+15	225
155	−30	900
170	−15	225
180	− 5	25
205	+20	400
190	+ 5	25
222	+37	1369
160	−25	625
184	− 1	1
172	−13	169
178	− 7	49
154	−31	961
198	+14	196
212	+27	729
210	+25	625
202	+17	289
164	−21	441
176	− 9	81
157	−28	784
211	+26	676

X	$X - \bar{X}$	$(X - \bar{X})^2$
175	-10	100
180	$- 5$	25
178	$- 7$	49
165	-20	400
190	$+ 5$	25
205	$+20$	400
195	$+10$	100
200	$+15$	225
180	$- 5$	25
175	-10	100
190	$+ 5$	25
192	$+ 7$	49
$\sum X = 592$		$\sum (X - \bar{X})^2 = 10{,}318$

$$\bar{X} = \frac{592}{32} = 185 \text{ mA}$$

$$s = \sqrt{\frac{(X - \bar{X})^2_{\text{tot}}}{n - 1}} = \sqrt{\frac{10{,}318}{32 - 1}} = \sqrt{332.8} = 18.25$$

$$Z_1 = \frac{X_1 - \bar{X}}{s} \qquad \text{where } X_1 = 225 \text{ mA (upper tolerance limit)}$$

$$Z_2 = \frac{X_2 - \bar{X}}{s} \qquad \text{where } X_2 = 150 \text{ mA (lower tolerance limit)}$$

$$Z_1 = \frac{225 - 185}{18.25} = 2.19$$

$$Z_2 = \frac{150 - 185}{18.25} = -1.92$$

Look in Table A-12 and find the percent area contained between the mean $\bar{X}$ and Z values of 2.19 and -1.92, respectively:

$$
\begin{aligned}
\text{area between } \bar{X} \text{ and } Z_1 &= 2.19 = 0.4857 \\
\text{area between } \bar{X} \text{ and } Z_2 &= -1.92 = \underline{0.4726} \\
\text{Total area} &= 0.9583
\end{aligned}
$$

Therefore, it may on the average be expected that 95.83% of the lot (or nearly 268 transformers) will fall within the limits of 150–225 mA. Twelve transformers among the lot of 280 are likely to be outside these limits. Expressed in percent, the lot of 280 transformers may, on the average, be expected to be

$$\frac{12}{280} \times 100 = 4.3\% \text{ defective}$$

Therefore, since the Ajax sampling plan can detect lots which are at least 3% defective in 9 of 10 cases, a 4.3% defective lot has a very good chance of being rejected.

It is interesting to note that the 32 transformers tested by the manufacturer were all within the limits of 150 to 225 yet the statistical analysis showed that they came from a lot which is very likely 4.3% defective. There is, of course, no certainty that Ajax's sampling scheme will detect that this lot is more than 3% defective, but the probability that this lot will be rejected, is at least 90%.

PROBLEMS

2-1 A shipment of 350 capacitors is tested on a 100% basis with the result that 28 are rejected. What is the percentage of acceptable units?

2-2 Twenty-one of every 200 television sets require some repair during their first year of operation. What are the chances that a set will play for this length of time without requiring repair?

2-3 Define mutually exclusive events and give the formula.

2-4 A stock bin holds a total of 60 chokes which bear no individual identification of their measured inductance value (inductance is measured in henries, H). The lot is known to contain 15 chokes measuring 50 H and 25 chokes measuring 40 H. If a single choke is issued from this bin, what is the probability that its inductance value is either 40 or 50 H?

2-5 A card is drawn from a regular pack of 52 playing cards. What is the probability that it is either a king of diamonds or a spade?

2-6 Suppose that an engineer is being interviewed for a job with an electronic company. The probability that his assignment will be in design is 0.55, the probability that he will work in research is 0.3, and the probability that he will be engaged in both design and research is 0.10. What is the probability that he will work in design or research, or in both?

2-7 Three cards are randomly drawn in succession from a regular pack of 52 playing cards. A drawn card is immediately replaced before the next is drawn. What is the probability that the first card drawn is a club, the second not a queen, and the third a spade?

2-8 The electronic parts on a circuit board operate independently of one another (the failure of one part does not affect the operation of other parts on the board). If the probability of survival of a choke is 0.75 and of a resistor 0.9, what is the probability of survival of two resistors and one choke connected in series?

2-9 A lot of 75 transistors contains 15 with low gain. What is the probability that the first transistor drawn has low gain and a second transistor drawn from the remaining 74 transistors also has a low gain?

2-10 Forty percent of all coils produced by a shop have the proper inductance. Fifty percent of the coils with proper inductance have the correct resistance. Fifty five percent of all coils have proper resistance. What is the probability of randomly selecting a coil with improper inductance and incorrect resistance?

2-11 A shipment of 250 refrigerators contains 4% defective units. In filling an order for 3 refrigerators, what are the chances that two of the three units chosen randomly are defective?

2-12 Two true coins are flipped.
(a) What is the probability of flipping one head and one tail?
(b) What is the probability that neither coin flipped turns up head?

2-13 Three true coins are flipped.
(a) What is the probability of flipping two heads and one tail?
(b) What is the probability of flipping no tails?

2-14 A lot of miniature transformers is 6% defective. A quality control directive for incoming inspection requires that the lot be accepted provided a five-unit sample, drawn randomly from this lot, contains no more than one defective unit. What is the probability that the lot will be accepted? Use the Poisson distribution as an approximation to the binomial to solve this problem.

2-15 Refer to Problem 2-14 and determine the probability that the sample drawn contains exactly three defective transformers if the lot is 20% defective.
(a) By using the binomial formula.
(b) By using Table A9(a), Appendix A.

2-16 Forty percent of a large quantity of fuses in a box have bad blow characteristics. The remainder are good. Three fuses are drawn randomly from this box.
(a) What is the probability that the 3 fuses are bad?
(b) What is the probability that the 3 fuses are good?
(c) What is the probability that 2 fuses are good and 1 is bad?
(d) What is the probability that at least 1 fuse is good?
(e) What is the probability that at least 2 fuses are good?
Use Tables A9(a) and A9(b), Appendix A.

2-17 Thirty percent of the light bulbs in a large shipment are marginal in performance. What is the probability that of a sample of six light bulbs, drawn randomly, at least two will be marginal? Use Table A9(b), Appendix A.

2-18 On the average, 10 of every 100 roller bearings produced by a production line are defective. Using the Poisson expression, determine the probability that a sample of eight bearings drawn randomly from production will contain the following number of defectives:
(a) Five defectives.
(b) Three defectives.
(c) One defective.
(d) No defectives.

2-19 An electronic system is to be operated for 400 h. The total failure rate for the equipment is 0.002 failure/hour. Using Table A10 in Appendix A, determine the probability of survival of the equipment. (*Note:* Probability of survival equals probability of zero occurrence of a failure.)

2-20 If the probability of failure of an equipment is not to be less than 80% and the operating time is 750 h, what is the maximum tolerable total equipment failure rate expressed in failures per 10^6 h?

2-21 What are the mean and the standard deviation for the following 48 load current readings obtained on an electric distribution panel over a period of 2 weeks?

Current (A)	*Number of Readings*
90	2
85	7
80	10
70	12
60	9
50	6
40	2

2-22 An incoming inspection department receives a shipment of 250 vacuum tubes. The criterion for acceptance is the value of plate current for a given plate voltage, bias, and grid excitation. This value is to range between 75 and 112 mA. Twenty-seven tubes are drawn randomly from the lot of 250 and tested for plate current, with the following results:

100	111	98	88	81
78	80	106	77	75
85	92	105	105	80
90	86	101	70	79
103	89	82	73	90
95	77			

What percentage of the total lot is likely to fall within the range of 75 and 112 mA? Assume that the normal distribution applies.

2-23 Assume that the activation pressure of a quantity of valves follows a normal distribution. The mean pressure value of the population was found to be 26 psi and the standard deviation 10 psi. What percentage of the valves produced would, on the average, fall within 12 and 55 psi?

3

STATISTICAL INFERENCE

3-1 BASIC CONCEPTS

Statistical inference is a process by which observations made on a random sample, selected from a population, are used for an estimation of the range within which the true population parameters lie.

Assume there are 10,000 recruits in an army training camp. The army wants to determine the average height of these recruits for the purpose of ordering fatigue uniforms. One method that could be employed would be to actually measure the height of each recruit, add these heights, and divide by the number of recruits. This method is tedious and much too time-consuming.

Another approach might be to pick a percentage of recruits from the whole population—say, a random sample of 100 recruits—and determine their mean height. Is it safe to assume that the mean height obtained from this sample represents the mean height of the 10,000 recruits?

What if we were to pick additional random samples of 100 recruits each and determined their mean height; would their mean be the same as in the first sample of 100 recruits? We would probably find this not to be the case. It can be shown that the means of the samples drawn from the population of recruits will cluster around the true mean of the population. The distribution of these sample means, for sample sizes of $n \geq 30$, will very closely represent a normal curve (see Figure 3-1). Every normal distribution is defined by its mean and by its standard deviation. This applies equally to the distribution of sample means depicted in Figure 3-1. It has a mean $\mu_{\bar{x}}$ and a standard deviation $\sigma_{\bar{x}}$. The mean of the distribution of sample means, $\mu_{\bar{x}}$ (assuming

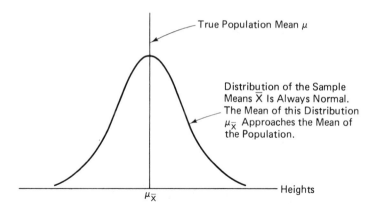

Figure 3-1 *Sampling distribution.*

a sufficiently large number of samples), approaches the mean of the population μ from which the samples were drawn.

The standard deviation of the distribution of sample means is referred to as the *standard error of the mean* and is expressed by equation (27):

$$\sigma_{\bar{x}} = \frac{\sigma}{\sqrt{n}} \qquad (27)$$

where

$\sigma_{\bar{x}}$ = standard deviation of the distribution of sample means (the standard error of the means)
σ = standard deviation of the population
n = sample size

Equation (27) theoretically assumes an infinite population or sampling with replacement.

EXAMPLE 3-1 The standard deviation of the resistance value of a population of 100,000 resistors is 0.56 Ω ($\sigma = 0.56$). The mean resistance value of the population is $\mu = 100$ Ω. If we draw a number of samples of 36 resistors each at random from this population, calculate the mean and the standard deviation of each sample, what could we expect to obtain for $\mu_{\bar{x}}$, the mean of this distribution, and for $\sigma_{\bar{x}}$, the standard deviation of this distribution (standard error of the means)?

Solution The mean of the distribution of sample means will equal, or will be very close to the mean of the population; therefore,

$$\mu_{\bar{x}} = \mu$$

or

$$\mu_{\bar{x}} = \underline{100 \ \Omega}$$

The standard deviation of the distribution of sample means (standard error of the means) equals

$$\sigma_{\bar{x}} = \frac{\sigma}{\sqrt{n}}$$

or

$$\sigma_{\bar{x}} = \frac{0.56}{\sqrt{36}} = \underline{0.093\ \Omega}$$

The nature of equation (27) suggests that as the sample size n increases, the standard error of the means decreases, and the sample means will cluster more closely around the true population mean. (see Figure 3-2).

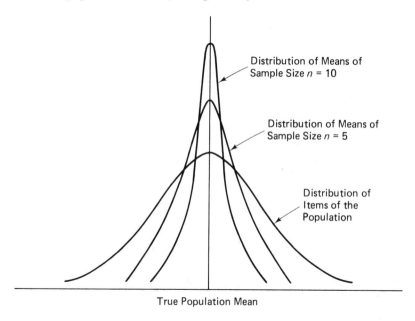

Distribution of Means of Sample Size n = 10

Distribution of Means of Sample Size n = 5

Distribution of Items of the Population

True Population Mean

Figure 3-2 *Illustration of distribution of population items and sample means of different sample sizes.*

3-2 ESTIMATE OF CONFIDENCE INTERVALS FOR POPULATION MEANS (LARGE SAMPLE IS ASSUMED, $n \geq 30$)

Knowledge of the value of the standard error $\sigma_{\bar{x}}$ of a sampling distribution furnishes information regarding the dispersion of random sample means around the mean of the population μ. In accordance with the normal-curve table, 68.27% of the sample means will fall within 1 standard deviation (standard error in this case), 95.45% will fall within 2 standard errors, and 99.73% will fall within 3 standard errors of the population mean. Expressed differently, the probability that the mean of any randomly chosen sample will not differ from the true mean of the population by more than 1, 2, or 3 standard errors is 68.26%, 95.45%, and 99.73%, respectively.

The relationship above may also be expressed in terms of a confidence interval. A 99.73% confidence interval refers to a range of ± 3 standard errors around the sample mean; 95.45% and 68.26% confidence intervals refer to ± 2 and ± 1 standard errors around the sample means, respectively. Therefore, the $\bar{X} \pm 3\sigma_{\bar{X}}$ confidence interval will contain the true population mean 99.73% of the time.

In general terms we may estimate a confidence interval for the population mean μ from the theoretical relation

$$\mu_{\bar{X}} \pm k\sigma_{\bar{X}} \qquad (28)$$

where

$\mu_{\bar{X}}$ = mean of the distribution of sample means
k = number of standard errors from $\mu_{\bar{X}}$ (for a 90% confidence
        interval, $k = 1.645$; for 95%, k is 1.96; and for 99%, k is 2.575)
$\sigma_{\bar{X}}$ = standard error of the means per equation (27)

In practice, a single random sample with a mean $\bar{X}$ and a standard deviation σ, drawn from a population suffices to form an estimate about the true mean of the population, provided the sample is at least $n \geq 30$ to assure a negligible sampling error. Thus, the mean of the population, μ, is estimated on the basis of the sample mean and the standard deviation of the population σ is estimated on the basis of the standard deviation of the sample. When using a sample to draw an inference about the population the σ in equations (27) and (28) is therefore replaced by an estimate of σ, namely, the sample statistic s, and $\mu_{\bar{X}}$ in equation (28) is replaced by $\bar{X}$. Example 3-2 illustrates the use of a single sample in establishing desired confidence limits about the population mean.

EXAMPLE 3-2   A test is to be administered to high school students nationwide. In order to assess the length of time to allow for this test, a random choice of 100 students was made and the test administered. The mean time for the test (in minutes, min) and the standard deviation were found to be

$$\bar{X} = 58 \text{ min}$$

$$s = 10 \text{ min}$$

Find the 95% confidence interval for the true mean test time.

Solution   A 95% confidence interval around a sample mean refers to a range of ± 1.96 standard deviations. Applying equations (27) and (28) to a single sample and substituting $\bar{X}$ and s for $\mu_{\bar{X}}$ and σ, gives

$$\bar{X} \pm 1.96 \cdot \frac{s}{\sqrt{n}}$$

$$= 58 \pm 1.96 \cdot \frac{10}{\sqrt{100}}$$

$$= 58 \pm 1.96 \text{ min}$$

Therefore, the interval 56.04 to 59.96 minutes will contain the true mean test time 95% of the time.

In the preceding discussions the assumption was made that we were dealing with populations of infinite size, or that items drawn from the population for a sample are replaced as drawn. The latter will assure that the population remains constant in size.

In practice, the population is seldom of infinite size, and sampling with replacement is not always done. For these cases, equation (27) for the standard error is modified as follows:

$$\sigma_{\bar{x}} = \frac{\sigma}{\sqrt{n}} \cdot \sqrt{\frac{N-n}{N-1}} \tag{29}$$

where N refers to the size of the population and the other symbols have the same meaning as before. The standard deviation of the sample, s, is substituted for σ when dealing with a sample. The correction factor $\sqrt{(N-n)/(N-1)}$ in equation (29) reduces $\sigma/\sqrt{n}$ in value, depending on the ratio of population size to sample size. If the size of the sample is $n = 30$ and the size of the population is $N = 1000$, the correction factor will be

$$\sqrt{\frac{N-n}{N-1}} = \sqrt{\frac{1000-30}{1000-1}}$$
$$= \sqrt{\frac{970}{999}}$$
$$= 0.98538$$

If the size of the population is 10,000, the correction factor is

$$\sqrt{\frac{10,000-30}{9999}} = 0.99855$$

As can be seen, when the population size N is large compared to the sample size, the value of the correction factor approaches 1.0, and therefore the reduction of the value of the standard error is negligible. In most instances the correction factor can be neglected.

EXAMPLE 3-3   A random sample of 225 steel rods was drawn (without replacement) from a production line that produced a total quantity of 1000 of these rods. The mean diameter of the sample was calculated to be 0.325 inch (in) and the standard deviation came to 0.06 in. Find the 95% and 99% confidence intervals for the true mean diameter of all the rods produced.

Solution   Since the sample is drawn without replacement, we use equation (29). (Refer to page 52 for the k-factor values.)

   (a) The 95% confidence limits are

$$\frac{0.06}{\sqrt{225}} \cdot \sqrt{\frac{1000-225}{1000-1}}$$
$$= 0.004 \cdot \sqrt{0.775}$$
$$= 0.00352 \text{ in}$$

The factor k for 95% confidence limits is 1.96; therefore, the 95% confidence limits are

$$0.325 \pm 1.96(0.00352)$$

or

$$0.325 \pm 0.0069 \text{ in}$$

(b) The 99% confidence limits are

$$k \text{ for } 99\% \text{ confidence limits} = 2.575$$

$$0.325 \pm 2.575 \,(0.00352)$$

or

$$0.325 \pm 0.0091 \text{ in}$$

3-3 ESTIMATE OF CONFIDENCE INTERVALS FOR POPULATION MEANS (SMALL SAMPLE IS ASSUMED, $n < 30$)

In Section 3.1 a point was made that when the sample sizes are 30 or larger, the distribution of the means of the samples can be very closely approximated by a normal curve. When the sample size decreases below $n = 30$, this approximation becomes less valid and we must turn to the Student t distribution to determine confidence limits.

The t distribution is similar in appearance to the normal distribution except it becomes flatter and spreads more for progressively smaller values of sample size.

Equation (30) is used to express confidence limits using the Student t distribution:

$$\bar{X} \pm t_c \cdot \frac{s}{\sqrt{n}} \tag{30}$$

where t_c is the confidence coefficient and the other symbols are as defined before. Values of t_c may be determined from Student t tables for the appropriate number of *degrees of freedom* and for the *level of significance* desired. The number of degrees of freedom equals the sample size n minus 1. The level of significance may be chosen as 0.01, 0.05, 0.10, etc., corresponding to 99%, 95%, 90%, etc., confidence intervals, respectively.

Figure 3-3 shows a Student t distribution for 95% confidence limits. The shaded areas together represent a value of $2 \times 0.025 = 0.05$. The remaining unshaded area is $1.00 - 0.05 = 0.95$. Applied to an actual example, and assuming 95% confidence limits, we look up in the Student t table and find that value of t_{c_1} for which the shaded area to the right of t_{c_1} is 0.025. It is necessary to enter the Student t table with the proper number of degrees of freedom. We select the applicable column: $t_{.025}$. The reason we look up t_{c_1} corresponding to a value of 0.025, rather than 0.05, is that the Student t table represents the shaded area in one tail of the distribution only. Therefore, when a confidence interval is sought, we use half the total shaded area.

Table 3-1 represents a short version of a Student t table and is useful for sample sizes of 5, 15, 20, 25, and 30.

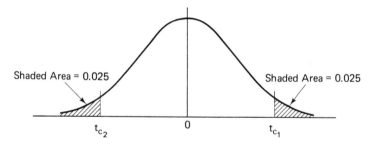

Figure 3-3 *Student* t *distribution for 95% confidence limits.*

Table 3-1 *Short Version of Student* t *Table.*

Degrees of Freedom	$t_{.005}$	$t_{.025}$	$t_{.05}$	$t_{.20}$
4	4.60	2.78	2.13	0.94
9	3.25	2.26	1.83	0.88
14	2.98	2.14	1.76	0.87
19	2.86	2.09	1.73	0.86
24	2.80	2.06	1.71	0.86
29	2.76	2.04	1.70	0.85

EXAMPLE 3-4 A sample of 10 items is drawn from a population and the mean weight and standard deviation determined.

$$n = 10, \bar{X} = 50 \text{ g}, s = 2.5 \text{ g}$$

Determine the 90% confidence interval for the true mean weight.

Solution

$$\bar{X} \pm t_{c_1} \cdot \frac{s}{\sqrt{n}}$$

$$50 \pm t_{c_1} \cdot \frac{2.5}{\sqrt{10}}$$

t_{c_1} is found in the Student t table (see Table 3-1) under 9 degrees of freedom and the vertical column headed $t_{.05}$

$$t_{c_1} \text{ is therefore } 1.83$$

$$50 \pm 1.83 \cdot \frac{2.5}{\sqrt{10}}$$

$$50 \pm 1.44 \text{ g}$$

Therefore, the interval 48.56 to 51.44 grams will contain the true mean weight 90% of the time.

Let us examine what changes are required in this type of calculation when we are not dealing with a confidence interval but with a one-sided confidence level. Figure 3-4 depicts such a situation for a 95% confidence level. In this case we enter the Student t table with the column headed $t_{.05}$ Applied to our previous example, we have

$$\bar{X} + 1.83 \cdot \frac{2.5}{\sqrt{10}}$$

$$50 + 1.44 = \underline{51.44 \text{ g}}$$

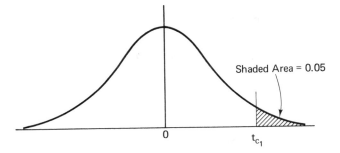

Shaded Area = 0.05

0 t_{c_1}

Figure 3-4 *Illustration of a one-sided confidence level of 95%.*

Hence, 51.44 grams constitutes the 95% one-sided upper confidence limit for the true mean weight.

3-4 CHI-SQUARE DISTRIBUTION

The *chi-square distribution* corresponding to various values of degrees of freedom dF is shown in Figure 3-5. The lower the value of dF is, the more asymmetrical the shape of the curve becomes.

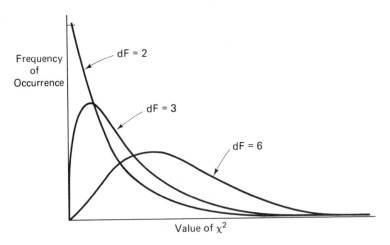

Frequency of Occurrence

dF = 2

dF = 3

dF = 6

Value of χ^2

Figure 3-5 *Chi-square distribution χ^2.*

In a manner similar to the t distribution, we can define confidence intervals (99%, 95%, etc.,) for χ^2 by using a chi-square table (see Table A13, Appendix A). For example, assume that we are interested in a 90% confidence interval; then we determine the critical values of χ^2, namely $\chi^2_{0.050}$ and $\chi^2_{0.950}$ for which 5% of the area under the curve lies in each tail. Let us further assume that the number of degrees of freedom is $dF = 4$; then the graphical interpretation is as shown in Figure 3-6.

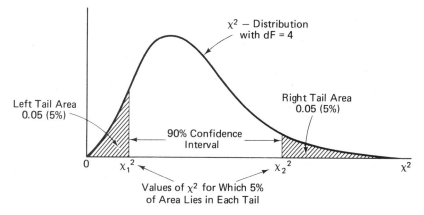

Figure 3-6 *Illustration of Chi-square distribution with dF = 4 and a 90% confidence interval.*

We will be specifically interested in the use of the χ^2 distribution in connection with establishing confidence intervals around estimates of failure rate or mean time between failures (MTBF). Referring to Figure 3-6, the true failure rate or MTBF would actually fall within the shown interval 90% of the time. (The failure rate of a piece of equipment equals the number of failures divided by the total operating hours. MTBF is the reciprocal of failure rate. Failure rate and MTBF are discussed in detail in Chapter 9).

The typical situation of interest may take on two variations:

1. A test is conducted on a group of parts (or equipment) and is terminated after a predetermined number of hours. At the end of this test time, we may, for example, obtain the following results:

$$\text{test time: } T = 1000 \text{ h}$$

$$\text{number of failures: } r = 3$$

2. Same as in (a), except that instead of terminating the test after a predetermined number of hours, the test is terminated after the rth failure has occurred.

In either of these cases, a confidence interval may be established which contains the true MTBF, or failure rate. The following formulas may be used to establish the confidence limits.

(a) Test Terminated at Predetermined Time

$$\text{lower confidence limit} = \frac{2T}{\chi^2_{2r+2;\,\frac{1-\phi}{2}}} \tag{31}$$

$$\text{upper confidence limit} = \frac{2T}{\chi^2_{2r;\,\frac{1+\phi}{2}}} \tag{32}$$

where

T = total number of hours of testing
r = number of failures
χ^2 = chi-square value found from the chi-square table for
    $2r + 2$ or $2r$ degrees of freedom.
ϕ = level of confidence selected (*Note*: ϕ equals
    1-total tail area; *see* Figure 3-6).

EXAMPLE 3-5 Twenty-six motors are tested for 200 hours. Three motors fail during this test. The failures occurred after the following test times:

Motor 1	50 h
Motor 2	61 h
Motor 3	146 h

(a) Determine the best estimate of the MTBF.
(b) What are the 90% confidence limits for the true MTBF?

Solution

	Operating Hours
23 motors at 200 h each	4600
1 motor at 50 h	50
1 motor at 61 h	61
1 motor at 146 h	146
Total	4857 h

$$\text{MTBF} = \frac{\text{total number of operating hours}}{\text{number of failures}}$$

$$= \frac{4857}{3} = 1619 \text{ h} \quad \text{(best estimate)}$$

Now let us use the chi-square distribution to determine the 90% confidence limits:

(a) Lower confidence limit $= \dfrac{2 \times 4857}{\chi^2_{0.05}}$ for 8 degrees of freedom where

$$\frac{1 - 0.90}{2} = 0.05$$

$$2r + 2 = 2(3) + 2 = 8$$

Look up in the chi-square table the χ^2 value for 8 degrees of freedom under the column heading 0.05. The χ^2 value is found to be 15.507. Hence,

$$\text{lower confidence limit of MTBF} = \frac{2 \times 4857}{15.507}$$

$$= 626 \text{ h}$$

(b) Upper confidence limit $= \dfrac{2 \times 4857}{\chi^2_{0.950}}$ for 6 degrees of freedom

where

$$\frac{1 + 0.90}{2} = 0.95$$

$$2r = 2(3) = 6$$

From the chi-square table: $\chi^2_{0.950}$ for 6 degrees of freedom is 1.635. Hence,

$$\text{upper confidence limit} = \frac{2 \times 4857}{1.635}$$

$$= 5941 \text{ h}$$

Therefore the interval of 626 to 5941 h will contain the true MTBF 90% of the time.

Frequently, when conducting a test, we are interested only in a lower confidence level. For example, we would be satisfied to know with 95% confidence that the estimated MTBF is above a certain level. We do not care how high it is—only that it does not fall below a minimum level. Equation (31) may be used for the calculation of the lower confidence level except that we use $1 - \phi$ instead of $(1 - \phi)/2$.

EXAMPLE 3-6 Referring to Example 3-5, what is the lower one-sided 95% confidence level of the MTBF?

Solution   MTBF $= \dfrac{2 \times 4857}{\chi^2_{0.05}}$ for 8 degrees of freedom

where

$$1 - \phi = 1 - 0.95 = 0.05$$
$$2r + 2 = 2(3) + 2 = 8$$

Look up in chi-square table for 8 degrees of freedom and the 0.05 area in the tail of the distribution. The χ^2 value is found to be 15.507.

$$\text{MTBF} \geq \frac{2 \times 4857}{15.507} = 626 \text{ h}$$

Hence, based on the test conducted, the 95% one-sided lower confidence limit for the true MTBF is at least 626 h.

EXAMPLE 3-7   A computer operated for 5000 h without a failure. Determine the 95% one-sided lower confidence limit for the true MTBF.

Solution   Even though no failure occurred, a lower confidence limit can be calculated using equation (31) and $1 - \phi$ instead of $(1 - \phi)/2$.

$$\text{Lower confidence limit} = \frac{2T}{\chi^2_{2r+2;\, 1-\phi}}$$

Since $r = 0$,

$$\text{lower confidence limit} = \frac{2T}{\chi^2_{2;\, 1-0.95}} = \frac{2T}{\chi^2_{2;\, 0.05}}$$

From the chi-square table for 2 degrees of freedom, under the column heading of 0.05, the χ^2 value is 5.991.

$$\text{Lower confidence limit} = \frac{2 \times 5000}{5.991} = \underline{1669 \text{ h}}$$

(b) Test Terminated When the *r*th Failure Has Occurred

So far we have dealt with situations in which a test is stopped at a predetermined number of hours, regardless of whether or not failures have occurred. Now we stop the test upon the occurrence of a specific failure. The formulas for lower and upper confidence limits of the MTBF for this case are as follows:

$$\text{lower confidence limit} = \frac{2T}{\chi^2_{2r;\, \frac{1-\phi}{2}}} \tag{33}$$

$$\text{upper confidence limit} = \frac{2T}{\chi^2_{2r;\, \frac{1+\phi}{2}}} \tag{34}$$

EXAMPLE 3-8   When the 10th failure occurred in an electronic equipment placed on a life test, the test was stopped. At this point the running time meter showed 7500 h of operation. Find the 90% confidence interval for the true failure rate for this system. What is the point estimate for the MTBF?

Solution   Since the test was stopped upon occurrence of a failure, formulas (33) and (34) apply:

(a)   Upper confidence limit $= \dfrac{2 \times 7500}{\chi^2_{20;\, 0.95}}$

where

$$\frac{1 + \phi}{2} = \frac{1 + 0.9}{2} = 0.95$$

$$2r = 2(10) = 20$$

From the chi-square table,

$$\chi^2_{20;\, 0.95} = 10.851$$

$$\text{upper confidence limit} = \frac{2 \times 7500}{10.851} = \underline{1382 \text{ h}}$$

(b) Lower confidence limit $= \dfrac{2 \times 7500}{\chi^2_{2r; \frac{1-\phi}{2}}}$

where

$$\frac{1 - \phi}{2} = \frac{1 - 0.9}{2} = 0.05$$

$$2r = 20$$

From the chi-square table,

$$\chi^2_{20; \, 0.05} = 31.41$$

$$\text{lower confidence limit} = \frac{2 \times 7500}{31.41} = 477 \text{ h}$$

Hence, 477 h and 1382 h constitute the time interval that contains the true MTBF of the equipment 90% of the time.

Without using the chi-square distribution for that computation, the best estimate we could have furnished of the MTBF would have been

$$\text{MTBF} = \frac{\text{total test time}}{\text{number of failures}}$$

$$= \frac{7500}{10} = 750 \text{ h}$$

This is a mere point estimate of MTBF and is not as useful as the statistically calculated confidence interval of MTBF. As may be expected, the value of 750 h falls within the range of the interval of 477 and 1382 h. Since the problem required failure rate (not MTBF) determination, we still have a small calculation left. Since failure rate $\lambda = 1/\text{MTBF}$, for MTBF $= 477$ h,

$$\lambda = \frac{1}{477} = 0.002097 \text{ failure/hour}$$

For MTBF $= 1382$ h,

$$\lambda = \frac{1}{1382} = 0.000723 \text{ failure/hour}$$

Hence, based on the results of the test, we may state that the interval of 0.002097 and 0.000723 failures per hour will contain the true failure rate 90% of the time.

EXAMPLE 3-9 One hundred amplifiers were tested for 1000 h each as part of a life-test program. No failures were experienced. Based on this test and a 90% one-sided confidence limit, what is the probability of survival of this amplifier for a 1000-h mission?

Note: The probability of survival is determined as follows (see Chapter 9 for a complete explanation):

$$P = e^{-t/\text{MTBF}}$$

where

$$e = \text{base of Naperian logarithm}$$

$$t = \text{operating time, hours}$$

Solution

$$100 \text{ amplifiers} \times 1000 \text{ h} = 10^5 \text{ unit hours}$$

Even though no failures occurred, we can establish a one-sided confidence limit for the MTBF. The lower confidence limit is:

$$\text{MTBF} \geq \frac{2T}{\chi^2_{2r+2;\, 1-\phi}}$$

$$\geq \frac{2 \times 10^5}{\chi^2_{2;\, 0.1}}$$

where

$$1 - \phi = 1 - 0.9 = 0.1$$

$$2r + 2 = 2(0) + 2 = 2$$

From the chi-square table, $\chi^2_{2;\, 0.1} = 4.605$,

$$\text{MTBF} \geq \frac{2 \times 10^5}{4.605}$$

$$\geq 43,431 \text{ h}$$

The probability of survival is

$$P = e^{-t/\text{MTBF}}$$

$$= e^{-1000/43,431}$$

$$= e^{-0.023}$$

From Table A10, Appendix A,

$$P = \underline{0.9773}$$

Hence, based on the test results obtained, we may state with 90% confidence that the probability of successful completion of a 1000-h mission for this amplifier is at least 97.73%.

PROBLEMS

3-1 What is statistical inference?

3-2 What is the minimal sample size for which the distribution of sample means closely approximates a normal curve?

3-3 What are the two parameters that define a normal distribution?

3-4 Explain the term "standard error of the mean" and state its formula.

3-5 From a continuous production line of inductors a sample of 30 pieces is randomly drawn on a daily basis. The population has a mean value of inductance of 8 henries and a standard deviation of 0.5 H.

    (a) What is the mean of the distribution of daily sample means?

    (b) Determine the standard error of the means.

3-6 What is the confidence interval that

    (a) Refers to a range of ± 1 standard errors around the sample means?

(b) Refers to a range of ± 2 standard errors around the sample means?

3-7   In an automatic bottling plant the volume of liquid is checked by a random sample of 100 bottles. The mean liquid volume was found to be 200 milliliters (ml). If we wanted to be 95% confident that the true mean liquid volume of the entire production quantity will fall between 190.2 and 209.8 ml, what is the maximum standard deviation that we could tolerate in the sample?

3-8   Assuming a normal distribution, determine the 90%, 95%, and 99% confidence limits for the voltage breakdown potential of individual toggle switches. To perform this computation, the quality-control inspector draws a sample from a large lot of switches. The mean voltage breakdown potential was found to be 745 volts (V) and the standard deviation 98 V.

3-9   (a) What are the circumstances that require the use of the Student t distribution?
       (b) How are the values for the confidence coefficient and the degrees of freedom determined?

3-10   A small sample ($n = 5$) is drawn from a production line and a designated characteristic measured. The values for the mean and the standard deviation are:

$$\bar{X} = 30$$

$$s = 1.5$$

(a) Determine the 90% confidence interval for the true mean of the population.
(b) Find the 80% one-sided upper confidence limit for the true mean of the population.

3-11   (a) What is the relationship between the asymmetry of the chi-square distribution and the degrees of freedom?
       (b) Sketch an approximate chi-square distribution for dF = 3 and identify a 90% confidence interval. Designate and shade the appropriate tail areas of the distribution.
       (c) Repeat step (b) except for a one-sided confidence level of 85%.

3-12   Twenty microcircuits are subjected to a 500-h life test. During this test, two microcircuits fail. The first failure occurs after 40 h and the second after 250 h of operation. What are the 80% confidence limits for the true MTBF?

3-13   A radar set was operated for 2000 h. Although one failure occurred, it has not been decided whether this failure is relevant or not (a nonrelevant failure, such as human error, should not count against equipment performance). Determine the 95% one-sided lower confidence limit for the true MTBF.
       (a) When the failure is relevant?
       (b) When the failure is not relevant?

3-14   A completed electronic equipment is life-tested under full power operating conditions. As failures occur, they are immediately diagnosed and repaired. Upon occurrence of the fifth relevant failure, the test is terminated. A total of 3850 continuous operating hours were logged at this point. Determine the 90% confidence interval for the true MTBF of the equipment.

3-15   During a 1000-h "shake-down" test of a control circuit, four failures caused short interruptions of the test. The cause for the four failures was attributed to operator error.
       (a) Should these failures be counted against the reliability of the device?
       (b) Utilizing the decision reached in part (a) and an 80% one-sided confidence limit, what is the upper limit for the failure rate?

3-16 A designer wants to use a certain electronic counter in the design of a piece of equipment. Results from a previous test on 50 counters showed that none failed during a
3000-hour test. Based on this test and a 90% one-sided confidence limit, is it realistic
to expect that the probability of survival of this counter will be at least 85% for a 1-year
period of operation?

4

PRODUCT QUALITY

4-1 BACKGROUND AND MEANING OF QUALITY CONTROL

The desire to control the quality of a produced item is probably as old as man's ability to produce it. The perfection of Egypt's pyramids, the flawlessness of the classical Greek masterworks and the endurance of Roman structures attest to a very conscious effort to control quality. In medieval times, the various trade guilds ensured quality by rigid craftsmanship training and by setting standards of workmanship. Their criteria for apprenticeship, performance, and master craftsmanship were solidly entrenched throughout Europe and the New World for a long period. With the advent of the Industrial Revolution, in the middle of the eighteenth century, production machinery and processes largely superseded the tools and know-how of the individual craftsman. Concurrent with the introduction of more complex production techniques, the general tempo of manufacturing increased, thereby greatly increasing the need for control of the quality of the end product.

All these early efforts to control quality, although effective in their own way, have, of course, little resemblance to today's statistical methods of control. It was not until the 1920s that the foundations for statistical quality control in its present meaning were laid. The meaning of the term *quality control* can best be understood by examining the meaning of the words *quality* and *control* individually. *Quality* has been defined as follows:

> *Quality is the extent to which a particular product satisfies the expectations of its consumer.*

Control means *to exercise restraint or direction over someone or something to assure the results desired.*

The meaning of *quality control* combines the above-stated meanings for the individual terms.

4-2 DEFINITION OF QUALITY CONTROL

There is a wide range of interpretation applied throughout industry to the term *quality control.* In a very narrow sense, some companies refer to their product inspection activity as quality control, since inspection is their sole assurance activity. In other companies, where a more elaborate quality organization is maintained, the term *quality control* encompasses not only inspection but also covers such activities as quality planning in production, process controls, incoming material control, analysis and corrective action in connection with production defects, and preparation of quality procedures and reports. Quality control should not be restricted to a narrow application of the term but should encompass all the activities necessary to ensure the desired level of product integrity. *Quality control is a system for verification and maintenance of a desired level of quality in a product or process by careful planning, the use of proper equipment, continuing inspection, and corrective action where required.*

4-3 NEED FOR QUALITY CONTROL

No two objects in the world around us, nor any two actions performed by the same or by different individuals, are exactly identical. Precision machine parts produced in quantity by the same operator using identical tools and equipment will, upon examination, show a definite variability. This variability may be caused by a variety of factors. For example, the raw material used may come from different lots with slightly different chemical composition or hardness or other significant characteristic; there are tolerances in connection with treatment or exposure of the material, such as pressure, temperature, electrical stress, and the duration of such factors. In addition, there is the variability of the human element, which further adds to the variability of the end product.

The best a manufacturer can do is recognize the causes for variability in his product and institute steps to control each contributing factor within appropriate limits, thereby controlling the variability of the end product. The complete elimination of variability in production is usually not feasible, and would be entirely uneconomical even if feasible. Instead, the manufacturer's philosophy is based on a tolerable, statistically predictable, level of imperfect product.

The need for quality control is thus well established. Its use enables us to ascertain sudden or gradual changes in product variability (or establish trends) to permit the institution of timely corrective action that will avoid production of costly scrap.

4-4   HISTORY OF QUALITY CONTROL

As noted in the introduction, statistical quality control began in 1926 with Shewart's statistical control charts. The influence of the military services on the adoption of sampling acceptance techniques is well established. Army Ordnance sampling inspection tables came out in 1942, but as early as 1940 American Standards Association (ASA), on the request of the War Department, initiated a project dealing with the application of statistical-quality-control methods to manufactured products. As a result, two standards were developed:

1. American War Standard AWS Z1.1, "Guide for Quality Control," May 1941.
2. American War Standard AWS Z1.2, "Control Chart Methods of Analyzing Data," May 1941.

These standards were widely used for educational programs and indoctrination by the War Production Board.

After minor modifications, the ASA approved the two revised quality-control standards as American Standards in November 1958. American Standards Z1.1—1958 and Z1.2—1958 gave detailed guidelines for handling problems concerning economic control of the quality of materials and manufactured products. They had particular reference to methods of collecting, arranging, and analyzing inspection and test records to detect lack of uniformity of quality and to apply the control-chart technique to ascertain lack of control.

The utilization of inspection sampling techniques was largely due to the 1944 published sampling inspection tables by H. F. Dodge and H. G. Romig.[1] In 1950 military sampling procedures and tables for inspection by attributes were consolidated in Military Standard MIL-STD-105,[2] a document which, in its updated form, is still in wide use today. In 1957, "Military Standard for Acceptance Sampling by Variables", MIL-STD-414[3] was introduced. The Department of Defense (DoD) also issued inspection and quality-control handbooks H107 and H108[4] for single- and multiple-level continuous-sampling procedures by attributes. The interest shown by the DoD in sampling procedures is understandable in light of the enormous expenditures for procurement which are authorized by the agencies of the DoD. These quality-control standards issued by the military were primarily concerned with the acceptability of products in procurement. They did not contain, nor did they impose upon the supplier, detailed quality program requirements or inspection techniques.

---

[1] H. F. Dodge and H. G. Romig, *Sampling Inspection Tables, Single and Double Sampling*, John Wiley & Sons, Inc., New York, 1944.

[2] Military Standard MIL-STD-105D, "Sampling Procedures and Tables for Inspection by Attributes," Government Printing Office, Washington, D.C., 1963.

[3] Military Standard MIL-STD-414, "Sampling Procedures and Tables for Inspection by Variables for Per Cent Defectives," Superintendent of Documents, Washington, D.C., June 1957.

[4] "Inspection and Quality Control Handbook (Interim) H107 and H108 for Single-Level and Multi-Level Continuous Sampling Procedures and Tables for Inspection by Attributes, Respectively," Superintendent of Documents, Washington, D.C., April 1959 and October 1958.

The need to impose very detailed requirements in these areas became quite pressing, however, resulting in comprehensive quality-control and quality-assurance programs. These programs, currently enforced by agencies of the DoD, include MIL-Q-9858A[5] and MIL-I-45208A[6] (discussed in a subsequent section). A comparable document, such as NHB5300.4(1B)[7], covers NASA quality assurance provisions.

In more recent years, the control of the integrity of the inherent design of the product has been given increased attention, resulting in a new discipline, reliability engineering. Its object is to assure that a design will perform its intended function with a specified probability for a given period of time. In 1946, a national organization, the American Society for Quality Control, was formed. This agency has continued to spearhead quality-control efforts in terms of organization, research, promotion, and publications.

4-5 DISTINCTION BETWEEN RELIABILITY AND QUALITY CONTROL

The definition of reliability that we have given involves the element of time. To say, for example, that the probability of survival of a piece of equipment is 95% for 1000 h of operation means that, on the average, 95 of 100 such pieces would operate without failure for 1000 h.

The incorporation of a time constraint in the reliability definition is not unusual. No product lasts forever; after its useful life period is over, it begins to show signs of wearing out, and a rapidly rising failure rate. The manufacturer therefore restricts his prediction of the reliability of the product to a time span that does not reach into the wear-out region of the life of the equipment.

Product quality, on the other hand, is not mathematically linked to a time period, nor is it stated in terms of a probability of performance. It may best be defined as the degree of conformance of the product to the applicable specifications, standards, and workmanship criteria.

It should be noted that the high quality of an item is not necessarily synonymous with high reliability. For instance, a component used in an assembly, or system, may have been exposed to all the necessary quality control and have met all of its specification requirements and yet run into unforeseen difficulties when working as part of the overall equipment. This difficulty may stem, for example, from a lack of recognition of adverse electrical or mechanical interactions between components. This is a reliability problem that should have been solved by a systems analysis, including a component application review. Such an analysis could well have resulted in a recom-

---

[5] Military Specification, "Quality Program Requirements," MIL-Q-9858A, Government Printing Office, Washington, D.C. 1963.

[6] Military Specification, "Inspection System Requirements," MIL-I-45208A, Government Printing Office, Washington, D.C., October 1961.

[7] NASA Reliability and Quality Assurance Publication NHB 5300. 4(1B), "Quality Program Provisions for Aeronautical and Space Systems Contractors," Government Printing Office, Washington, D.C., April 1969.

mendation for substitution of a more suitable component for the application. Clearly, the most stringent quality-control measures applied to the unsuitable component would not have altered the situation.

Although the separate functions of quality control and reliability engineering are well recognized, there are at times organizational considerations associated with the assignment of responsibilities. This situation is mainly encountered where organizations have combined quality control and reliability engineering in one department. Where separate departments are maintained, the lines of demarcation of duties and responsibilities are quite clear.

4-6 ACHIEVING PRODUCT QUALITY

(a) Establishing Product Quality Policies and Objectives

B. L. Hansen, in his book on quality control, cites three broad areas of quality assurance in the operation of a manufacturing business:[8]

1. Quality of design.
2. Quality of conformance.
3. Quality of performance.

These areas are by necessity linked to the cost and economics of quality.

1. Quality of Design

Two products may have the same functional use but be very different in design. One product may be made of cheap plastic with ill-fitting parts and an obviously short life expectancy in use. The other may be well designed, using strong, durable materials; have parts machined to close tolerances; and give all indications of being able to withstand extended use. The quality of design of the second product is clearly superior.

The quality of design is greatly influenced by the application, cost considerations and market demand. The quality (as well as reliability) that is demanded of electronic equipment for a moon probe is obviously many times more stringent than that which is employed (and which can be afforded) for an ordinary commercial pocket transistor radio. As a matter of fact, it is not infrequent that the quality of mass-produced items is purposely made marginal in order to stimulate both the repair and spare-parts activities of the business as well as replacement sales of the product. It is, of course, vital to the health of the business that any relaxation in the level of quality of the product be consistent with the advertising claims for the product. On the other hand, care must be employed in the design effort to avoid needless overdesign of the product.

[8] Bertrand L. Hansen, *Quality Control: Theory and Applications*, Prentice-Hall, Inc., Englewood Cliffs, N.J., 1963, p. 2.

For example, when a tolerance of ± 0.01 in suffices for a machined part, there is no need to specify the much more costly tolerance of ± 0.001 in for the part.

The factors that influence product quality policies depend on the above-cited considerations, as well as the position that the company aims to maintain in a particular field. For example, the company may enjoy a leadership role in the field and may be anxious to maintain its reputation.

Accordingly, a list of the major factors influencing a company's quality of design policy must include recognition of the general objectives of the company relative to its standing in the field, the demands imposed by the customer, the competitive situation in the market place, the profit policy of the firm, the availability and nature of parts and materials that could be used in the manufacture of the product, and, last but not least, the legal responsibility borne by the company in connection with product safety.

2. Quality of Conformance

Quality of conformance refers to the extent to which the product complies with the specifications, standards, and workmanship criteria imposed upon its manufacture. A product manufactured to specification and in conformance with the control limits of the production processes should satisfy the customer provided that the specifications have correctly translated the customer's requirements (and that pertinent reliability aspects were considered in the design). The difference between quality of design and quality of conformance is illustrated by the following example: Two electrical freezers are made to the same design, specifications, procedures, and standards—they both have the same quality of design. One of them is unable to carry its cooling load as advertised; it therefore does not conform to the specification, and the two units differ in their quality of conformance.

In general terms, quality of conformance deals with three broad areas of control:

1. Defect prevention.
2. Defect finding.
3. Defect analysis and correction.

DEFECT PREVENTION

Defect prevention is dedicated to the prevention of failures before they occur. This includes such activities as devising and implementing process controls, developing quality standards, determining the extent, type, and frequency of inspections for incoming materials and product in process, and stipulating final acceptance criteria.

DEFECT FINDING

Defect finding involves the application and use of the specific methods of inspection, test, and statistical analysis instituted by quality engineering to detect defects in production runs or to spot unusual trends which foretell impending trouble in production. It should be noted, however, that the confirmation of the presence of a number of defectives in a production run which does not exceed the aspired quality level is perfectly acceptable.

Defect Analysis and Correction

Defect analysis and corrective action is an important factor in the quality-of-conformance cycle. Merely removing the defects from the production line without analyzing the failed items to determine the cause of failure would not give the necessary insight into the problem and would not permit the institution of appropriate corrective action. Such corrective action may be found to require tightening of incoming inspection limits or in-process tolerances, the issuance of additional assembly or workmanship specifications or criteria, or added controls, or simply better training of operators.

Defect analysis during production is usually the responsibility of the quality-control department. Failures of parts or components operating in a system and failing during specific system or subsystem testing are normally analyzed by the reliability engineering department as part of its failure-analysis and corrective-action activity.

3. Quality of Performance

The *quality of performance* of a product is a function of its quality of design and quality of conformance. Maintaining both at a high level will render a high quality of performance. If the quality of design is poor or if there is a lack of conformance to applicable specifications, the quality of performance will be adversely affected. Obviously, the highest level of conformance to specifications will be useless if the design is poor.

(b) Quality Control Engineering

The quality control department is responsible for the overall quality planning of the company. To produce a viable quality plan, the quality-control engineer has to translate customer or market requirements into quality objectives and goals which are consistent with the company's competitive posture in the marketplace and its general financial and operating principles.

Following this initial activity, the quality-control engineer has to collaborate with the design engineers to obtain a firsthand understanding of the details of the proposed design of the product and the processes and equipment that will be required for production. It is during this early review that the quality-control engineer may be in a position to influence the design or the choice of an intended process in a manner that might prevent quality problems. In companies, where new designs are subject to a design review, the quality-control engineer usually attends this review along with representatives of the production department and therefore has the advantage of a formal forum to present his views and recommendations for the quality aspects. When formal design reviews are not held, the quality-control engineer has to interface with the design and manufacturing group on an independent basis, with disputes settled by management.

With the quality objectives firmly established and the review of the basic design and the proposed processes behind him, the quality-control engineer is now ready to plan the necessary quality inspections, controls, and tests on parts, materials, and processes, as well as on subassemblies and on the final product.

The plan for the quality inspections, controls, and tests involves the quality control engineer in the following activities:

1. Provide detailed quality procedures for controlling product quality and process quality.
2. Establish flow charts that identify quality-control checkpoints.
3. Establish criteria for in-flow product quality inspections, including detailed information for testers and inspectors.
4. Collaborate with design engineering to reflect quality requirements in the purchase specifications and drawings.
5. Issue instructions to the receiving department regarding inspection methods and criteria for purchased parts and materials.
6. Examine capability of proposed machines, tools, dies, fixtures, gages, and meters to assure the required product quality.
7. Devise control charts and other statistical methods to permit continuous monitoring of the quality of the manufactured product.
8. Set up or arrange for laboratory facility to examine and analyze defects and and failures.

In addition to the above, the quality-control engineer has responsibilities as follows:

1. He helps to define the quality characteristics of the finished product—usually in conjunction with design engineering—to enable the marketing function to commit the company to definitive quality standards of performance, appearance, life, and so on, in the marketing literature or in contractual agreements.
2. He diagnoses manufacturing quality problems and proposes appropriate remedies.
3. He provides reports on quality status and quality problems to management.
4. He actively participates in the quality indoctrination and training activities of the company.
5. He conducts quality audits including those of vendors and subcontractors.

(c) Planning for Quality

A viable quality program must be both effective and economical. It must consider the company's quality objectives, market commitments, production facilities, and process capabilities; it must, furthermore, look at personnel qualifications, financial and budgetary constraints, and so on, and attempt to strike an optimum balance between all these factors.

The extent of the planning depends on whether an already existing quality-control program is to be adopted to a new product line, or whether a quality-control program is to be first initiated. In the former case, company orientation, personnel training, quality-control methods and procedures, inspection test and reporting requirements, etc., have already been devised and used, and it is merely necessary to adopt the existing capability to the new product. This may, for example, necessitate the addition of more suitable or more modern equipment in the factory in cases where the existing equipment is incapable of providing the necessary quality. As an illustration, the available tools may be unable to hold the required tolerances of machined parts. The use of these tools would produce a vast number of defective part rejects in the subsequent quality-control inspection. It is therefore necessary that quality planning include an analysis of the available tools, equipments, and processes.

When a formal quality-control activity is first to be instituted in a company, the planning that must precede such a step must by necessity start at the beginning and be more comprehensive. Generally, the following quality aspects should be considered:

1. Ascertain or help develop company quality objectives.
2. Determine quality requirements and characteristics of the product(s) to be produced.
3. Assess capability of existing machines, tools, equipments, and processes to fulfill the quality goal.
4. Assess adequacy of existing inspection and surveillance personnel and develop training plans.
5. Define characteristics to be controlled and the extent of control.
6. Define quality level, number of defectives permitted, and average outgoing quality limit (AOQL).
7. Develop directives, procedures, forms, and documentation.
8. Define inspection, monitoring, and surveillance responsibilities.
9. Plan details of statistical control of specific quality characteristics by control charts.
10. Define procedures for handling nonconforming material (Material Review Board).
11. Define controls for the storage and use of age-sensitive materials.

12. Provide for laboratory facility and technical personnel to perform failure analyses, study trends, and recommend corrective action.

Failure of quality-control programs to provide for adequate coverage in the above-cited general areas of control require additional quality planning. Section(i) which deals with quality program and inspection system requirements contains information that is useful in the formulation of quality plans (see page 81).

(d) Organizing for Quality

Companies create organizational structures which best achieve company objectives. The organization and function of quality-control departments has, over the years, developed from a mere inspection activity of the final product reporting to the production manager, to a sophisticated organization reporting directly to top management. Organizationally, the quality-control manager is now on an equal level in the company hierarchy with the production manager, or the manager of purchasing and similar executives. The reason for this shift in the line of command of quality-control organization was the desire to remove the control over acceptance and rejection of manufactured products from the very organization that is responsible for its production. A quality-control organization that reports directly to top management is clearly in a better position to remain unbiased and is therefore able to fulfill its quality assurance function more effectively.

The modern quality-control organization employs, in addition to the traditional inspectors, quality control engineers, technicians, and specialists. The emphasis is now placed not only on acceptance testing the final product, but, to an increasing extent, on assuring quality of design and on prevention of defectives in production. This is achieved by a variety of planned activities which include quality-level decisions, process controls, incoming material inspection, in-process and final inspections, defective material review, analysis and corrective action, and so on. It is the responsibility of the quality-control engineers, technicians, and specialists to plan and implement these activities under the leadership of the quality-control manager. Since they deal with problems in production, it is desirable that quality-control personnel have some degree of shop and production background.

The responsibilities of the quality-control manager also extend to participation in the planning of new products and deciding on methods of production and plant facilitation. He must, furthermore, assess the impact of quality assurance decisions on company policy and market posture and provide his expert views and council to top management. Because of variations in organizational structure, the role and duties of the quality assurance manager may be combined with the duties of the reliability manager; the corresponding job title would be "Quality and Reliability Assurance Manager." In some companies the all-encompassing department name for the various assurance functions is "Product Assurance." In large companies the product assurance manager may have several department managers reporting to him, heading up the major assurance functions such as reliability/maintainability, quality assurance, and integrated logistic support. Each of these functions is broken down into further functional subdivisions and may be headed up by a supervisor. The ILS activity is concerned with the maintenance and support aspect of the equipment produced and includes the development of repair procedures, preventive maintenance, spares planning and provisioning, training, and technical publications.

The organizational structure of a company's assurance function and the type and size of the subordinate functions is largely dependent on the size of the company and the type of business it is in. Large companies engaged in major aerospace and/or military contract work may well have need for a comprehensive organizational plan as illustrated in Figures 4-1 and 4-2.

Smaller companies may not require quite as elaborate an organization; for example, there may be no need for an ILS activity, or the reliability/maintainability and quality-control functions may be combined in one activity. If the company supports the commercial market, there is frequently less emphasis on the reliability/maintainability aspect and more concentration on quality control. A possible exception may be commercial industries producing highly complex pieces of equipment, such as computers. Then it simply constitutes good business and is justifiable profitwise to maintain a substantial reliability/maintainability activity and a group responsible for the maintenance engineering and spares functions, although the latter may not be known as ILS.

Figure 4-1 is an organizational structure showing the relationship of the product assurance (or equivalent) activity to other activities and to top management. Figure 4-2 shows a very comprehensive product assurance organization (military and aerospace) for a large company. Figures 4-3 and 4-4 show progressively smaller organizational assurance structures.

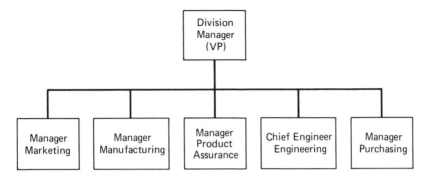

Figure 4-1 *Organizational structure showing relationship of the product assurance activity (or equivalent) to other activities and top management.*

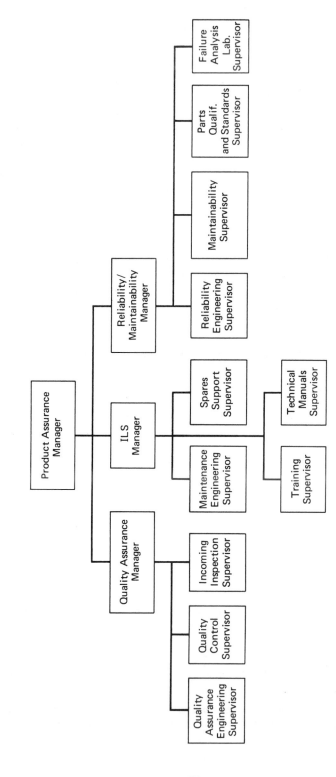

Figure 4-2 *Organizational structure of assurance functions in a large company engaged in aerospace and military contracts.*

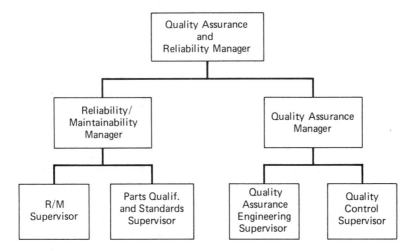

Figure 4-3 *Organizational structure of assurance functions in a medium size company.*

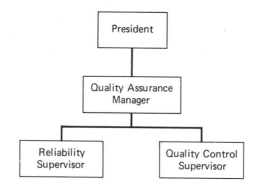

Figure 4-4 *Organizational structure of the assurance functions of a small company.*

(e) Motivating for Quality

The pressures experienced by the production manager to meet cost, schedule, quality, and so on, are by necessity carried down to the level of the production worker. The latter frequently faces conflicting requirements: he is asked to speed up production, reduce the number of defectives, and often work tiresome periods of overtime.

In order to fulfill these requirements and responsibilities the employee needs motivation. If overly fatigued and lacking motivation, he is likely to take trouble-producing shortcuts. A few individuals might even be tempted to purposely produce defective material or damage production tools to force an interruption of the frustrating tasks. There are, of course, situations in which the worker lacks the necessary skill to produce a quality product. If this situation prevails, he is simply not in a position to eliminate defects.

In some instances the difficulty experienced by production workers to maintain an acceptable quality level can be traced to causes which are beyond their control. For example, the design of the item, or the processes used in production, may be faulty, or the drawings could be ambiguous. These types of problems are management problems that require speedy rectification to restore production quality.

The motivation of production personnel requires planned action by management. During wartime, when materials needed for defense are produced, motivation of employees in the factory is spurred on naturally as a result of patriotic feelings; also, there is no employee indifference due to job insecurity.

In peacetime, however, job security is often lacking, as is the patriotic drive of the individual. The production worker often finds himself assigned to a repetitive and uninteresting operation and very likely will adopt the philosophy that "errors are human." Without management's special attention to this problem, the quality of the end product may suffer. Management's duty is to try to provide an environment in which self-motivation of employees for conscientious performance is encouraged.

The reason it is necessary to show the worker the effect of his contribution on the end product stems from the fact that modern technology tends to have a dehumanizing effect; no longer does a worker contribute to the product throughout its manufacture, but he is likely to be involved in a singular, isolated operation which, by itself, may be downright boring.

Motivating production employees to be quality conscious requires that management attempt to make the job challenging by explaining how each task, well done, contributes measurably to the success of the whole job. Management must further stress the importance that each operator strive for perfection and must create certain standards of performance against which each employee can be measured. Fulfilling an expected level of performance individually and as a member of a production group or team will give the employee a sense of accomplishment, especially if rewarded through appropriate recognition. Recognition may include simply praise and encouragement or salary increases and advancement.

Management must stimulate motivation for product quality not only in the production department but in all other departments of the company; this includes marketing, engineering, and purchasing. One of the most effective methods management can use to make employees quality conscious is a company-wide quality-improvement employee suggestion system with appropriate prizes and publicity. In addition to the suggestion system, there are many other methods of creating a climate of quality consciousness, such as citations for excellence for professional employees who have made a significant contribution to product quality. Each company is a unique organization, and the specific plan that would work best in one company may not be as successful in another; however, the basic principles underlying employee motivation, recognition, and reward are certainly applicable in all cases.

(f) Quality Indoctrination

The consistency and effectiveness with which a manufacturer is able to produce a quality product depends, to a significant degree, not only on the motivation of his employees, but also upon their background and acceptance of quality concepts, principles, and practices. It is difficult to motivate employees to adopt a sound quality attitude and observe strict quality rules and practices in their daily activities when they lack the necessary understanding for such rules and practices. To start, each employee, on his own level, needs indoctrination as to the company objective relative to product quality. The impact which poor quality may have on the company's reputation, on its competitive posture in the marketplace, and, ultimately, on the job security of the employees should be stressed. With the stage thus set, formalized training in quality-control techniques may then be instituted.

The content of quality-control courses should be tailored to the level, occupation, and education of the participants. Courses tailored to assemblers on the production line need not be concerned with the statistical aspects of quality control, but should stress the solution to quality problems, methods of failure prevention, and general quality awareness. Quality-control courses for inspectors and testers should include technological and basic statistical techniques. Courses for professional men, such as engineers and designers, should cover more advanced techniques of control and statistical methodology.

Quality indoctrination provided for the various levels of management should include topics dealing with general quality-control fundamentals and emphasize the planning and control aspects of quality programs. Also, the managerial courses should include topics dealing with motivational principles and training needs of personnel.

In-plant training courses in quality control are the most desirable, since quality topics of special interest to the company may be included. When the need exists, outside instructors and consultants are available. When in-plant training courses are not practical, consideration may be given to courses in quality control offered in a number of colleges or to courses offered by the American Society for Quality Control.

(g) Cost and Economics of Quality

Quality assurance activities are a financial burden for a company; the higher the quality of design and the quality of conformance, the greater this burden. In return, the company will enjoy a greater share of success and goodwill in the marketplace. It is obviously necessary to strike a balance between the costs and benefits of a company's assurance program. The costs of a quality program are made up of failure costs (rework, retest, scrap), preventive costs (operator training, proper equipment, and instructions), and inspection costs. Of the three cost groups, the inspection costs account for approximately one third of the total quality costs. The remaining costs are divided in varying proportion among preventive costs and failure costs.

Not easily determined numerically are the benefits that accrue from a quality assurance program in terms of goodwill in the marketplace or in terms of additional sales. In the case of aerospace or military contracts, the level of quality assurance is specified by the contract and must be complied with. For consumer goods, the cost allotted to the quality assurance function must be closely related to the type of market that the company is aiming to capture and the competitive nature of the product line. From the customer's point of view, there are several factors which help him choose the product of one manufacturer over that of another; these include price, delivery, reputation of the manufacturer, and quality. Reputation and quality aspects are, of course, closely related.

Three broad areas of quality assurance have been defined: the qualities of design, conformance, and performance; only the first two represent cost elements. The costs incurred in connection with quality of design involve the identification of key product requirements and the preparation of a quality plan designed to satisfy these requirements.

Quality-of-conformance costs are incurred for the purpose of preventing failures by proper quality planning of the quality program, by carrying out inspection and test activities, and by analyzing the causes of defective parts or equipment removed from production and the implementation of corrective action.

The allotment of quality funds should be based on the prevailing quality of production and should be flexible. When the defect level rises, it may, for example, be advisable to spend more money on preventive costs rather than on added inspection. Similarly, if failure causes are not readily determinable, more extensive failure-analysis activities need to be funded.

(h) Quality Survey Team

Some companies use the team approach to perform a quality survey.[9] The team is normally composed of representatives of quality engineering, manufacturing, and design engineering; members of other company activities may be brought in for consultation on an as-required basis.

The quality survey team is charged with conducting an in-depth review of all factors that have bearing on the quality of a planned product. Each team member contributes his individual expertise and experience to the findings of the group. The survey will concern itself with a complete analysis review and assessment of the design requirements, quality considerations, manufacturing specifications and procedures, test and inspection methods and criteria, past performance and quality history of similar items produced, and other topics that influence the quality of the product. Where a design model and/or prototype production unit exists, the survey team will be interested in careful examinination of this unit for compliance with, and deviation from, the specified requirements. The activity of the survey team will include a physical examination of proposed production equipment, tools, and test and inspection facilities to assess their adequacy for the job.

---

[9] Hansen, *Quality Control*, p. 260.

The findings and recommendations of the survey team are covered by a report. Corrective actions are recommended and the applicable company activities are charged with the responsibility of implementing these recommendations. Follow-up action by the team members assures compliance with the implementation of the recommended corrective actions. Management is kept informed of the progress and eventual completion of all action items. The effective functioning of a quality survey team depends to a large extent on the support it receives for its actions from management. If the support is half-hearted, the success of the team will be in question.

The use of a quality survey team is costly and can be justified only by large design and manufacturing organizations. Smaller organizations may find modified or simplified versions of the type of quality survey team described above adequate for their purposes.

(i) Specifications for Quality Program and Inspection System Requirements

The best examples of comprehensive quality program and inspection system requirements may be found among the military quality assurance specifications. The following two military quality documents are the most widely used for government procurement.

1. "Quality Program Requirements," MIL-Q-9858A, December 16, 1963.
2. "Inspection System Requirements," MIL-I-45208A, December 16, 1963.

MIL-Q-9858A covers a comprehensive quality program for producers of equipment for the military services. MIL-I-45208A is a quality specification that deals primarily with inspection requirements for contractors. Compliance with MIL-Q-9858A automatically covers compliance with MIL-I-45208A. These documents are reproduced in Appendix B.

(j) Quality-Control Manuals

The quality requirements for a product are defined in quality specifications. The latter may constitute complete quality program specifications, as has been mentioned previously (MIL-Q-9858A and MIL-I-45208A), or they may involve anything from parts and material quality or inspection requirements, process control criteria, and defective material review to storage, packing, handling, and shipping.

Manufacturers who are awarded contracts invoking quality program specifications such as MIL-Q-9858A will usually generate in-house quality-control manuals which comply with and implement the various requirements of MIL-Q-9858A. Figure 4-5 is a comprehensive listing of quality-control areas typically covered by these manuals.

1. APPLICABLE DOCUMENTS

   (a) Specifications
   (b) Publications
   (c) Handbooks

2. MANAGEMENT, ORGANIZATION,
   AND PERSONNEL

   (a) Organization Chart—General
   (b) Organization Chart—Quality Assurance
   (c) Plant Layout
   (d) List and Location of Quality Control Inspection Stations

3. QUALITY-CONTROL ENGINEERING

   (a) Quality-Control Policy
   (b) Preparation and Issuance of Quality-Control Procedures
   (c) Quality Inspection System
   (d) Acceptance Sampling Plans
   (e) Quality-Control Charts
   (f) Quality-Control Audit Program

4. QUALITY-CONTROL FORMS AND
   DOCUMENTS

   (a) "Approved" Tag
   (b) "Hold" Tag
   (c) "Reject Part" Tag
   (d) Nonconformance Report
   (e) Failure Report
   (f) Corrective Action Request
   (g) Test Discrepancy Report
   (h) Receiving Report
   (i) "Age-Sensitive Material" Tag

5. DATA REPORTING AND RETENTION

   (a) Quality-Control Records
   (b) Storage of Quality Records

Figure 4-5 *Contents of typical quality-control manual.*

6. TRAINING PROGRAMS
   (a) Quality-Control Course for Supervisory Personnel
   (b) Training of Operators
   (c) Training of Inspectors
   (d) Certification of Operator/Inspection Personnel

7. ENGINEERING CONTROL
   (a) Drawings, Specifications, and Standards
   (b) Test Specifications
   (c) Drafting Procedures
   (d) Engineering Change Orders
   (e) Drawing and Specification Review

8. PURCHASING CONTROL
   (a) Purchasing Procedures
   (b) Selection of Procurement Sources
   (c) Review and Approval of Purchase Orders

9. PURCHASED MATERIALS
   (a) Procurement Documents
   (b) Vendor Quality Rating
   (c) Incoming Inspection
   (d) Control and Storage of Purchased Material Parts and Assemblies
   (e) Vendor Corrective Action
   (f) Surveillance of Subcontractors
   (g) Failure and Deficiency Feedback

10. MANUFACTURING AND PROCESS CONTROLS
    (a) Shop Operations
    (b) Materials and Material Control
    (c) Material Flow
    (d) Workmanship Standards
    (e) Production Processing and Fabrication
    (f) Measuring and Testing Equipment
    (g) Standard Repairs
    (h) In-Process Inspection
    (i) Identification and Storage of Limited-Life Items
    (j) Special Processes

Figure 4-5 *Contents of typical quality-control manual* (continued).

11. PRODUCT ACCEPTANCE
    (a) Quality Assurance Testing
    (b) Lot Acceptance
    (c) Inspection and Test Procedures
    (d) Inspection Stamps
    (e) Returned Products (from field)
    (f) Test Review Board

12. DISCREPANT MATERIAL CONTROL
    (a) Material Review Board
    (b) Reject, Rework and/or Scrap Procedure
    (c) Control of Nonconforming Material

13. FAILURE ANALYSIS AND CORRECTIVE ACTION
    (a) Methods and Techniques of Failure Analysis and Corrective Action

14. EQUIPMENT CALIBRATION AND STANDARDS
    (a) Electrical Measuring Instruments
    (b) Calibration and Evaluation of Test Equipment
    (c) Tool and Gage Control

15. PACKAGING AND SHIPPING CONTROLS
    (a) Packaging, Shipping, and Storage Procedure
    (b) Preservation, Handling

16. GLOSSARY OF QUALITY-CONTROL TERMS
    (a) Definitions of Common Quality-Control Terms

Figure 4-5 *Contents of typical quality-control manual* (continued).

The following examples of quality specifications are presented for illustrative purposes:

Figure 4-6. Tool and Gage Control (item 14c in Figure 4-5).
Figure 4-7. Incoming Inspection—Diodes (part of item 9c in Figure 4-5).
Figure 4-8. Material Review Board (item 12a in Figure 4-5).

1. <u>PURPOSE</u>   To provide a tool and gage control system which shall ensure that all tools and gages are calibrated at periodic intervals to verify their accuracy and reliability.

2. <u>SCOPE</u>   This procedure applies to the inspection and calibration of tools and gages used in the manufacture of electro-mechanical devices on contracts which specify Quality Program Requirements MIL-Q-9858A.

3. <u>REFERENCE</u>   Military Specifications MIL-C-45662A Calibration System Requirements and MIL-I-45607B Inspection Equipment, Acquisition, Maintenance and Disposition, MIL-STD-120 Gage Inspection.

4. <u>PROCEDURE</u>

a. <u>Types of Tools and Gages</u>
This procedure covers tools and gages employed in the manufacture of electro-mechanical devices. They include: Calipers, Indicators, Gage Blocks, Micrometers, Comparators, and similar devices.

b. <u>Calibration</u>
Tools and gages shall be calibrated periodically (6 months interval, unless otherwise prescribed) in accordance with Section 8 of Gage Inspection Manual MIL-STD-120, unless an individual calibration procedure applies.

c. <u>Marking and Identification of Tools and Gages</u>
All tools and gages shall be permanently marked with serial number.

d. <u>Calibration Cards</u>
A calibration card shall be prepared for each tool and gage which shall fully identify the item and list the serial number, Calibration frequency code and date of result of calibration.

e. <u>Calibration Frequency Code</u>
Calibration frequency shall be identified by a color code on the calibration card.
The color code for standard calibration (6 months) is green (see Para 4b). Other calibration frequencies are identified by the following color codes:

Daily calibration	— red
Weekly calibration	— blue
Monthly calibration	— yellow
Three months calibration	— brown
Yearly calibration	— black

f. <u>Calibration Identification</u>
Tools and Gages shall be identified by labels or stamps to indicate the most recent date of calibration, the name of the technician responsible for the calibration and the next due date for calibration.

g. <u>Tool and Gage Storage</u>
Tools and gages shall be stored in specially designated areas where they are protected from deterioration and damage.

h. <u>Tool and Gage Inspection Records</u>
Tool and gage inspection records shall be kept on file for five (5) years and shall be available for review by quality assurance personnel.

Figure 4-6 *Tool and gage control.*

<u>1. PURPOSE</u>   To provide criteria for inspection and testing of diodes and rectifiers during incoming inspection.

<u>2. SCOPE</u>.   This procedure applies to the incoming inspection of diodes and rectifiers procured to military or commercial specifications.

<u>3. APPLICABLE
DOCUMENTS</u>
a. For Military Devices: MIL-S-19500 with appropriate slash-number spec.
b. For Commercial Devices: Commercial Data Sheet.
c. Quality Control Procedure, Incoming Inspection: General Procedure.

<u>4. PROCEDURE</u>
a. Visual Inspection
  1. <u>Configuration</u>: As shown on drawings or specifications.
  2. <u>Finish, Workmanship</u>: Free from damage and flaws in material, finish, and workmanship.
  3. <u>Markings</u>: All markings shall be clear, legible, and permanent, and shall conform to the applicable specifications.

b. Mechanical Inspection
  1. <u>Dimensions</u>: Dimensions, overall, mounting, stud length, threads, etc., shall conform to the applicable specification.
  2. <u>Hardware</u>: Quantity, type, and finish of studs, screws, standoffs, bushings, etc., shall be per specification.

c. Electrical Tests
  1. Forward Voltage Drop
     Using the circuit shown in Figure (i), increase the applied DC voltage gradually until the specified forward current is measured. The forward voltage drop is measured on the VTVM across the diode under test. Measurement is performed at $25°$C room ambient temperature.
  2. Reverse Current
     Using the circuit as in Figure (i), except with reversed polarity of the applied DC voltage, increase the DC voltage gradually to the value specified and read the reverse current. Measurement is performed at $25°$C room-ambient conditions.
  3. Zener Voltage
     Use circuit as delineated for reverse current. Increase DC voltage gradually until the Zener current of specified value is reached. Read the Zener voltage on the VTVM. The value read must fall within specified Zener voltage limits. Measurement is performed at $25°$C room-ambient conditions.
  4. Zener Impedance
     Use circuit as per Figure (ii). Increase DC voltage until the specified DC current is obtained through the diode (Zener current). Now increase the applied 60 Hz voltage to transformer T_1 until the RMS value of the AC current is 10% of the DC current. The RMS value of the AC current is calculated as follows:

$$I_{RMS} = \frac{E_R}{R}$$

The Zener impedance is calculated as:

$$Z_{ZENER} = \frac{E_{AC}}{I_{RMS}}$$

Figure 4-7 *Incoming inspection—diodes.*

4. <u>PROCEDURE</u>
(continued)

5. Silicon Controlled Rectifier Tests
The following parameters shall be verified:

V_{GF} — Gate voltage to fire SCR

I_{GF} — Gate current to fire SCR

V_{BO} — Anode-Cathode forward break over voltage

PFV — Anode-Cathode repetitive peak forward blocking voltage

I_S — Anode-Cathode forward leakage current

PRV — Cathode repetitive peak reverse blocking voltage

6. <u>Internal Contamination — Glass Diodes</u>
Use 20 power magnification and examine the inside cavity of the diode. There shall be no contamination such as extraneous matter (solder balls, pieces of wire, etc.)

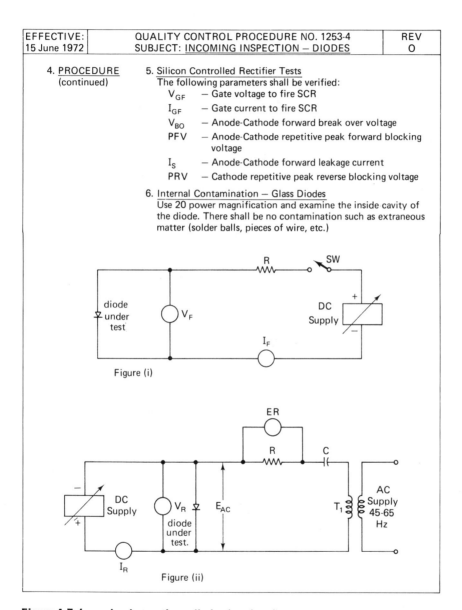

Figure (i)

Figure (ii)

Figure 4-7 *Incoming inspection—diodes* (continued).

EFFECTIVE: 4 June 1972	QUALITY CONTROL PROCEDURE NO. 1210 SUBJECT: MATERIAL REVIEW BOARD (MRB)	REV O

1. <u>PURPOSE</u>   The purpose of the Material Review Board is the control, review and disposition of non-conforming material including corrective action related thereto.

2. <u>SCOPE</u>   This procedure outlines the policy for reporting, reviewing, controlling, and dispositioning of non-conforming materials for programs which are subject to MIL-Q-9858A control.

3. <u>PROCEDURE</u>   a. The MRB will be appointed on each program involving deliverable hardware.

b. The MRB will consist of one member each from Design Engineering and Quality Assurance and, when so directed by contract, one member representing the Government/Customer.

c. The MRB may be convened at the request of any member of the board.

d. The MRB will have the following duties and responsibilities:
1. Determine whether material submitted for material review action is within the scope of the jurisdiction of the board.
2. The MRB shall determine the cause and then decide on the disposition of the material submitted for review. Material review action consists of choosing one of the following possible actions:
   a. accept, as is
   b. rework (per MRB)
   c. scrap
   d. return to vendor
3. The MRB shall assure that appropriate records be kept covering all MRB actions and decisions.

4. Acceptance of non-conforming materials requires unanimous agreement of the MRB.

5. The MRB shall be responsible for follow-up action to assure implementation of the MRB decision.

Figure 4-8 *Material Review Board (MRB).*

PROBLEMS

4-1   Summarize the background and meaning of quality control.

4-2   Define quality control.

4-3   Discuss the need for quality control.

4-4   Who is credited with the introduction of
(a) Statistical control charts?
(b) Sampling inspection tables?

4-5   What were the American Standards? When were they adopted?

4-6   Name the most popular military document for inspection by attributes.

4-7   Which are the two military documents covering inspection system requirements and quality program requirements?

4-8   Discuss the distinction between reliability and quality control.

4-9   Discuss the concept "quality of design."

4-10 Name and discuss the three major areas of control covered by quality of conformance.

4-11 Whom should a quality-control organization report to? Name some of the type of personnel of a modern quality-control organization.

4-12 What are some of the duties of a quality-control manager?

4-13 In large organizations the overall assurance function is called *Product Assurance*. Name the disciplines that are covered by this title.

4-14 Draw a typical organizational chart showing the relationship of product assurance to other company functions and to top management.

4-15 Draw an example of an organizational chart showing the breakdown of the assurance function for a
    (a) medium-sized company.
    (b) small company.

4-16 Discuss some of the aspects of motivating employees for quality. What is an effective method?

4-17 After employees have been motivated to produce a quality product, they have to be indoctrinated. Discuss some of the methods used for indoctrination.

4-18 What are the duties and responsibilities of the quality-control engineer?

4-19 List 10 areas of control with which quality planning is concerned.

4-20 What is the purpose of a quality survey team? What are the methods of its operation?

4-21 What are the contractor responsibilities as delineated in MIL-I-45208A?

4-22 Discuss the scope of MIL-Q-9858A.

4-23 What are the quality program management requirements imposed by MIL-Q-9858A?

4-24 Prepare a quality-control procedure for a product of your choice.

4-25 List the major areas of control which a typical quality-control manual would cover.

5

CONTROLLING THE PROCESS

In Section 4.3 it was stated that no two objects in the world around us, nor any two actions performed by the same or by different individuals, are exactly identical. In this section statistical methods will be discussed which will permit determination of whether the variability of a product is due to chance variations or whether there are assignable causes.

5-1 VARIABILITY IN MATERIALS, MACHINES, AND OPERATORS

Despite the fact that a machine operator uses the same precision methods and machines and endeavors to produce identical parts, the finished products will show a definite variability.

The variability of a product can more broadly be attributed to one of two causes:

1. Variability due to chance causes.
2. Variability due to assignable causes.

Chance variations are seldom produced by one cause only but are usually the result of a number of minor interacting causes. Chance variations are random in nature; they do not occur in repetitive cycles and their occurrence is not predictable.

Variability of the product due to assignable causes results from discrepancies in one or more of the following areas: raw materials, processes and machines, as well as from conscious or unconscious actions by the operators. Once the cause for an assignable variation in a product is determined, further occurrences of this type of variability are, of course, predictable until corrective action is implemented. The latter will restore the process to chance variation. As long as the variability of a product is only due to chance effects, the manufacture of the product is said to be "under control."

Assignable causes for product variability usually signal an "out-of-control" condition in the manufacture of the product.

The quality-control chart is a statistical tool which superimposes an upper and lower sigma control limit ($\pm 3\sigma$) on readings taken on the product. The space bounded by these control limits is considered as the locus of nearly all chance variations of the product coming off the line.

To achieve greatest sensitivity, control limits are best applied to sample averages or sample ranges (a range is the arithmetic difference between the maximum and minimum value in the data), rather than to individual values. The use of a statistic, such as the sample averages, has the additional advantage that its distribution is normal regardless of the fact that the distribution of individual values is nonnormal. The normal distribution of the sample averages complies with the statistical laws of chance, and the standard deviation is therefore applicable.

5-2 QUALITY-CONTROL-CHART TECHNIQUE

(a) General Principles

In order to exercise control over a process or over a physical characteristic of an item produced, we set limits on variability due to chance effects. Since $\pm 3\sigma$ represents 99.73% of the total population, all readings that fall within this range are considered to have a variability due to chance causes. This means that in fewer than 3 cases in 1000 (2.7 to be exact), a measured physical characteristic will exceed the $\pm 3\sigma$ range by falling in the extreme tail(s) of the distribution.

If the process goes out of control due to an assignable cause, there will soon be evidence of a substantial deviation from the established $\pm 3\sigma$ limits. In the latter case, immediate corrective action is indicated to regain control of the process. By limiting ourselves to $\pm 3\sigma$ limits of variability, we have, in fact, created a gate that permits passage of 99.73% of the population. The remainder of the population, amounting to 0.27%, can be viewed in the role of a barometer as to the continuing control of the process. A rising trend in numbers in that portion of the population warrants investigation.

As long as the inspection tolerances coincide with the $\pm 3\sigma$ spread in variability production yield will be at a maximum. Significant shifts in the process due to assignable causes will result in a shift of the frequency curve of the particular characteristic of interest in relation to the inspection tolerance range. This will decrease the portion of the population which is acceptable, even though its variability is due to chance. Figure 5-1a depicts the situation where 99.73% of the population falls within the $\pm 3\sigma$ limits and the inspection tolerance limits coincide with these limits. Figure 5-1b illustrates an upward shift in the mean value of the frequency curve, showing the

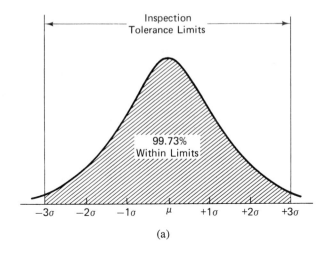

(a)

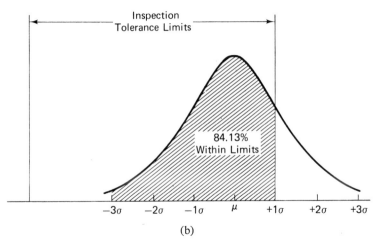

(b)

Figure 5-1 *(a) Process is under control. The ±3σ control limits coincide with the inspection tolerance limits. (b) The mean of the curve has shifted upward and the standard deviation remains unchanged. The area under the curve remaining between the inspection tolerance limits has substantially decreased.*

relative displacement of this curve with respect to the unchanged inspection tolerance range. As a result, the portion of the area under the curve which falls within the tolerance range is reduced and production yield is reduced from the original 99.73% to 84.13%.

In order to be able to assess whether a process is under control and to spot any alarming trends, it is necessary to draw sample quantities from the production line at regular intervals and measure the particular characteristic of interest. The results of these sample checks are plotted in the form of quality-control charts to permit continuous monitoring of production and timing of corrective action, when required.

(b) Control Limits and Specification Limits

Even though industrial processes do not always conform strictly with the normal curve, the assumption is made that it applies. Accordingly, since $\pm3\sigma$ (a total of 6 standard deviations) encompass 99.73% of the population, nearly all the chance parameter variations will fall within these limits. What this means in a practical instance is that if, for example, a manufacturing line produces 10,000 steel rods, a total of 9973 rods will, on the average, fall within the acceptable $\pm3\sigma$ limits.

Table 5-1 lists corresponding quantities for a range of σ values. In each case the inspection tolerance limits coincide with the σ values. From Table 5-1, if a range in variability of $\pm4\sigma$ is considered, 99.94% of the population would be included. This means that 9994 of the 10,000 rods would be expected to fall within this range strictly on the basis of chance alone. If we restrict ourselves to a range of variability of $\pm2.5\sigma$, only 9876 of the 10,000 rods would fall within these limits. In order to cover the entire population, that is, all 10,000 rods, we would have to go beyond the $\pm4\sigma$ limits.

Table 5-1 *Number of Acceptable Rods for a Range of σ Values (The inspection tolerance limits coincide with the σ values).*

σ Values	Number of Acceptable Rods From a Lot of 10,000
$\pm2.5\sigma$	9876
$\pm3.0\sigma$	9973
$\pm3.29\sigma$	9990
$\pm4.0\sigma$	9994

If the inspection tolerance limits for a characteristic are chosen so that they agree with the $\pm3\sigma$ variability of this parameter, we would be assured that, as long as the process remains under control (no significant change in the mean and the standard deviation), only 27 of 10,000 rods would fall outside this range.

If the required inspection tolerance limits greatly exceed the $\pm3\sigma$ control limits, that is, if 100% of the product is easily passed through inspection without even coming close to the limiting tolerance values, then the manufacturing process is better than needed and therefore too costly. If, on the other hand, the inspection tolerance limits are less than the $\pm3\sigma$ control limits, either the process variability needs to be reduced (increased cost!), or, if possible, the inspection tolerance limits have to be widened to improve production yield. In either case, specification requirements and cost need to be considered.

(c) Types of Control Charts

Control charts can be categorized in two general classes:

1. Control charts for inspection by attributes (count).
2. Control charts for inspection by variables (measurement).

The method of inspection by attributes is based on a simple "go, no–go" evaluation of the product. For example, the final test in a vacuum-tube production line is a continuity test for the filament. Either the filament is intact and is acceptable, or it is broken, in which case the tube is judged defective.

Obviously, there is no way to tell whether the defective item is only slightly or considerably over the limit when using inspection by attributes. Attribute control charts include:

1. *p* charts.
2. *np* charts.
3. *c* charts.
4. *u* charts.

When the quality of a manufactured item is most effectively determined by one or more variable parameters, and individual readings of these parameters will reveal whether the process is stable or shows shifting trends, the variable control chart is used. The variable chosen for control charts must be a parameter that is measurable and that can be expressed in numbers. Examples of such parameters are weight, physical dimensions, concentration, pH factor, hardness, and others. Variable control charts include:

1. $\bar{x}$ charts.
2. *R* charts.
3. Moving-range charts.

Attribute and variable control charts will be discussed in detail in the remainder of this chapter.

1. Control Charts for Inspection by Attributes

p CHARTS

The *p* charts deal with fraction defectives. Fraction defective, *p*, is defined as the ratio of the number of defective items to the total number of items inspected.

> **EXAMPLE 5-1** Determine *p* for the following case. A manufacturer draws 10 items hourly from his production line and subjects them to an attribute inspection. By the end of the day a total of 80 units were inspected and 4 defective units found. What is the fraction defective *p* for the day?
>
> **Solution**
>
> $$p = \frac{\text{number of defectives}}{\text{total number inspected}} = \frac{4}{80} = 0.05$$

Instead of fraction defective, we may also express the defect rate of a product in terms of *percent defective*. Percent defective is 100 times fraction defective. In the preceding example the percent defective is

$$0.05 \times 100 = 5\%$$

Control charts for attributes are based on the binomial distribution. The reason for the use of a discrete probability distribution is that there are only two classifications of product: defective or non-defective; we can count the number of defectives in a given number inspected and the probability of occurrence of a defective can be assumed to be constant.

The upper and lower control limits for the attribute control chart are spaced from the average fraction defective $\bar{p}$ by $\pm 3\sigma$.[1]

$$\text{UCL} = \bar{p} + 3\sigma \qquad (35)$$

$$\text{LCL} = \bar{p} - 3\sigma \qquad (36)$$

where

$\quad$ UCL = upper control limit
$\quad$ LCL = lower control limit
$\quad\quad \bar{p}$ = average fraction defective of a number of sample lots
$\quad\quad \sigma$ = standard deviation for the binomial distribution

$$= \sqrt{\frac{\bar{p}(1-\bar{p})}{n}} \qquad (37)$$

$\quad\quad n$ = number of items in a sample

The following example will illustrate the use of the attribute control chart.

EXAMPLE 5-2 A manufacturer produces a quantity of 10,000 special valves during a 10-day period. Twice each day he draws a random sample of 50 valves and determines how many defectives are among them. Since this is the first order for this item, the manufacturer wishes to determine if the process is under control and establish control limits for future use.

$\quad$ Table 5-2 lists the number of defectives found in each of the 20 sample lots of 50 valves each. The last column in this table gives the fraction defective calculated by dividing the respective number of defectives by 50.

Solution The average fraction defective $\bar{p}$ is

$$\bar{p} = \frac{180}{20 \times 50} = 0.18$$

The standard deviation of the binomial distribution is

$$\sigma = \sqrt{\frac{0.18(0.82)}{50}} = 0.054$$

$$\text{UCL} = 0.18 + 3(0.054) = 0.342$$

$$\text{LCL} = 0.18 - 3(0.054) = 0.018$$

The next step is to plot the attribute control chart (see Figure 5-2). The horizontal axis identifies the sample lots; the vertical axis, fraction defective. Horizontal lines

---

[1] It is customary to use σ for the standard deviation in control-chart work.

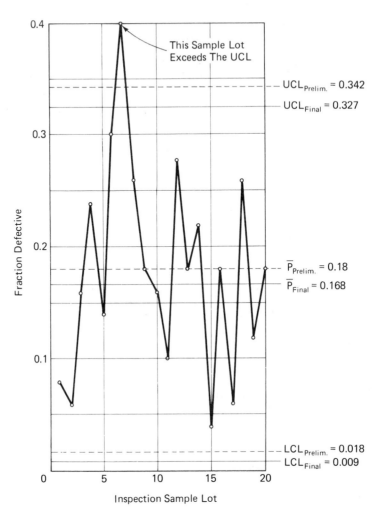

Figure 5-2 *Attribute control chart.*

representing average fraction defective $\bar{p}$, UCL, and LCL are drawn, followed by the individual fraction-defective values from Table 5-2 for each of the 20 sample lots.

Inspection of the control chart reveals that sample lot 7 exceeds the UCL. All other sample lots produced fraction-defective values well within the UCL and LCL. Since the manufacturer wants to establish realistic control limits for the new product, he will be interested to determine whether the excessive number of defects in sample lot 7 could be due to an assignable cause. If an assignable cause is established, then, after corrective action is implemented, there is justification to exclude the sample lot 7 readings. Suppose that an investigation is initiated and the review of the shop records indicate that a new operator started this particular shift and that he misunderstood a torque specification for bolts; after a brief indoctrination of the correct torquing method, this cause for production rejects is eliminated.

Table 5-2 *Record of Defects in Sample Lots.*

Sample Lot Number (n = 50)	Number of Defectives	Fraction Defective
1	4	0.08
2	3	0.06
3	8	0.16
4	12	0.24
5	7	0.14
6	15	0.30
7	20	0.40
8	13	0.26
9	9	0.18
10	8	0.16
11	5	0.10
12	14	0.28
13	9	0.18
14	11	0.22
15	2	0.04
16	9	0.18
17	3	0.06
18	13	0.26
19	6	0.12
20	9	0.18
Total 1000	Total 180	

Since the reason for the high defect count in sample lot 7 is now removed, we may recalculate the final parameters for the attribute control chart.

$$\bar{p}_{final} = \frac{160}{19 \times 50} = 0.168$$

$$\sigma = \sqrt{\frac{0.168(0.832)}{50}} = 0.053$$

$$UCL_{final} = 0.168 + 3(0.053) = 0.327$$

$$LCL_{final} = 0.168 - 3(0.053) = 0.009$$

The new lines for $\bar{p}_{final}$, UCL_{final}, and LCL_{final} are now drawn in the control chart and the earlier corresponding lines are labeled $\bar{p}_{prelim}$, UCL_{prelim}, and LCL_{prelim}.

As is now apparent, all remaining 19 sample lots fall between UCL_{final} and LCL_{final} and are not clustered either above or below the $\bar{p}$ line. The process is therefore considered under control. The final control limits may be used for future production control.

np CHARTS

In Example 5-2 we dealt with fraction defective. Since the sample quantity remained the same throughout the production run, we could just as easily have plotted the number of defectives instead of the fraction defective. The number of defectives is the product of the sample size and the fraction defective.

$$\text{number of defectives} = np \tag{38}$$

where

$$n = \text{size of sample}$$
$$p = \text{fraction defective}$$

If we divide the number of defectives by n, the fraction defective p is obtained:

$$\frac{np}{n} = p \tag{39}$$

The standard deviation of the binomial distribution for the case using number of defectives is

$$\sigma = \sqrt{n\bar{p}(1 - \bar{p})} \tag{40}$$

where $n\bar{p}$ is the average number of defectives.

EXAMPLE 5-3   What is the number of defectives of a sample lot of 30 known to be 10% defective?

Solution

$$n = 30$$
$$p = 0.1$$
$$\text{number of defectives } np = 30 \times 0.1 = 3$$

EXAMPLE 5-4   The number of defectives in a sample lot of 35 is 7. What is the fraction defective?

Solution

$$np = 7$$
$$p = \frac{7}{n} = \frac{7}{35} = 0.2$$

EXAMPLE 5-5   A manufacturer uses injection molding to produce a plastic insulation barrier. He inspects 100 barriers daily, picked randomly from production, and determines the number of defectives by visual examination. He wishes to use the data accumulated during a 10-day period to construct an attribute control chart for number of defectives. The results of the daily inspection of 100 barriers is shown in Table 5-3. Does the attribute control chart reveal that the process is under control?

Solution   The average number of defectives $n\bar{p}$ is

$$n\bar{p} = \frac{98}{10} = 9.8$$

The average fraction defective $\bar{p}$ is

$$\bar{p} = \frac{98}{1000} = 0.098$$

Table 5-3 *Record of Defects in Sample Lots.*

Sample Lot Number	Sample Size	Number of Defectives np
1	100	8
2	100	7
3	100	12
4	100	5
5	100	18
6	100	2
7	100	10
8	100	16
9	100	14
10	100	6
	Total 1000	Total 98

The standard deviation of the binomial distribution is

$$\sigma = \sqrt{9.8(1 - 0.098)}$$
$$= \sqrt{8.839} = 2.98$$
$$\text{UCL} = n\bar{p} + 3\sigma = 9.8 + 3(2.98) = 18.74$$
$$\text{LCL} = n\bar{p} - 3\sigma = 9.8 - 3(2.98) = 0.86$$

The attribute control chart for number of defectives is shown in Figure 5-3.

All points are found to be within the $\pm 3\sigma$ limit. There is, furthermore, an equal number of points above and below the center line indicative of a reasonably random distribution. Therefore, the process appears to be well under control.

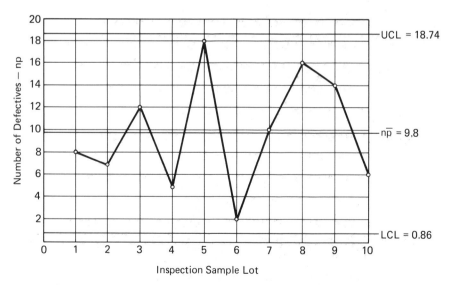

Figure 5-3 *Attribute control chart for number of defectives.*

c CHARTS

The c chart is applied in instances where we are concerned with the number of defects per unit of product. The unit of a product may involve a single unit or it may consist of a subassembly of units. The important thing is that the inspection unit remain of constant size.

Representative examples of instances when the c chart is applied are as follows:

1. The number of pinholes in equal size sheets of insulation.
2. The number of defective welds in airplane wings.
3. The number of surface scratches on printed circuit boards.
4. The number of mechanical defects per lots of a given quantity of units.

The c chart is based on the Poisson distribution on the assumption that the frequency of defects follows the Poisson law. The average and the standard deviation of the Poisson distribution are as follows:

$$\text{average of Poisson distribution} = \bar{c} = \frac{\text{Total Number of Defects}}{\text{Total Number of Units}} \quad (41)$$

$$\text{standard deviation of Poisson distribution } \sigma = \sqrt{\bar{c}} \quad (42)$$

EXAMPLE 5-6 Ten airplane wings were inspected for defective welds. The number of defects per wing ranged from 26 to 55. The total number of defects for the 10 units was 360. Determine the average and the standard deviation of the Poisson distribution as required for a c chart.

Solution The average number of defects per wing is

$$\bar{c} = \frac{360}{10} = 36$$

The standard deviation of the Poisson distribution is

$$\sigma = \sqrt{\bar{c}} = \sqrt{36} = 6$$

The UCL and LCL for the c chart are the customary $\pm 3\sigma$ away from $\bar{c}$. Hence,

$$\text{UCL} = \bar{c} + 3\sqrt{\bar{c}} \quad (43)$$

$$\text{LCL} = \bar{c} - 3\sqrt{\bar{c}} \quad (44)$$

$$\text{UCL} = 36 + 3\sqrt{36} = 54$$

$$\text{LCL} = 36 - 3\sqrt{36} = 18$$

EXAMPLE 5-7 Table 5-4 represents a listing of the number of mechanical defects observed on lots of 10 miniature worm-gear drives drawn daily from the line. This is a pilot production run lasting 20 days. Construct a c chart.

Table 5-4 *Record of Defects in Sample Lots.*

Lot Number (10 units each)	Number of Defects per Lot
1	13
2	15
3	19
4	8
5	6
6	17
7	7
8	9
9	3
10	23
11	17
12	11
13	7
14	11
15	14
16	6
17	16
18	10
19	2
20	6
20 lots (200 units)	220 total number of defects

Solution   From Table 5-4,

$$\bar{c} = \frac{220}{20} = 11$$

$$\sigma = \sqrt{11} = 3.32$$

$$\text{UCL} = 11 + 3(3.32) = 20.96$$

$$\text{LCL} = 11 - 3(3.32) = 1.04$$

See Figure 5-4 for the c chart. Examination of this chart reveals that lot 10 is outside the control limits. The manufacturer should examine the records for lot 10 to ascertain if an assignable reason for this out-of-control condition prevailed. If this can be established, the constants for the c chart should be recalculated, as demonstrated in Example 5-2.

u CHARTS

There are cases where the constant sample lot sizes, as used for the c chart, are not feasible. In those instances the u chart is used. It furnishes a constant central line but varies in control limits from sample lot to sample lot. In setting up a u chart, it is necessary to decide on a common unit which is to be used as the basis for the chart. For sample inspection lots composed, for example, of 45, 50, 40, 64, 55, and 70 pieces,

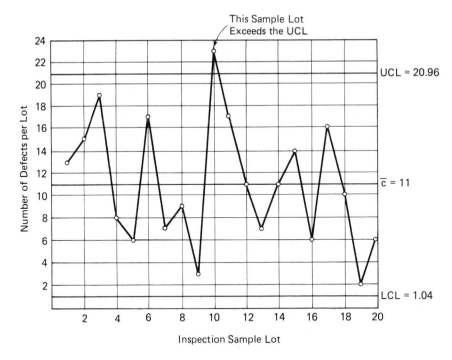

This Sample Lot
Exceeds the UCL

UCL = 20.96

$\bar{c}$ = 11

LCL = 1.04

Number of Defects per Lot

Inspection Sample Lot

Figure 5-4 *Attribute control chart for defects per lot (c – Chart).*

respectively, the common unit (referred to henceforth as the "common basic lot size") chosen may be 50 or 55, or, for that matter, any other convenient number appropriate for the range of sample sizes. The purpose of introducing the common basic lot size is for expressing the defects observed in the unequal-size sample inspection lots in terms of a common yardstick. The inspection sample lots do not necessarily have to consist of n individual units but could, for example, refer to n square yards of cloth within a continuous roll.

If, for example, the size of the inspection sample lot is 70 and the common basic lot size is 55, the inspection sample lot is equivalent to $k = 70/55 = 1.272$ common basic lots. We thus express the inspection sample lots in terms of k, the number of equivalent common basic lot sizes. In general terms k is expressed as

$$k = \frac{\text{size of inspection sample lot}}{\text{size of common basic lot}} \tag{45}$$

The values plotted in the u chart are the calculated values of u for each inspection sample lot, where u refers to the number of defects per inspection sample lot converted to equivalent common basic lots.

If c identifies the number of defects found in the inspection sample lot, u may be calculated using equation (46):

$$u = \frac{c}{k} \tag{46}$$

where

$$c = \text{number of defects in inspection sample lot}$$
$$k \text{ is as defined in equation (45).}$$

EXAMPLE 5-8 The number of defects found in one of a group of inspection sample lots is $c = 5$. The size of the lot is $n = 20$. The common basic lot size chosen was $n = 30$. Determine the value of u.

Solution

$$c = 5$$

$$k = \frac{20}{30} = 0.667 \qquad \text{[per equation (45)]}$$

$$u = \frac{5}{0.667} = 7.49 \qquad \text{[per equation (46)]}$$

The average number of defects per common basic lot may be determined from all sample inspection lots in accordance with equation (47):

$$\bar{u} = \frac{\text{total number of defects } (\sum c)}{\text{sum of } k \text{ values } (\sum k)} \tag{47}$$

Knowing k and $\bar{u}$, we are now in a position to define the control limits:

$$\text{UCL} = \bar{u} + 3\sqrt{\frac{\bar{u}}{k}} \tag{48}$$

$$\text{LCL} = \bar{u} - 3\sqrt{\frac{\bar{u}}{k}} \tag{49}$$

EXAMPLE 5-9 The data taken on three sample inspection lots are shown in Table 5-5 (in practice, a much larger number of lots would be used). A common basic lot size of $n = 100$ has been chosen. Determine the following parameters:

(a) The u values.
(b) $\bar{u}$.
(c) UCL and LCL.

Table 5-5 *Tabulation of Inspection Data.*

Inspection Sample Lot Number	Number of Parts per Inspection Sample Lot	Number of Defects per Inspection Sample Lot (c)
1	100	4
2	90	3
3	110	5
		Total 12

Solution Determine the k and u values per equations (45) and (46), respectively. Following this, $\bar{u}$ is determined. Then, for convenience, calculate the values for $3\sqrt{\bar{u}/k}$ for each lot to facilitate the determination of the control limits (see Table 5-6 for a listing of the results of all these calculations).

Table 5-6 *Tabulation of Calculated Data.*

Inspection Sample Lot Number	k	u	$3\sqrt{\dfrac{\bar{u}}{k}}$	UCL: $\bar{u} + 3\sqrt{\dfrac{\bar{u}}{k}}$	LCL: $\bar{u} - 3\sqrt{\dfrac{\bar{u}}{k}}$
1	1.0	4.00	6.00	10.00	—
2	0.9	3.33	6.33	10.33	—
3	1.1	4.55	5.73	9.73	—
	$\sum k = 3.0$				

A sample calculation using inspection sample lot 2 follows:

$$k = \frac{90}{100} = 0.9 \qquad \text{[per equation (45)]}$$

$$u = \frac{3}{0.9} = 3.33 \qquad \text{[per equation (46)]}$$

$$\bar{u} = \frac{12}{3} = 4 \qquad \text{[per equation (47)]}$$

$$3\sqrt{\frac{\bar{u}}{k}} = 3\sqrt{\frac{4}{0.9}} = 6.33$$

$$\text{UCL} = 4 + 6.33 = 10.33 \qquad \text{[per equation (48)]}$$

$$\text{LCL} = 4 - 6.33 = \text{not a positive number} \qquad \text{[per equation (49)]}$$

Note: Since the number of defects corresponding to the lower control limit (LCL) is not a positive number, the overall control range for this particular inspection sample lot extends from 0 to 10.33.

Figure 5-5 depicts the u chart that is based on the calculations above.

EXAMPLE 5-10 Referring to Example 5-9, would a fourth inspection sample lot with $n = 80$ and $c = 16$ defectives fall within the control limits?

Solution

$$k = \frac{80}{100} = 0.8$$

$$u = \frac{16}{0.8} = 20$$

$$\bar{u} = \frac{4 + 3 + 5 + 16}{1 + 0.9 + 1.1 + 0.8} = 7.36$$

$$3\sqrt{\frac{\bar{u}}{k}} = 3\sqrt{\frac{7.36}{0.8}} = 9.09$$

$$\text{UCL} = 7.36 + 9.09 = 16.45$$

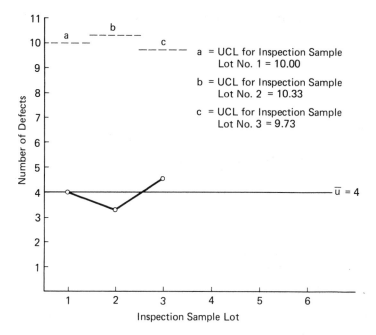

Figure 5-5 *Plot of u-Chart for Example 5-9.*

Since the value of $u = 20$ calculated for this inspection sample lot exceeds UCL = 16.45, the inspection sample lot is outside the control limits.

2. Control Charts for Inspection by Variables

Attribute control charts, discussed in the previous section, are based on the presence or absence of a quality characteristic. The actual value of this characteristic is not ascertained.

Inspection by variables requires that one or more variable quality characteristics of the product be measured for actual value. Variable characteristics include pressure, temperature, and concentration.

The variable control charts most commonly used are the $\bar{X}$ and R charts. They are based on measures of central tendency and spread. The $\bar{X}$ chart controls the mean of a process, while the R chart is concerned with its range (the range is equal to the difference between the largest and smallest value measured). When a process is under control, both of the sample statistics, the mean and the range, should fall within the control limits. When the mean, or the range, or both, are found to have shifted significantly, undesirable changes in the process may be assumed to have occurred.

Control charts for variables are usually based on sampling distributions of the mean and the range of a group of sample lots. As will be recalled, sampling distributions of means approach normality regardless of the nonnormality of the actual measurements. Therefore, the use of standard deviation in the determination of statistical control limit of means is justified.

In addition to $\bar{X}$ and R charts there are moving-average and moving-range charts. These are particularly suited for monitoring continuous processes, such as in chemical processing plants. These types of control charts are beyond the scope of this chapter.

$\bar{X}$ Charts

This type of control chart depicts variations in the means of sample lots drawn from production. The variations are shown around a central line which represents $\bar{\bar{X}}$, the grand mean (the mean of the sample means $\bar{X}$). The UCL and LCL are $3\bar{\sigma}$ limits either side of $\bar{\bar{X}}$, where $\bar{\sigma}$ is the grand average of σ, the standard deviation of the sample lots.

$$\text{UCL} = \bar{\bar{X}} + 3\bar{\sigma} \tag{50}$$

$$\text{LCL} = \bar{\bar{X}} - 3\bar{\sigma} \tag{51}$$

Calculation of the $\bar{\bar{X}}$ from a group of sample lots is not very time-consuming, but the computation of σ for each sample lot and of the overall $\bar{\sigma}$ would be very lengthy.

Fortunately, these lengthy calculations are not necessary, since a tabulation may be used which permits determination of the 3σ control limits for $\bar{X}$ charts using the value of $\bar{R}$. Table 5-7 furnishes factor A_2 for a series of sample sizes for use in $\bar{X}$-chart design. This factor is then substituted in equations (52) and (53) to obtain the desired control limits. The table also contains similar factors for constructing the R-chart.

Table 5-7 *Tabulation of Factors A_2 for $\bar{X}$ Charts and D_3 and D_4 for R Charts.*

Size of Inspection Sample Lot n	Factor for $\bar{X}$ Chart: A_2	Factors for R Chart	
		D_3	D_4
2	1.88	0	3.27
3	1.02	0	2.57
4	0.73	0	2.28
5	0.58	0	2.11
6	0.48	0	2.00
7	0.42	0.08	1.92
8	0.37	0.14	1.86
9	0.34	0.18	1.82
10	0.31	0.22	1.78
11	0.29	0.26	1.73
12	0.27	0.28	1.72
13	0.25	0.31	1.69
14	0.24	0.33	1.67
15	0.23	0.35	1.65

This table was adapted from Eugene L. Grant, *Statistical Quality Control*, 2nd ed. McGraw-Hill Book Company, N. Y., 1952. Used by permission.

The upper and lower control limits for the $\bar{X}$ chart are

$$\text{UCL} = \bar{\bar{X}} + A_2\bar{R} \tag{52}$$

$$\text{LCL} = \bar{\bar{X}} - A_2\bar{R} \tag{53}$$

where the center line of the control chart is $\bar{\bar{X}}$.

R CHART

The R chart depicts variations in the range of sample lots drawn from production. As in the $\bar{X}$ chart, control limits are usually set at 3 standard deviations above and below the central value.

The center line for the R chart is $\bar{R}$. The upper and lower control limits are:

$$\text{UCL} = D_4\bar{R} \tag{54}$$

$$\text{LCL} = D_3\bar{R} \tag{55}$$

where factors D_4 and D_3 may be determined from Table 5-7. The following example will illustrate the use of Table 5-7 and the calculation of the control limits for a $\bar{X}$ chart and R chart. In practice the R chart is constructed first to assure that $\bar{R}$ used in the calculation for the $\bar{X}$ chart is controlled. In order to confirm that a production process is fully under control both the R chart and $\bar{X}$ chart must show no points outside the control limits.

EXAMPLE 5-11   Six inspection sample lots of five amplifiers each are drawn from production. Table 5-8 lists the power output obtained for each amplifier. Plot an $\bar{X}$ chart.

Table 5-8 *Tabulation of Power Outputs for Audio Amplifiers.*

Inspection Sample Lot Number	Amplifiers					$\bar{X}$	R
	1	2	3	4	5		
1	12	8	11	9	10	10.0	4
2	9	10	12	11	13	11.0	4
3	15	12	9	10	13	11.8	6
4	11	13	14	12	15	13.0	4
5	10	12	16	13	12	12.6	6
6	14	12	13	11	9	11.8	5
					Total	70.2	29

Solution   First, $\bar{X}$ and R are calculated for each inspection sample lot (see the last two columns in Table 5-8). Sample calculations for inspection sample lot 3 in Table 5-8 follow:

$$\bar{X} = \frac{15 + 12 + 9 + 10 + 13}{5} = 11.8$$

$R = 15 - 9 = 6$ (the difference between the highest and lowest values in the group)

Next, $\bar{\bar{X}}$ and $\bar{R}$ are calculated.

$$\bar{\bar{X}} = \frac{\text{sum of all } \bar{X}\text{'s}}{\text{number of inspection sample lots}} = \frac{70.2}{6} = 11.7$$

$$\bar{R} = \frac{\text{sum of all } R\text{'s}}{\text{number of inspection sample lots}} = \frac{29}{6} = 4.83$$

The UCL and LCL for the R chart are determined using equations (54) and (55). Factors D_4 and D_3 from Table 5-7, for $n = 5$, are

$$D_4 = 2.11$$
$$D_3 = 0$$
$$\bar{R} = 4.83 \text{ as determined above}$$
$$\text{UCL} = 2.11\,(4.83) = 10.19$$
$$\text{LCL} = 0(4.83) = 0$$

See Figure 5-6 for a plot of the R chart. An examination of Figure 5-6 reveals that all points are below the upper control limit. It should be noted that no lower control limit exists for this case. The UCL and LCL for the $\bar{X}$ chart are now determined using equations (52) and (53). From Table 5-7 we determine the value of A_2 for $n = 5$.

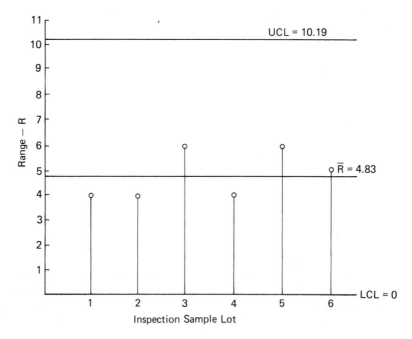

Figure 5-6 *R-chart for Example 5-11.*

$$A_2 = 0.58$$

$$\bar{\bar{X}} = 11.7, \text{ as determined earlier.}$$

$$\text{UCL} = 11.7 + 0.58(4.83) = 14.50$$

$$\text{LCL} = 11.7 - 0.58(4.83) = 8.90$$

The center line of the $\bar{X}$ chart is $\bar{\bar{X}} = 11.7$. With these calculations completed, the $\bar{X}$ chart can now be constructed (see Figure 5-7). An examination of Figure 5-7 shows no point outside the control limits.

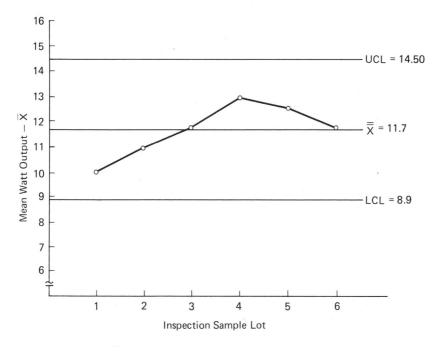

Figure 5-7 $\bar{X}$-*chart for Example 5-11.*

(d) Interpretation of Control Chart Patterns

When there is evidence of points falling outside the control limits, there is very good reason to believe that the process is out of control. Certainly, if more than one point is outside the control limits, this will be the case. If one point falls outside the control limits, there could still be a reasonable chance that there is no assignable cause for this variation. The reason for this is that the $\pm 3\sigma$ control limits do not encompass 100% of the universe. On the average, we may expect 27 in 10,000 instances where points in a normal distribution fall outside the $\pm 3\sigma$ limits. The chance that a point in a control chart will fall outside the $\pm 3\sigma$ control limits without assignable cause is therefore 1 in 370. The chance that two or more points will fall outside the control limits (for no assignable reasons) is very small indeed.

When an out-of-control point shows up in a control chart, it should be investigated for an assignable cause; but even if such a cause does not immediately become apparent, hasty action, such as shutting down production, is not always indicated. It will be wise to look first for tell-tale trends, such as shifts in process average, with more and more points falling on one side of the central line, or an increased spread of the process. To complicate matters, these two conditions could occur simultaneously.

A properly controlled process is always discernible on a control chart by the location of an approximately equal number of points on either side of the central line. A further indication of continued control is the persistence of this pattern. Abrupt changes in the pattern demand attention and analysis. When the points start crowding toward both the upper and lower control lines (or, for that matter, toward the center), an increase in process spread with potential loss of control is indicated. Concentration of the majority of points on one side of the central line is indicative of a shift in process average. Shifts in process average and/or increases in process spread decrease the available margin provided by the fixed specification tolerance limits and are therefore to be watched. The prudent manufacturer will see to it that trends indicating potential process problems are promptly analyzed and that corrective action is implemented before the control limits are exceeded.

(e) Evaluation of Process Capability

Prior to starting a large-scale production run it is desirable to make an analysis of the capability of the intended process to comply with the imposed design tolerances. An analysis of process capability can be performed by comparing the total natural variability of the process to the design tolerance spread.

For example, in order to study the specific process characteristic of interest, such as the length or diameter of a rod, measurements are made on a sufficiently large sample of units. A quantity of 100 pieces, for example, drawn from an initial production trial run will normally suffice. The next step is to calculate the standard deviation, σ, of the observed readings. Since $\pm 3\sigma$ represents 99.73% of the total population, the manufacturer would be quite safe if 6σ, which nearly represents the total variability of the process, does not exceed the total specification tolerance spread.

It should not be overlooked, of course, that some differences between the initial trial run and subsequent production are possible and must be anticipated, despite all efforts to make the trial run as representative of production conditions as feasible. Consequently, the total variability of the process should preferably be a good deal less than the total specification tolerance spread. A criterion that the total variability be about 75% of the total tolerance spread is considered reasonable.

As an illustration, let us assume that the specification calls for a steel rod with a diameter of 0.400 $\pm$ 0.005 in. The total specification tolerance range is therefore:

$$0.005 \times 2 = 0.01 \text{ in}$$

The total variability of the process based on the 75% criterion should therefore not exceed $6\sigma = 0.75 \times 0.01 = 0.0075$ in. and the maximum standard deviation should be no more than $\sigma = 0.0075/6 = 0.00125$. A trial run could verify whether the mean of the observed characteristic is sufficiently close to the desired central value, 0.400, and whether the total variability of the process is close to the calculated 6σ value.

PROBLEMS

5-1 (a) What two possible causes may variability in products be attributed to?
　　(b) Discuss the two causes of variability.

5-2 (a) What is the advantage of using sample averages as a measure of a population characteristic?
　　(b) Define a sample range.

5-3 (a) When a population parameter is normally distributed, what percentage of the total population falls between ± 2.5 standard deviations?
　　(b) What level of variability is usually acceptable when a process is under control?

5-4 (a) What are assignable and nonassignable causes of product variability?
　　(b) It is decided to accept a product with a variability of $\pm 3\sigma$. Assuming a normally distributed, controlled process, how many rejects may, on the average, be expected in a population of 100,000?

5-5 (a) Discuss the relationship between design tolerance limits and variability due to chance causes.
　　(b) In a normally distributed, controlled population, what percentage of the population is not included in $\pm 4\sigma$?

5-6 A normally distributed population of capacitors has a mean capacitance value of $\bar{X} = 100$ microfarads (μF) and a standard deviation of $\sigma = 5$ μF. The process is under control. The advertised nominal value and tolerance of this particular design is 100 μF $\pm 10\%$.
　　(a) Is the variability of the process compatible with the acceptable tolerance spread for capacitance?
　　(b) If a population of 10,000 capacitors is tested, how many will, on the average, be rejected considering the $\pm 10\%$ tolerance?
　　(c) How many more capacitors could be accepted if the tolerance in part (b) is increased to $\pm 15\%$?

5-7 (a) Name four types of attribute control charts and three types of variables control charts.
　　(b) Which control chart deals with fraction defectives?

5-8 (a) A manufacturer finds, on the average, 6 rejects in lots of 100 units of his product. Determine the fraction and percent defectives.
　　(b) If the standard deviation in part (a) is $\sigma = 0.015$, what are the upper and lower control limits for an attribute control chart?

5-9   The following tabulation lists the result of inspection of 10 sample lots of 40 units each:

Sample Lot Number	Number of Defects
1	11
2	2
3	10
4	18
5	5
6	12
7	6
8	3
9	11
10	8
	Total 86

Determine the upper and lower control limits and plot the attribute control chart (p chart).

5-10   Refer to Problem 5-9. Determine if any of the sample lots fall outside the control limits. (Should this be the case, assume that the problem was due to assignable causes and, after censuring the readings, add the final control limits to the chart.)

5-11   (a) For a sample size of 100 units and a fraction defective value of 0.05, what is the number of defectives and the percent defective?
(b) For the sample given in part (a), what is the standard deviation of the binomial distribution for 10 samples and for $\bar{p} = 0.05$?
(c) A lot of 50 units contains 5 defectives. What is the fraction-defective value?

5-12   The average number of defectives of a number of sample lots is $n\bar{p} = 7.8$. The average fraction defectives is $\bar{p} = 0.075$. Determine the upper and lower control limits of the attribute control chart.

5-13   (a) When are c charts used? Give typical examples.
(b) What probability distribution is used as a basis for c charts?
(c) What is the formula for the standard deviation of this distribution?

5-14   Twenty sample sheets of Mylar insulation were examined for the presence of pinholes. The number of pinholes per sheet ranged from 5 to 16 for 17 sheets and from 24 to 28 for the remainder. The total number of pinholes for the 20 sheets was 180.
(a) Determine the average and the standard deviation of the Poisson distribution as required for a c chart.
(b) How many sample sheets fall outside the control limits?

5-15   (a) When are u charts used?
(b) What is the purpose of using the common basic lot size?
(c) If the size of the inspection sample lot is 100 and the common basic lot size is 50, how many equivalent common basic lot sizes k does the inspection sample represent?

5-16   A manufacturer draws a daily inspection sample lot of 30 units from the production line. The first day he finds 6 defects.

(a) Based on a common basic lot size of $n = 50$, what is the value for u (the number of defects per inspection sample lot converted to equivalent common basic lots)?

(b) What are the upper and lower control limits for the first day's sample? Assume that $\bar{u} = 9.6$.

5-17 The following tabulation lists inspection data for five sample inspection lots. The common basic lot size is $n = 50$.

Inspection Sample Lot Number	Number of Parts per Inspection Sample Lot	Number of Defectives per Inspection Sample Lot
1	90	6
2	80	3
3	100	5
4	85	2
5	105	4

(a) Determine the upper and lower control limits for a u chart.

(b) Plot the u chart and identify the individual u values.

5-18 (a) State the difference between attribute control charts and variables control charts.

(b) Give examples of characteristics that are suitable for variables control charts.

(c) Which are the two most commonly used variables control charts.

5-19 (a) State the general UCL and LCL formulas for the $\bar{X}$ chart.

(b) State the UCL and LCL formulas using factor A_2.

(c) What determines the center line for the $\bar{X}$ chart?

5-20 (a) Ten inspection sample lots of five power solenoids each are drawn from production and measured for pull-in current. The readings obtained range from 8 to 34 amperes (A). The following tabulation lists the readings obtained for each sample. Compute the necessary parameters and construct the $\bar{R}$ chart.

Inspection Sample Lot Number	Solenoid				
	1	2	3	4	5
1	18	16	18	21	18
2	17	18	21	19	19
3	16	17	16	19	20
4	19	19	17	18	20
5	32	21	30	22	34
6	17	22	21	16	17
7	16	16	17	18	19
8	23	24	16	17	21
9	8	9	9	10	8
10	20	20	22	21	16

(b) Examine the control chart and determine if any points fall outside the control limits. If no more than one point is found outside the control limits assume that this point is due to a nonrelevant failure.

5-21 Refer to Problem 5-20.
   (a) Plot the $\bar{X}$ chart.
   (b) Determine whether any points fall outside the control limits.   *5 and 9*
   (c) Having plotted the $\bar{X}$ and R charts, what conclusions may be drawn about the process?   *Out of control*

5-22 A small engine plant produces pistons whose lengths must be within 2.125 ± 0.007 in. If the total variability of piston lengths is limited to 75% of the specification tolerance range, what is the maximum tolerable standard deviation?   *.00175*

5-23 Discuss control-chart patterns that are indicative of loss of process control.

6

PRINCIPLES OF ACCEPTANCE CONTROL

Acceptance control of manufactured goods is a major function of statistical quality control. No production line is perfect nor are checks and controls—there will be defects. Quality-control techniques and in-process inspections will help tremendously, but completed production lots will, in all likelihood, still contain a small percentage of unacceptable parts. It is simply not practical, nor economically feasible, to prevent all defects from reaching the shipping platform.

It is, however, possible to control the quality of lots accepted by statistical sampling schemes. The latter define the risk of accepting production lots based on limiting the number of defects permitted in sample lots drawn randomly from production.

6-1 OBJECTIVES OF ACCEPTANCE CONTROL

Acceptance control encompasses the procedures and tests applied to completed product by either the producer or the receiver of the goods, or by both. The objective of acceptance control is to ensure that the final product, submitted for acceptance, is consistent with the quality specified.

6-2 TYPES OF ACCEPTANCE CONTROLS

The following general types of acceptance controls are practiced:

1. Spot checks.
2. 100% inspection.

3. Certification.
4. Acceptance sampling.

Each of these methods will be discussed briefly.

(a) Acceptance Control by Spot-Check

Excepting the method of certification, the spot-check method is probably the least scientific of the cited methods. Even the name suggests a certain arbitrary aspect. Spot-check procedures frequently call for testing of 10% samples of each production lot on a go–no go basis, with the acceptance of only those lots whose samples were failure-free. The method of selection of these samples is often not defined, and the choice of a 10% sample with a zero-defects criterion may often not be realistic. This method also gives different levels of protection for different lot sizes.

Another approach to spot-checking involves the periodic removal of a finished production unit for an acceptance test. For example, every tenth, twentieth, fiftieth, etc., unit may be drawn for this test. The choice of frequency is arbitrary and is subject to later adjustment if needed. Should a defective part be drawn, then all the parts between the previously drawn part and this defective part are tested on a 100% basis and any defective units removed.

For example, assume a production lot of 1000 pieces and a spot-check frequency of every fiftieth unit. This means that a total of 20 sample units will be drawn from production, assuming none of these is found to be defective. If, for example, the third sample is defective, then units 101 through 149 (a total of 49 units) are individually checked. If no defects are found, the defective unit (No. 150) is replaced and the spot-check continues for the rest of the production lot. If there are defects among the 49 units, they are also replaced.

The problem with the spot-check method is that it is not based on the laws of probability and therefore does not relate to sampling risks. The 100% inspection imposed between a defective sample and the previous sample, as described above, is costly and statistically unwarranted.

Last but not least, whereas scientific sampling inspection can discriminate to a desired degree between good and bad lots, the spot-check method does not have this ability. In practice, therefore, a statistical sampling method is superior to the spot-check method.

(b) Acceptance Control by 100% Inspection

Inspection of finished production on a 100% inspection basis is a technique which, with proper controls, theoretically should be one of the surest ways to eliminate defective products from being shipped. In practice, the 100% inspection is not as foolproof as may be expected. Experience has shown that the monotony and repetition inherent in 100% inspection tends to create boredom and fatigue, with the result that not all defective units are eliminated. Other undesirable features of 100% inspection are the very high cost and the tendency to eliminate inspection of some of the characteristics to reduce cost. Automated 100% inspection is frequently not feasible or too costly and is subject to mechanical or electrical breakdown.

(c) Certification

Certification as a means of acceptance control involves the receipt of a certificate from the vendor stating that certain tests and inspection have been implemented. This certificate is usually signed by a responsible management official. If the company is considered reliable and has established a good past history of quality, such certificates are sometimes used as evidence of product quality in lieu of acceptance control tests. Obviously, the buyer who uses this method assumes a certain risk.

(d) Acceptance Sampling

In contrast to the stated disadvantages of spot-checks, 100% inspection, and certification, the method of acceptance sampling has several desirable features. It is relatively inexpensive, less time consuming, not fatiguing, and it is based on well-established principles of probability theory. Acceptance sampling by attributes (go–no go) has been widely accepted as the preferred method of acceptance control. Military procurement would today be unthinkable without acceptance sampling.

Tables, charts, and graphs have been developed to facilitate the implementation of acceptance sampling and the selection of specific sampling plans. MIL-STD-105D[1] and MIL-STD-414[2], to be discussed later, are sources of widely used military acceptance control information. MIL-STD-105D and MIL-STD-414 define a range of acceptable quality levels (AQL). These quality levels limit the percentage of allowable defects in the production lot.

The charts developed for acceptance sampling define criteria for the size of the sample to be tested based on the size of the production lot, as well as the maximum number of defectives permitted in the sample. When these criteria are met, the probability of accepting lots with this defective rate is specified by the applicable plan. There is, to be sure, an inherent risk in all acceptance sampling plans of rejecting lots that should be accepted, and vice versa. The level of protection and the risk to producer and consumer as afforded by the various sampling plans are a matter of negotiation between the parties concerned. Economic factors relating to the cost of inplementing various inspection plans, as well as the consequence of passing unacceptable parts or material, will enter this decision-making process.

---

[1] MIL-STD-105D, "Sampling Procedures and Tables for Inspection by Attributes," Department of Defense, April 29, 1963.

[2] MIL-STD-414, "Sampling Procedures and Tables for Inspection by Variables," Department of Defense, June 11, 1957.

6-3 FUNDAMENTALS OF ACCEPTANCE SAMPLING

(a) Nature of Acceptance Sampling

Acceptance sampling is the method of evaluating a group of units, drawn from a production lot, in order to assess the acceptability of the whole lot. Inasmuch as only a portion of the whole lot is inspected, there is always a risk that the quality of the sample will not reflect the quality of the lot. As a result, production lots may, at times, undeservedly be rejected and faulty lots accepted. The former situation will be mainly of interest to the producer. He wants to minimize the risk of having good lots rejected. This risk is identified as the producer's risk. The risk of accepting bad lots in a sampling scheme is, of course, the primary concern of the consumer. This risk is called consumer's risk. The method by which both the producer's risk and the consumer's risk are quantified statistically is by constructing an operating characteristic curve.

Before proceeding, let us define the several sampling terms used above:

(b) Producer's Risk

The *producer's risk*, α, is the probability of having production lots rejected as a result of a sampling plan, even though they meet the specified acceptable quality requirements. The most frequently used value for the producer's risk is $\alpha = 0.05$. This means that the producer assumes a 5% risk of having production lots, which in reality comply with the specified AQL, rejected by the sampling plan. Conversely, he has a $1 - 0.05 = 0.95$, or 95%, chance of having production lots accepted which meet the specified AQL.

(c) Consumer's Risk

The consumer's risk, β, is the probability of accepting production lots as a result of a sampling plan which do not meet the poorest tolerable quality requirements.

The risk assumed by the consumer is usually set at $\beta = 0.1$, or 10%. This means that the consumer assumes a 10% risk of accepting production lots so poor in quality that they are no longer tolerable by him. This quality level (which may be many times the percent-defective value of the AQL) is designated as the lot tolerance percent defective. At the same time, the consumer has a $1 - 0.1 = 0.9$, or 90%, chance that the sampling plan will reject production lots of LTPD quality.

(d) Acceptable Quality Level

The *acceptable quality level*, or AQL, is the maximum percent defective that for purposes of sampling inspection can be considered satisfactory as a process average (lot has a high probability of acceptance, usually 95%).

(e) Lot Tolerance Percent Defective

The *lot tolerance percent defective*, or LTPD, is that quality level (in percent defectives) which is so poor that is is at the point of being intolerable and therefore has a low probability of acceptance (usually 10%). The usual sampling plan will therefore accept good lots (lot quality = AQL) 95 out of 100 times and reject the not-quite-tolerable lots (lot quality = LTPD) 90 out of 100 times. Lot quality levels between the AQL and LTPD values have probabilities of acceptance as defined by the OC curve.

(f) Operating Characteristic Curve

The *operating characteristics curve*, or OC curve, is a graph depicting the relationship between process defect percentage and the corresponding probability of acceptance of such lots by the sampling plan.

1. Power of Discrimination

An ideal OC curve which provides perfect discrimination between good lots and bad lots is shown in Figure 6-1. As can be seen, when the quality of submitted lots does

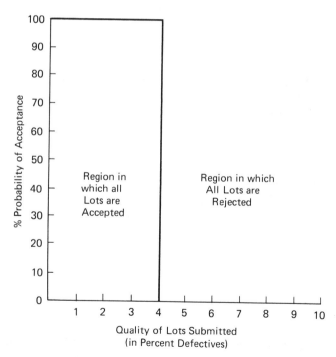

Figure 6-1 *Operating characteristic curve for an ideal sampling plan.*

not exceed 4% defectives, there is a 100% chance of acceptance; above 4% defectives, there is 0% chance of acceptance. In practice, no OC curve can be made to be that discriminatory; the power of discrimination of sampling plans varies depending on sample size and on the assigned accept/reject criterion of the plan.

A sampling plan and its associated OC curve are fully defined by the sample size and the acceptance number. For example, a sampling plan may be specified as follows:

Sample size n	50
Acceptance number (A_c)	3
Reject number (R_e)	4

This is interpreted that when a random sample of 50 units is drawn from the production lot and 100% inspected, the production lot will be accepted if no more than three defectives are found in the sample lot. The production lot will be rejected if four or more defectives are present. The reject number R_e is always equal to $A_c + 1$.

The fact that an OC curve does not discriminate good from bad lots perfectly is due to the inherent sampling risks of the plan. No matter how carefully we design an acceptance sampling plan, there always remains some risk that good lots will be rejected and bad lots accepted. However, the larger the sample size, the better the ability to discriminate good from bad lots.

2. Typical OC Curve

A more realistic OC curve than that shown in Figure 6-1 is the one depicted in Figure 6-2. This OC plot is curved. It indicates a 100% probability of acceptance of submitted lots when the quality of the lots is 0% defectives (which is to be expected), and a near 0% probability of acceptance for 12% defective lots.

Points A and B on the OC curve refer, respectively, to the acceptable quality level, AQL, and the lot tolerance percent defective, LTPD. Point A is situated on the OC curve at the intersection of a horizontal drawn through the 95% probability point, while point B is at the intersection of the 10% horizontal line and the curve. These probability percentages agree with the AQL and LTPD acceptance probabilities that are common.

The corresponding quality of lots submitted from inspection of Figure 6-2 is

$$AQL = 1\%$$
$$LTPD = 8\%$$

The OC curve permits the determination of the probability of acceptance for lots of varying quality. For example, in Figure 6-2 the probability of acceptance can be determined for any lot defective rate between 0 and 12%. For a submitted lot that is 3% defective, the probability of acceptance will be 70% (see point P).

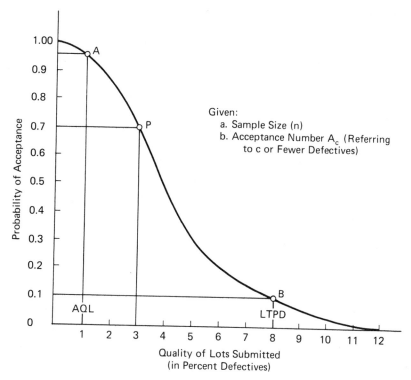

Figure 6-2 *Typical OC—curve for sampling plan.*

3. Calculation of OC Curve

The OC curve depicts the graphical relationship between a range of lot quality levels and the corresponding probability of acceptance and is based on the Poisson distribution (see Section 2.6).

The following two parameters are needed to calculate and plot an OC curve;

1. The sample size, n
2. The acceptance number, A_c (referring to c or fewer defectives).

We assume a range of values of percent defective for the lot submitted for inspection and calculate the expected number of defects by multiplying by the sample size n. Example 2-21 (p. 32) illustrates the determination of the expected number of defects for a random production sample of 50 resistors when, on the average, 10% of the produced resistors are defective ($p = 0.1$).

The expected number of defectives for this case was determined as follows:

$$np = 50(0.1) = 5$$

In the case of an OC curve calculation, the product quality is not restricted to one value but is varied over a range. The following example illustrates an assumed range of percent defective from 2% to 20%, together with the corresponding expected number of defectives.

EXAMPLE 6-1 The size of a sample drawn from production is $n = 50$. Calculate the expected number of defectives for 2%, 5%, 10%, 15%, and 20% defective lots.

Solution

Percent Defective	Expected Number of Defectives ,np
2	$np = 50(0.02) = 1.0$
5	$np = 50(0.05) = 2.5$
10	$np = 50(0.10) = 5.0$
15	$np = 50(0.15) = 7.5$
20	$np = 50(0.20) = 10.0$

It will be noticed that the expected number of defectives for 5% and 15% defective lots renders values for expected number of defectives which are not whole numbers. Although it is difficult to visualize 2.5 or 7.5 defective parts in a sample, this is mathematically acceptable.

Once the expected number of defectives (np) for an assumed range of percent-defective lots has been determined and an acceptance number c is known, we may employ the cumulative Poisson tables (Table A11, Appendix A) to determine the probability of c or fewer defectives occurring for each value of np.

If an acceptance number of $A_c = 4$, for instance, is assumed in Example 6-1, the calculation of the probabilities using the Poisson distribution or the use of the Poisson tables would be based on the following proposition: for each of the assumed 2%, 5%, 10%, 15%, and 20% defective production lots, what is the probability of 4 or less defectives occurring in an inspection sample of $n = 50$?

The cumulative probability of 4 or less defectives is the sum of five discrete probabilities:

1. Probability of 0 defectives
2. Probability of 1 defective
3. Probability of 2 defectives
4. Probability of 3 defectives
5. Probability of 4 defectives

Sum: the probability of 0 to 4 defectives

The calculation of individual Poisson terms has been shown in Table 2-7 (p. 34). In that table the probability of k occurrences of an event, when the expected number of occurrences is a, was given as

$$P = \frac{a^k \cdot e^{-a}}{k!}$$

Applied to the calculation of probabilities for the individual Poisson terms of a cumulative probability expression, the basic formula above is expressed as

$$P_{(n, p, r)} = \frac{(np)^r \cdot e^{-np}}{r!} \tag{56}$$

where

$$n = \text{size of sample}$$
$$p = \text{percent defectives}$$
$$r = \text{number of defectives}$$

The use of equation (56) is demonstrated in the following example.

EXAMPLE 6-2   Production of a line of ceramic capacitors contains 3% defectives. What are the chances of finding 4 or less defectives in a sample lot of 100, drawn from production, when this sample lot is tested 100%? The answer is to be determined by the addition of individually calculated Poisson terms and confirmed by the use of the cumulative Poisson Table A11, Appendix A.

Solution   (a) Calculate the Poisson term for the probability of exactly 4 defectives occurring in the lot of 100.

$$n = 100$$
$$p = 0.03$$
$$r = 4$$

Per equation (56),

$$P(100, 0.03, 4) = \frac{(100 \times 0.03)^4 \cdot e^{-(100 \times 0.03)}}{4!}$$

$$= \frac{(3^4) \cdot e^{-3}}{4 \times 3 \times 2 \times 1} = \frac{(81) \cdot e^{-3}}{24}$$

The value of e^{-3} may be determined from Table 6-1. Substituting for e^{-3} from Table 6-1 and simplifying,

$$P(100, 0.03, 4) = \frac{(81)0.049787}{24} = 0.168031$$

Table 6-1 *Partial Tabulation of Exponential Factors.*

e^{-1}	0.367879
e^{-2}	0.135335
e^{-3}	0.049787
e^{-4}	0.018316
e^{-5}	0.006738
e^{-6}	0.002479
e^{-7}	0.000912
e^{-8}	0.000335
e^{-9}	0.000123
e^{-10}	0.000045

(b) The Poisson term for the probability of finding exactly 3 defectives in the lot of 100:

$$P(100, 0.03, 3) = \frac{3^3 \cdot e^{-3}}{3!}$$

$$= \frac{(27)0.049787}{6} = 0.224041$$

(c) The Poisson term for the probability of finding exactly 2, 1, and 0 defectives in the lot of 100 by the same method is

$$P(100, 0.03, 2) = \frac{3^2 \cdot e^{-3}}{2!} = 0.224041$$

$$P(100, 0.03, 1) = \frac{3^1 \cdot e^{-3}}{1!} = 0.149361$$

$$P(100, 0.03, 0) = \frac{3^0 \cdot e^{-3}}{0!} = 0.049787$$

(d) The summation of the Poisson terms calculated above is

$$
\begin{array}{r}
0.168031 \\
0.224041 \\
0.224041 \\
0.149361 \\
0.049787 \\
\hline
\text{Total} \quad 0.815261
\end{array}
$$

Accordingly, the probability of finding up to 4 failures in a lot of 100, which is 3% defective, is 0.815261.

(e) Using the summation of terms of the Poisson exponential distribution per Table A11, Appendix A, renders a probability value of 0.815 for $a = np = 3.0$ in the first column under $c = 4$ (4 or less defectives). This checks with the above-calculated value (the table only furnishes three significant digits).

As was mentioned earlier, OC curves are calculated by assuming a range of product quality, such as in Example 6-1. For each of the $a = np$ values in this range, the probability of acceptance is determined from Table A11, Appendix A, for a given value of the acceptance number A_c. Values of A_c depend on the particular level of quality desired and the amount of protection afforded by the plan. MIL-STD-105D contains a large number of inspection plans which define sample size and accept/reject numbers, as related to acceptable quality levels and OC curves.

4. Effects of Sampling Plan Parameter Changes

The protection provided by a sampling plan can be altered by

1. Changing the acceptance number.
2. Changing the sample size.
3. Changing both acceptance number and sample size.

5. Effects of Changing Acceptance Number

In Example 6-3 we will study the effect which a variation in acceptance number c has on the shape of the OC curve, and hence on the protection offered by the inspection plans.

> **EXAMPLE 6-3**   A sample of $n = 110$ parts is drawn from production. The assumed range of product quality is from 0% defectives to 22% defectives.
>
> (a) Calculate and construct OC curves for the following three acceptance criteria:
> 1. $A_c = 0$, no defectives permitted.
> 2. $A_c = 4$, 4 or less defectives permitted.
> 3. $A_c = 12$, 12 or less defectives permitted.
> (b) Determine the AQL and LTPD values for each OC curve graphically, assuming 95% and 10% probability values, respectively.
>
> **Solution**   The following tabulation lists the percent defective, the expected number of defectives, and the probability of acceptance for each of the three acceptance numbers. The three OC curves are plotted in Figure 6-3. Curve A is the case for 0 defective, curve B for 4 defectives or less, and curve C for 12 defectives or less. A horizontal line drawn through the 0.95 probability point intersects the OC curves at points A_1, A_2, and A_3. These points constitute the 95% probability of acceptance for the corresponding percent defectives for each OC curve and represents the AQL

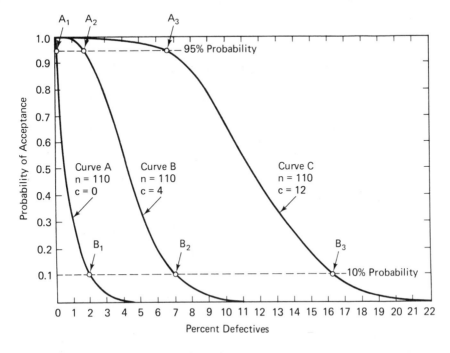

Figure 6-3 *OC—curves for Example 6-3; variation in acceptance number A_c for c or fewer rejects.*

for each plan. In a similar manner, a horizontal line through the 0.1 probability point intersects the OC curves at B_1, B_2, and B_3, thus defining the LTPD values for each plan.

Range of Product Quality, Defectives (%)	Expected Number of Defectives, np	PROBABILITY OF FINDING IN A SAMPLE LOT OF 110:		
		0-Defectives	4 Defectives or Less	12 Defectives or Less
0	$0 \times 110 = 0$	1.00	1.00	1.00
0.5	$0.005 \times 110 = 0.55$	0.577	0.999	0.999
1.0	$0.01 \times 110 = 1.1$	0.333	0.995	0.999
2	$0.02 \times 110 = 2.2$	0.111	0.928	0.999
3	$0.03 \times 110 = 3.3$	0.037	0.762	0.999
4	$0.04 \times 110 = 4.4$	0.012	0.551	0.999
6	$0.06 \times 110 = 6.6$	0.001	0.213	0.982
8	$0.08 \times 110 = 8.8$		0.052	0.885
10	$0.10 \times 110 = 11.0$		0.015	0.689
12	$0.12 \times 110 = 13.2$		0.004	0.445
14	$0.14 \times 110 = 15.4$		0.001	0.268
16	$0.16 \times 110 = 17.6$			0.118
18	$0.18 \times 110 = 19.8$			0.040
20	$0.20 \times 110 = 22.0$			0.015
22	$0.22 \times 110 = 24.2$			0.005
		Curve A	Curve B	Curve C

In the following tabulation the specific AQL and LTPD values are extracted from Figure 6-3. Accordingly, the inspection plan represented by OC curve A ($n = 110$, acceptance number $A_c = 0$) has an AQL $= 0.1\%$ and an LTPD $= 2.0\%$. This means that if the product contains 0.1% defectives, there is a 95% chance of finding zero defectives in inspection lots of $n = 110$ units. Under this plan, these lots are to be accepted. The LTPD $= 2.0\%$ means that if the product contains 2.0% defectives, there is a 10% chance of finding zero defectives in submitted inspection lots of size $n = 110$, and therefore 1 in 10 lots will, on the average, be accepted.

		% Defectives
AQL	A_1	0.1
	A_2	1.8
	A_3	6.7
LTPD	B_1	2.0
	B_2	7.0
	B_3	16.2

In the case of OC curve B ($n = 110$, $A_c = 4$), inspection lots drawn from production which contain 1.8% defectives have a 95% chance of acceptance. Similarly, if production contains 7.0% defectives, there is a 10% chance of lot acceptance.

In the case of OC curve C ($n = 110$, $A_c = 12$), lots containing 6.7% defectives are accepted 95% of the time, while lots containing 16.2% defectives are accepted 10% of the time.

Curve A, with an acceptance number $A_c = 0$, assumes a concave shape that is typical for zero acceptance OC curves. As the acceptance number increases, the OC curves (see curves B and C in Figure 6-3) lose their original concave shape, bulge toward the top, and lose some of their steepness. This means that inspection lots with a larger percent defective will now be accepted by the plan.

It is evident that the protection afforded by the plan changes as the acceptance number changes. For a given sample size, the higher the acceptance number, the lower the protection.

6. Effects of Changing the Sample Size n

The effects of changing the sample size n while keeping the acceptance number constant are examined in the following example.

EXAMPLE 6-4 Refer to Example 6-3. Assume that the acceptance number $A_c = 0$ remains unchanged but that the sample size n varies from 110 to 35 and 15. Calculate and plot the OC curves and determine the AQL and LTPD values.

Solution The following tabulation lists the percent defectives, the expected number of defectives, and the probability of finding zero defectives. The three OC curves are plotted in Figure 6-4. Curve A is the $n = 110$, curve B is the $n = 35$, and curve C is the $n = 15$ case. The acceptance number is $A_c = 0$ for each case.

Range of Product Quality, Defectives (%)	EXPECTED NUMBER OF DEFECTIVES			PROBABILITY OF FINDING ZERO DEFECTIVES IN AN INSPECTION LOT OF:		
	$n = 110$ np	$n = 35$ np	$n = 15$ np	Sample Size $n = 110$	Sample Size $n = 35$	Sample Size $n = 15$
0	0	0	0	1.00	1.00	1.00
0.5	0.55	0.175	0.075	0.577	0.840	0.933
1.0	1.1	0.35	0.15	0.333	0.705	0.861
2	2.2	0.70	0.30	0.111	0.497	0.741
3	3.3	1.05	0.45	0.037	0.350	0.638
4	4.4	1.40	0.60	0.012	0.247	0.549
6	6.6	2.10	0.90	0.001	0.123	0.407
8	8.8	2.80	1.20		0.061	0.301
10	11.0	3.50	1.50		0.030	0.223
12	13.2	4.20	1.80		0.015	0.165
14	15.4	4.90	2.10		0.0075	0.123
16	17.6	5.60	2.40		0.004	0.091
18	19.8	6.30	2.70		0.002	0.067
20	22.0	7.00	3.00		0.001	0.050
22	24.2	7.70	3.30			0.037
				Curve A	Curve B	Curve C

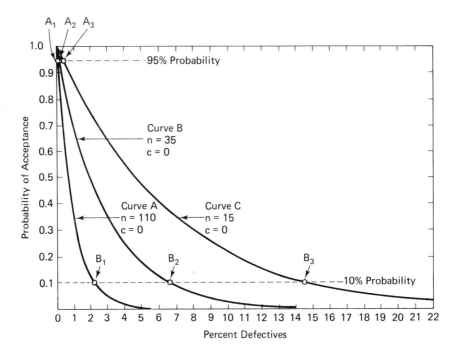

Figure 6-4 *OC—curves for Example 6-4; variation in sample size n; A_c = 0 for c or fewer defectives.*

The following tabulation lists the AQL and LTPD values extracted from the OC curves in Figure 6-4. Accordingly, OC curve *A* has an AQL = 0.1%, LTPD = 2.2%; OC curve *B* has an AQL = 0.2%, LTPD = 6.5%; and OC curve *C* has an AQL = 0.5%, LTPD = 14.5%.

		% Defectives
AQL	A_1	0.1
	A_2	0.2
	A_3	0.5
LTPD	B_1	2.2
	B_2	6.5
	B_3	14.5

Figure 6-4 demonstrates that an increase in sample size steepens the OC curve and improves the discriminatory power of the plan (small changes in product quality will result in more noticeable changes in the probability of acceptance).

7. Simultaneous Change of Acceptance Number Ac and Sample Size n

In practice, it is often necessary to adjust a sampling plan to give the desired AQL and LTPD percentages based on existing product quality. In trying to match the OC curve to the desired risk conditions, both sample size and acceptance number may have to be adjusted simultaneously. The effects of changes in these parameters are illustrated in the following example.

EXAMPLE 6-5   Refer to Examples 6-3 and 6-4. Assume that the sample size n and acceptance number c change simultaneously as follows:

Sample Size, n	Acceptance Number, A_c
110	4.0
35	1.0
15	0

The following tabulation lists all the necessary values for the plotting of the OC curves (see Figure 6-5).

Range of Product Quality, Defectives (%)	EXPECTED NUMBER OF DEFECTIVES			PROBABILITY OF FINDING c OR FEWER DEFECTIVES IN AN INSPECTION LOT OF SIZE n		
	$n = 110$ np	$n = 35$ np	$n = 15$ np	$c = 4$, $n = 110$	$c = 1.0$, $n = 35$	$c = 0$, $n = 15$
.0	0	0	0	1.00	1.00	1.00
0.5	0.55	0.175	0.075	0.999	0.987	0.93
1.0	1.1	0.35	0.15	0.995	0.951	0.861
2	2.2	0.70	0.30	0.928	0.844	0.741
3	3.3	1.05	0.45	0.762	0.718	0.638
4	4.4	1.40	0.60	0.551	0.592	0.549
6	6.6	2.10	0.90	0.213	0.431	0.407
8	8.8	2.80	1.20	0.052	0.231	0.301
10	11.0	3.50	1.50	0.015	0.147	0.223
12	13.2	4.20	1.80	0.008	0.078	0.165
14	15.4	4.90	2.10	0.002	0.044	0.123
16	17.6	5.60	2.40		0.024	0.091
18	19.8	6.30	2.70		0.013	0.067
20	22.0	7.00	3.00		0.007	0.050
22	24.2	7.70	3.30		0.004	0.037
				Curve *A*	Curve *B*	Curve *C*

The following tabulation lists the AQL and LTPD values extracted from the OC curves in Figure 6-5. OC-curve *A* has an AQL = 1.8%, LTPD = 7.2%; OC curve *B* has an AQL = 1%, LTPD = 11.0%; and OC curve *C* has an AQL = 0.4%, LTPD = 15.0%.

		% Defectives
AQL	A_1	1.8
	A_2	1.0
	A_3	0.4
LTPD	B_1	7.2
	B_2	11.0
	B_3	15.0

Examination of Figure 6-5 shows several intersections of the OC curves. This did not occur in the two previous examples, where only one parameter was changed at a time. Also, it is evident that when both sample size and acceptance number increase (see curve A in Figure 6-5), the perfect OC curve depicted in Figure 6-1 is more closely resembled.

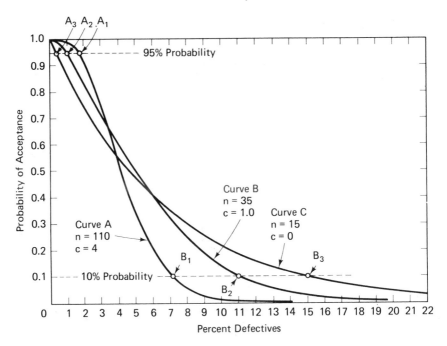

Figure 6-5 *OC—curves for Example 6-5; variation in sample size n and acceptance number c for c or fewer defectives.*

The three preceding examples demonstrate the enormous flexibility of sampling plans, in terms of producer's risk (and corresponding AQL) and consumer's risk (and corresponding LTPD), which may be achieved by appropriate choice of sample sizes and acceptance numbers.

(g) Selection of Sampling Plans

Each sampling plan is unique and provides its own characteristic measure of protection for both producer and consumer. The OC curves associated with the sampling plans provide a convenient way of ascertaining effects of changes in product quality, sample size, acceptance number, and risks. By examination of the OC curves the relative effectiveness of several sampling plans may be compared and the one chosen that best fits the particular situation.

Compromise is often necessary when the size of the inspection sample, chosen for the desired degree of discrimination, would result in prohibitive testing costs. In general, the AQL selected represents a compromise among the quality needs of the consumer, the quality capability of the producer, and the prevailing industry experience. Not infrequently, vendor performance data are not available and AQL values have to be set by estimation, subject to future verification or change, as factual data become available.

LTPD sampling plans, also known as LQ plans (limiting quality plans), limit the consumer's risk to accepting, on the average, no more than 1 in 10 lots, when production runs so defective as to be no longer considered tolerable. Published sampling plans contained in MIL-STD-105D, MIL-STD-414, and the Dodge–Roming Inspection tables[3] permit a wide choice of sampling schemes. In the Dodge–Roming tables the sampling plans are cataloged by *average outgoing quality limit* (AOQL) or by LQ plans (LTPD). Some LQ plans are also contained in MIL-STD-105D under the heading "Limiting Quality Protection."

The AOQL plan is another frequently employed plan. The AOQL is the average quality of the outgoing product based on accepted lots and those which were initially rejected but subsequently 100% screened to weed out the defective items. AOQL plans will be discussed later.

(h) Nonscientific Sampling Plans

Nonscientific sampling plans are generally not based on defined risks assumed by the producer and consumer and use no statistical tools. Figure 6-4 presents the opportunity to study the effectiveness of a not infrequently employed, nonscientific sampling scheme. This scheme provides for a 10% inspection sample to be drawn from any completed production lot. The acceptance number is usually $A_c = 0$, which means that the entire production lot is rejected if the 10% inspection sample contains one or more defectives.

On the surface this would appear to be a reasonable and effective sampling scheme but its shortcoming becomes quickly apparent when examining Figure 6-4. The sample sizes of $n = 110, 35,$ and 15, used in this figure, may be thought of being 10% inspection samples, drawn from production lots of size 1110, 350, and 150, respectively.

---

[3] H. F. Dodge and H. G. Romig, *Sampling Inspection Tables*, John Wiley & Sons, Inc., New York, 1959.

Assuming production normally runs 2% defective, the chances of accepting the 10% inspection lots (and thereby the production lots from which these samples were drawn) as determined from Figure 6-4 are as follows:

Sample Size, n	Probability of Acceptance
110	0.111
35	0.497
15	0.741

The probability of acceptance of these three sample lots is seen to depend on the size of these lots. That is, a production lot of size $n = 1100$ which contains 2% defectives has a 11.1% chance of acceptance using the 10% inspection sample scheme, while an equally defective lot of size $n = 150$ would have a 74.1% chance of acceptance. Using this plan, a producer of a product, which runs 2% defective, would obviously prefer to submit lots of size $n = 150$ for acceptance inspection.

In the case of scientific sampling, no such disparity in risk is present, as the very selection of sampling plans is based on a primary choice of risk levels.

6-4 TYPES OF SAMPLING PLANS

Probably the most commonly used sampling plans are those provided by MIL-STD-105D and MIL-STD-414.[4]

(a) Plans Provided by MIL-STD-105D

MIL-STD-105D provides sampling inspection plans by attribute. The plans provided by this standard fall into two classes:

1. Those that ensure lot quality.
2. Those that ensure average quality.

1. Sampling Plans which Ensure Lot Quality

When the product is submitted for inspection in discrete lots, AQL plans are usually used. When the lots are of various sizes, or mixed, or when there is no identity of discrete lots, as in the case of a continuous production flow, average-quality protection plans are used.

MIL-STD-105D provides two types of sampling plans which protect lot quality:

1. AQL plans.
2. LTPD plans.

---

[4] Selected portions from the texts and tables from MIL-STD-105D and MIL-STD-414 are contained in Chapter 7.

AQL Plans

As stated earlier, AQL, by definition, represents that quality level which, for purposes of sampling inspection, can be considered satisfactory as a process average. The probability of acceptance of lots with a quality level equal to AQL is usually 95%.

AQL levels are normally specified by the procurer. For example, the government may specify a 1.0% AQL inspection plan for a contracted item. If the manufacturer, based on previous production experience, or from intimate knowledge of his process capability, has established that his production indeed runs 1% defective, he may anticipate acceptance of 95 of 100 production lots. Knowledge of how many lots will, in all likelihood, be rejected, enables the manufacturer to compute his anticipated cost of manufacture under this plan.

It should be noted that the government, or industrial procurer, will usually specify an AQL plan without regard to the process quality of the prospective vendor. In fact, procurement is usually done on a competitive basis. The manufacturer who is unable to come close to the specified AQL stands a less chance of having his product accepted by the sampling plan. He must provide for the anticipated losses by raising the price for his product, or bring his production line up to the necessary level of quality. The latter usually requires more modern equipment and/or better training or supervision, all of which tends to increase his production cost and will be reflected in the pricing of the product. For bids to be competitive, production costs must be kept down. At the same time, product quality must not be compromised in order to avoid rejected lots during acceptance tests. In using AQL sampling plans, the government procurement agency, or commercial buyer, assures that if the product is good, it will be accepted by the plan 95% of the time; defective product, on the other hand, has a low chance of being accepted.

AQL plans are basically producer-oriented; they keep the latter's risk of having lots rejected, even though they meet the specification sampling criteria, to 5% (or to 1 lot in 20). The consumer protection afforded by AQL plans in MIL-STD-105D varies considerably from plan to plan. Unlike the producer's protection, it is not maintained at any specific value.

LTPD Plans

Depending on the steepness of the OC curve, the LTPD value may vary considerably, even though the AQL remains relatively unchanged. Thus, in specifying an AQL plan, the aspect of consumer's risk, related to acceptance of a product with a substantial percentage of defectives, has not been defined. When it is desirable to protect the consumer from accepting submitted lots whose quality is poorer than he is willing to tolerate, an LTPD plan is used.

As stated earlier, LTPD plans provide a 10% probability of accepting lots having a quality level of equal to the specified LTPD value. The Dodge–Roming inspection tables contain a selection of LTPD plans. MIL-STD-105D also includes LTPD plans with a 5% and a 10% acceptance criterion. Only the LTPD tables contained in MIL-STD-105D will be considered here.

2. Sampling Plans That Ensure Average Quality

Sampling plans that ensure average quality of the outgoing product are referred to as *average outgoing quality* (AOQ) plans. AOQ plans require the 100% inspection of those lots which do not meet the acceptance criterion of the sampling plan. Any defective items discovered during the 100% test have to be replaced. Plans of this type cannot be used when the inspection tests used to ascertain conformance are destructive.

AOQ plans afford protection to the consumer at a known risk. When submitted lots exceed the required AOQ, there is a low probability that the product will be accepted by the plan. AOQ plans have an upper limit of quality of the outgoing product (called average outgoing quality limit, AOQL).

Regardless of the percent defectives of the product submitted for inspection, the quality of the outgoing product will not exceed this limit. The AOQL is calculated in the following manner:

$$AOQL = (\text{factor})\left(1 - \frac{\text{sample size}}{\text{lot size}}\right) \tag{57}$$

where the factor is determined from Table 7-9 (p. 161).

3. Level of Inspection

The inspection level determines the relationship between the production lot size and the inspection sample size. A code letter designates the latter. MIL-STD-105D contains three general inspection levels: I, II, and III (see Table 7-1, p. 153). Unless otherwise directed, inspection level II is used. Inspection level I provides less discrimination (60% less than level II) because it uses smaller sample sizes, while level III increases the discrimination 1.6 times that of level II inspection because of larger sample sizes for the same lot size.

When dealing with relatively small sample sizes, Table 7-1 provides four special inspection levels: S_1, S_2, S_3, and S_4. Thus, there are a total of seven levels of inspection which permit the balancing of the cost of inspection against the amount of inspection desired. General inspection levels I, II, and III are generally used for nondestructive types of inspection. Special levels S_1 to S_4 are used when inspection involves destruction of the sample or when the cost of inspection is very high.

4. Inspection Severity

The acceptance sampling plans in MIL-STD-105D provide for a change in the severity of inspection based on the experience gained from the inspection of immediately preceding sample lots.

There are three levels of inspection.

1. Normal inspection.
2. Tightened inspection.
3. Reduced inspection.

The inspection procedure is initiated with normal inspection. If two out of five consecutive lots are rejected, normal inspection is shifted to tightened inspection. In changing to tightened inspection, there is a reduction in the values of accept/reject numbers, but the AQL value is retained. For example, in the case of a single sampling plan for sample-size code letter K with a sample size 125 and AQL $= 2.5$, the change in accept/reject numbers is as follows:

	Accept Number	Reject Number
Normal inspection	7	8
Tightened inspection	5	6

After tightened inspection has been in effect and five consecutive sample lots have been found to be acceptable, normal inspection may be resumed.

Normal inspection may be changed to reduced inspection provided the preceding 10 sample lots have been through normal inspection without a rejection and the following additional conditions are met:

1. The total number of defectives from the preceding 10 sample lots tested is equal to or less than the applicable number given in Table 7-12 (p. 164).
2. Production is at a steady rate.
3. Reduced inspection is considered desirable by the responsible authority.

Reduced inspection reverts to normal inspection if any of the following conditions occurs during original inspection:

1. A sample lot is rejected.
2. The reduced inspection was terminated without an acceptance or rejection criterion having been reached.
3. Production becomes irregular.
4. Other conditions warrant the change.

5. Single, Double, and Multiple Sampling

MIL-STD-105D provides three basic types of sampling plans: simple, double, and multiple.

SINGLE SAMPLING

In single sampling plans, the inspection sample size is as given by the plan. The production lot is considered acceptable if the number of defectives found in the inspection sample is equal to or is less than the acceptance number. If the number of defectives exceeds the acceptance number, the production lot is rejected. Single sampling plans are found in Table 7-2 (p. 154).

DOUBLE SAMPLING

Inspection plans for double sampling give two sample sizes per sample-size code letter, with the corresponding acceptance and rejection number. Inspection starts with the first sample size for the particular code letter and the applicable accept/reject criterion. As in the single-sampling plan, if the number of defectives found is equal to or is less than the accept number, the production lot is accepted. If the accept number is exceeded and the number of defectives is equal to or greater than the reject number, the production lot is rejected.

If the number of defectives found in the first sample is between the first acceptance and rejection numbers, a second sample of the size given by the plan is inspected. If the sum of the number of defectives found in the first and second inspection samples equals or is less than the second acceptance number, the production lot is considered acceptable. If the sum of the number of defectives in the first and second inspection samples is equal to or exceeds the second rejection number, the production lot is rejected. Double-sampling plans for normal inspection are found in Table 7-5 (p. 157).

MULTIPLE-SAMPLING PLANS

Multiple-sampling plans are an extension of the double-sampling plans. The difference is that more than two successive samples may be required to reach a decision regarding the acceptance of the production lot. Multiple-sampling plans for normal inspection are found in Table 7-8 (p. 160).

6. Tables and Operating Characteristic Curves Related to Sample-Size Code Letters

MIL-STD-105D contains operating characteristic curves for each sample-size code letter (code letters A through R) together with tabulated values of the OC curves and sampling plans for single-, double-, and multiple-sampling schemes. For illustrative purposes, these data have been reproduced for sample-size code letters J and K (see Tables 7-14 and 7-15, pp. 167–168).

7. Average Sample Number (ASN)

The ASN is the average sample size inspected per sampling lot and applies to single-, double-, and multiple-sampling plans. In single-sampling plans, the ASN equals the inspection sample lot size n. In double and multiple sampling, an accept decision may not be reached with the first sample, and one or more sample lots may have to be inspected.

The particular ASN values may be found from Table 7-13 (see p. 165). The curves depicted in Table 7-13 assume no curtailment of inspection, and there is no 100% inspection requirement of rejected lots. Furthermore, the curves are approximate to the extent that they are based upon the Poisson distribution, and that the sample sizes for double and multiple sampling are assumed to be $0.631n$ and $0.25n$, respectively, where n is the equivalent single-sample size.

As is evident from the curves, double and multiple sampling are almost always more economical, since fewer units require inspection. This is especially true when process quality is either very low or very high. A number of typical examples, applying the methods of MIL-STD-105D, are presented in Chapter 7.

(b) Plans Provided by MIL-STD-414

1. Inspection by Variables

Thus far we have discussed acceptance sampling by attributes. The latter is based on the classification of an inspected item as either acceptable or defective. If the characteristic inspected can be judged on a variable scale and not merely as good or bad, acceptance sampling by variables may be used. The latter requires the reading of actual values of the characteristic.

Examples of characteristics that can only be inspected by attributes are:

1. The presence or absence of a required marking on a fabricated item.
2. The diameter of a metal rod using a MIN-MAX gage.
3. The weight of an article meeting or exceeding a maximum limit.

Examples of characteristics that are inspected on a variable scale are:

1. The temperature rise of an electrical component.
2. The leakage current through insulation.
3. The fuel economy of an automobile engine.

Variable inspection furnishes information as to the extent of the variation in the sample. This information is very useful in assessing how good or how poor the sample really is. Based on this information, the need for corrective action may be assessed. If the variable readings are just very slightly outside the limits, the consumer may still deem them acceptable; with attribute sampling, such an assessment would not be possible.

Variable sampling plans apply to a single characteristic that can be measured on a continuous scale and for which quality is expressed in terms of percent defective. In a practical application, we may, for example, have a device whose maximum operating temperature must not exceed a given value. We draw a sample from production and measure and record the individual temperature values observed. As few as five readings may suffice to apply variable sampling procedures to arrive at an accept/ reject decision for the lot. In a similar manner we may be interested in a lowest temperature or in a range of allowable temperatures with an upper and a lower limit. Here, too, variable sampling may be used.

There are instances where both attribute and variable sampling methods could be employed. The use of either of these two sampling techniques has advantages and disadvantages.

The advantages of inspection by variables compared to attribute sampling are:

1. Smaller sample sizes will furnish equal-quality protection for a given percent defective lot.
2. When the rejection rate is high, actual readings are more informative for devising process changes than go–no go readings.
3. Variables sampling is very effective when the focus is on a critical quality characteristic.

The disadvantages of variables sampling are:

1. Variable sampling plans apply to a single quality characteristic. A separate variable sampling plan is required for each additional quality characteristic inspected. With attribute sampling, only one plan is needed to reach an accept/reject decision.
2. Variable sampling is more costly than attribute sampling, because it usually requires greater inspection skill and involves the recording of more data.

The theory underlying the development of the MIL-STD-414 variable sampling plans, including operating characteristic curves, assumes that measurements of the quality characteristic are independent, identically distributed normal random variables. This standard is divided into four sections. Section A describes general procedures of the sampling plans. Sections B and C describe specific procedures and applications of the sampling plans when variability is unknown. In Section B the estimate of lot standard deviation is used as the basis for an estimate of the unknown variability, and in Section C the average range of the sample is used. Section D describes the plans when variability is known.

Each of sections B, C, and D is divided into three parts: (I) Sampling Plans for the Single Specification Limit Case, (II) Sampling Plans for the Double Specification Limit Case, and (III) Procedures for Estimation of Process Average and Criteria for Tightened and Reduced Inspection. For the single specification limit case, the acceptability criterion is given in two forms: Form 1 and Form 2. Either of the forms may be used, since they are identical as to sample size and decision for lot acceptability or rejectability. In deciding whether to use Form 1 or Form 2, the following points should be borne in mind. Form 1 provides the lot acceptability criterion without estimating lot percent defective. The Form 2 lot acceptability criterion requires estimates of lot percent defective. These estimates also are required for estimation of the process average. Form 1 will be used in the subsequent examples.

2. Quality Characteristic

The quality characteristic may be expressed in terms of an upper or a lower specification limit (single specification limit), or both upper and lower specification limits (double specification limits).

3. Acceptable Quality Level

The particular AQL value to be used for a single quality characteristic of a given product must be specified in terms of percent defectives. Table 7-16 (p. 179) lists AQL values to be used with MIL-STD-414 and delineates the ranges of AQL for which the listed values are applicable.

4. Sample Size

Sample sizes are designated by code letters. The sample-size code letter depends on the inspection level and the lot size. There are five inspection levels: I, II, III, IV, and V. Unless otherwise specified, inspection level IV is used. Table 7-17 (p. 179) lists the sample sizes and code letters applicable to the particular inspection level and size of lots.

5. Operating Characteristic Curves

MIL-STD-414 contains operating characteristic curves for each of the sample-size code letters listed in Table 7-17. A representative OC curve is shown in Table 7-18. These OC curves show the relationship between the quality of inspected lots and the probability of acceptance for a series of AQL values. The OC curves are based on the assumption that measurements are selected at random from a normal distribution. The OC curves for sample-size code letter D, for example, cover an AQL range from 0.65% to 15%. Assuming that the quality of submitted lots is, for instance, 8% and the specified AQL $= 4\%$, the probability of acceptance of these lots is 75%.

6. Lot Acceptability

MIL-STD-414 provides sampling plans based on unknown and known variability. The latter case occurs very infrequently and will not be discussed here. In the case of sampling plans based on unknown variability, two alternative methods are provided to reach an accept/reject decision, one based on the estimate of lot standard deviation and the other on the average range of the sample readings. These are referred to as the *standard deviation method* and the *range method*. Both methods may be used for single or double specification limits. As an example, a single specification limit would apply to a one-sided quality characteristic, such as a maximum or a minimum temperature value. A double specification limit applies to a range of temperatures bounded by an upper and a lower limit. In the latter case, MIL-STD-414 includes the option of using different AQL values for the upper and lower limits of the quality characteristic. Most commonly the same AQL value is used, and we will confine ourselves thereto.

The following is a further attempt to keep the presentation of the MIL-STD-414 methodology for lot acceptance uncomplicated. The standard provides methods for the estimation of process average and criteria for tightened and reduced inspection. The determination of process average is required for the earlier-mentioned Form 2 lot-acceptability criterion. Because of our decision to use the equally valid Form 1 acceptability criterion, a discussion of the estimation of process average is not included. For the application of tightened and reduced inspection, which is based on the determination of the process average, the reader is referred to Part III of MIL-STD-414.

STANDARD DEVIATION METHOD—SINGLE SPECIFICATION LIMIT; VARIABILITY UNKNOWN, FORM 1 ACCEPTABILITY CRITERION

1. Determine the sample-size code letter from Table 7-17 (p. 179) by using the lot size and inspection level.
2. Obtain a plan from Table 7-19 (p. 181) by selecting the sample size n and the acceptability constant k. The latter is indicated in the column of the table corresponding to the applicable AQL value.
3. Select at random the sample of n units from the lot; inspect and record the measurement of the quality characteristic for each unit of the sample.
4. Compute the sample mean $\bar{X}$ and an estimate of the lot standard deviation s.
5. Compute the quantity

$$Q_U = \frac{U - \bar{X}}{s}$$

for an upper specification limit U, and the quantity

$$Q_L = \frac{\bar{X} - L}{s}$$

for a lower specification limit L.
6. If the quantity

$$\frac{U - \bar{X}}{s} \quad \text{or} \quad \frac{\bar{X} - L}{s}$$

is in value equal to or greater than k, the lot meets the acceptibility criterion. If k is greater in value, or if the calculated value is negative, the lot does not meet the acceptability criterion.

STANDARD DEVIATION METHOD—DOUBLE SPECIFICATION LIMIT; VARIABILITY UNKNOWN

1. Determine the sample-size code letter from Table 7-17 (p. 179) by using the lot size and inspection level.
2. Select a plan from Table 7-20 (p. 182). Obtain the sample size n and the maximum allowable percent defective M.
3. Select at random the sample of n units from the lot; inspect and record the measurement of the quality characteristic on each unit of the sample.
4. Compute the sample mean $\bar{X}$ and estimate of lot standard deviation s.
5. Compute the quality indices

$$Q_U = \frac{U - \bar{X}}{s} \quad \text{and} \quad Q_L = \frac{\bar{X} - L}{s}$$

6. Determine the estimated lot percent defective $p = p_U + p_L$ from Table 7-21 (p. 183).

7. If the estimated lot percent defective p is equal to or less than the maximum allowable percent defective M, the lot meets the acceptability criterion; if p is greater than M or if Q_U, Q_L, or both are negative, the lot does not meet the acceptability criterion.

RANGE METHOD—SINGLE SPECIFICATION LIMIT; VARIABILITY UNKNOWN, FORM 1 ACCEPTABILITY CRITERION

1. Determine the sample-size code letter from Table 7-17 (p. 179) by using the lot size and the inspection level.

2. Obtain a plan from Table 7-22 (p. 192) by selecting the sample size n and the acceptability constant k.

3. Select at random the sample of n units from the lot; inspect and record the measurement of the quality characteristic for each unit of the sample.

4. Compute the sample mean $\bar{X}$ and the average range of the sample, $\bar{R}$. In this method, $\bar{R}$ is the average range of the subgroup ranges. Each of the subgroups consists of 5 measurements, except for those plans with sample size 3, 4 or 7 in which case the subgroup size is the same as the sample size. In computing $\bar{R}$, the order of the sample measurements as made must be retained, subgroups of consecutive measurements must be formed and the range of each subgroup obtained. R is the average of the individual subgroup ranges.

5. Compute the quantity

$$Q_U = \frac{U - \bar{X}}{\bar{R}}$$

for an upper specification limit U or the quantity

$$Q_L = \frac{\bar{X} - L}{\bar{R}}$$

for a lower specification limit L.

6. If the quantity

$$\frac{U - \bar{X}}{\bar{R}} \quad \text{or} \quad \frac{\bar{X} - L}{\bar{R}}$$

renders a value equal to or greater than k, the lot meets the acceptability criterion; if k is greater in value or if the calculated value is negative, the lot does not meet the acceptability criterion.

RANGE METHOD—DOUBLE SPECIFICATION LIMIT; VARIABILITY UNKNOWN

1. Determine the sample-size code letter from Table 7-17 (p. 179) by using the lot size and the inspection level.

2. Select a plan from Table 7-23 (p. 193) Obtain the sample size n, the factor c, and the maximum allowable percent defective M.

3. Select at random the sample of n units from the lot; inspect and record the measurement of the quality characteristic on each unit of the sample.
4. Compute the sample mean $\bar{X}$ and average range of the sample, $\bar{R}$. (See explanatory remarks given under Range Method-Single Specification Limit).
5. Compute the quality indices

$$Q_U = \frac{(U - \bar{X})c}{\bar{R}} \quad \text{and} \quad Q_L = \frac{(\bar{X} - L)c}{\bar{R}}$$

6. Determine the estimated lot percent defective $p = p_U + p_L$ from Table 7-24 (p. 194).
7. If the estimated lot percent defective p is equal to or less than the maximum allowable percent defective M, the lot meets the acceptability criterion; if p is greater than M or if either Q_U or Q_L or both are negative, the lot does not meet the acceptability criterion.

The application of the standard deviation method and the range method is demonstrated in a number of typical examples in Chapter 7.

PROBLEMS

6-1 (a) State the objectives of acceptance control.
  (b) What are the four types of acceptance control methods?
  (c) What are the main disadvantages of the spot-check technique compared to statistical sample schemes?
6-2 (a) Why is 100% acceptance inspection not necessarily foolproof?
  (b) What is the risk of acceptance of a product by certification?
  (c) State four advantages of acceptance sampling.
6-3 (a) What is acceptance sampling by attributes?
  (b) What is a commonly used military document for acceptance sampling?
  (c) Define AQL.
6-4 (a) Discuss the terms "producer's risk" and "consumer's risk" and give commonly used values for these risks.
  (b) What is an operating characteristic (OC) curve?
  (c) What parameters are needed to calculate an OC curve?
6-5 (a) Define LTPD. What is its usually chosen value?
  (b) What are the accept and reject numbers?
6-6 (a) Draw an OC curve that has very good discrimination and then superimpose one that has very poor discrimination.
  (b) What is the mathematical relationship between accept and reject numbers?
6-7 (a) Draw an OC curve to cover a range of quality of submitted lots from 0 to 15%. Identify the AQL and LTPD points (AQL at 95% probability, LTPD at 10% probability of acceptance).
  (b) What is the expected number of defectives when a sample of 38 pieces is drawn from a production lot that is 5% defective?
  (c) What probability distribution is used for calculating OC curves?

6-8 State the general formula for calculating individual Poisson terms, making use of the symbol for expected number of defectives. Explain the meaning of each symbol used.

6-9 What are the chances of finding 3 or less defectives in a sample lot of 50 bearings when production is running 10% defective? Use the method of summation of the individual Poisson terms.

6-10 Repeat Example 6-9 but use the summation of terms of the Poisson distribution.

6-11 Calculate and draw a family of three OC curves for the following conditions:

> sample size: 50 parts
> acceptance criteria: 0 defectives
>   5 defectives or less
>   9 defectives or less

Identify the AQL and LTPD points (probability of acceptance 95% and 10%, respectively). Assume a 0 to 30% range of percent defectives.

6-12 Calculate and draw a family of four OC curves, keeping the acceptance criterion of 3 defectives or less constant and varying the sample size from 25 to 100 in increments of 25. Identify AQL and LTPD points (probability of acceptance 95% and 10%, respectively).

6-13 (a) What is an AOQL plan?
      (b) What are LQ plans?

6-14 (a) Are AQL plans consumer- or producer-oriented?
      (b) What are sampling plans that ensure average quality called?
      (c) What is AOQL?

6-15 (a) A manufacturer produces a lot of 5000 switches. General inspection level II applies and the AQL = 1.0%. Determine the AOQL value.
      (b) What inspection levels of MIL-STD-105D are used when the sample sizes are very small?

6-16 (a) Name the three levels of inspection severity.
      (b) What is the criterion for shifting from normal inspection to tightened inspection?
      (c) When is normal inspection changed to reduced inspection? State all applicable conditions.

6-17 (a) State the differences among single, double, and multiple sampling.
      (b) What is the average sample number (ASN)?
      (c) What is the ASN for single-sampling plans?

6-18 (a) Which military specification covers inspection by variables?
      (b) What is the difference between inspection by attributes and inspection by variables.
      (c) List several characteristics that may be judged by inspection by variables.
      (d) Give some examples of characteristics suitable for inspection by attributes.

6-19 (a) Give the advantages and disadvantages of inspection by variables.
      (b) Two types of sampling plans are provided by MIL-STD-414 for single-sampling plans based on an unknown variability. What are they?
      (c) Which inspection level is generally used in connection with MIL-STD-414?

6-20 Describe the steps you would use in applying the standard deviation method–single-specification limit using the Form 1 acceptability criterion. Variability is assumed unknown.

7

MILITARY STANDARDS
FOR ACCEPTANCE SAMPLING

7-1 GENERAL BACKGROUND

Earlier we traced the beginnings of organized statistical-quality-control activities to the World War II period, with the Armed Services providing the primary motivational drive. As a result, sampling procedures for inspection by attributes gained in acceptance and began to be used extensively in military and commercial procurements. With several earlier versions preceding it, MIL-STD-105A was finally introduced in 1950. During the next 13 years two revisions to the standard were issued (-105B and -105C). The present verison, MIL-STD-105D, was published in April 1963 and is approved by the Department of Defense for mandatory use by the Department of the Army, the Navy, the Air Force, and the Defense Supply Agency. Working groups representing the military services of Canada and the United Kingdom as well as European organizations for quality control collaborated with their counterparts in the United States in the development of this standard, thereby assuring its acceptance outside the United States. The international designation of this document is ABC-STD-105.

MIL-STD-414, the military standard for sampling procedures and tables for inspection by variables for percent defective was approved by the Department of Defense and has been mandatory for use by the Armed Services since June 1957. This document supersedes an earlier ordnance document, ORD-M608-10.

Chapter 6 gave general descriptions of the plans contained in MIL-STD-105D and MIL-STD-414 and described their applications, advantages, and disadvantages. No description or summary of the narrative contents of MIL-STD-105D and MIL-STD-414 can be as effective as the original texts. Accordingly, this chapter is devoted to pertinent portions of the text and a selected number of tables and figures. This is followed by a number of typical sampling inspection problems with solutions.

7-2 TEXT OF MIL-STD-105D*

SAMPLING PROCEDURES AND TABLES
FOR INSPECTION BY ATTRIBUTES

1. SCOPE

1.1 PURPOSE. This publication establishes sampling plans and procedures for inspection by attributes. When specified by the responsible authority, this publication shall be referenced in the specification, contract, inspection instructions, or other documents and the provisions set forth herein shall govern. The "responsible authority" shall be designated in one of the above documents.

1.2 APPLICATION. Sampling plans designated in this publication are applicable, but not limited, to inspection of the following:

   a. End items.

   b. Components and raw materials.

   c. Operations.

   d. Materials in process.

   e. Supplies in storage.

   f. Maintenance operations.

   g. Data or records.

   h. Administrative procedures.

These plans are intended primarily to be used for a continuing series of lots or batches.

The plans may also be used for the inspection of isolated lots or batches, but, in this latter case, the user is cautioned to consult the operating characteristic curves to find a plan which will yield the desired protection (see 11.6).

1.3 INSPECTION. Inspection is the process of measuring, examining, testing, or otherwise comparing the unit of product (see 1.5) with the requirements.

1.4 INSPECTION BY ATTRIBUTES. Inspection by attributes is inspection whereby either the unit of product is classified simply as defective or nondefective, or the number of defects in the unit of product is counted, with respect to a given requirement or set of requirements.

1.5 UNIT OF PRODUCT. The unit of product is the thing inspected in order to determine its classification as defective or nondefective or to count the number of defects. It may be a single article, a pair, a set, a length, an area, an operation, a volume, a component of an end product, or the end product itself. The unit of product may or may not be the same as the unit of purchase, supply, production, or shipment.

* The correlation between Table Numbers in the MIL-STD-105D and this book is as follows:

MIL-STD-105D	This book	MIL-STD-105D	This book
I	7-1	VI	7-10, 7-11
II	7-2, 7-3, 7-4	VII	Not included
III	7-5, 7-6, 7-7	VIII	7-12
IV†	7-8	IX	7-13
V A	7-9	X†	7-14, 7-15
V B	Not included		

†Table not reproduced in its entirety.

2. CLASSIFICATION OF DEFECTS AND DEFECTIVES

2.1 METHOD OF CLASSIFYING DEFECTS.
A classification of defects is the enumeration of possible defects of the unit of product classified according to their seriousness. A defect is any nonconformance of the unit of product with specified requirements. Defects will normally be grouped into one or more of the following classes; however, defects may be grouped into other classes, or into subclasses within these classes.

2.1.1 CRITICAL DEFECT.
A critical defect is a defect that judgment and experience indicate is likely to result in hazardous or unsafe conditions f o r individuals using, maintaining, or depending upon the product; or a defect that judgment and experience indicate is likely to prevent performance of the tactical function of a major end item such as a ship, aircraft, tank, missile or space vehicle. NOTE: For a special provision relating to critical defects, see 6.3.

2.1.2 MAJOR DEFECT.
A major defect is a defect, other than critical, that is likely to result in failure, or to reduce materially the usability of the unit of product for its intended purpose.

2.1.3 MINOR DEFECT.
A minor defect is a defect that is not likely to reduce materially the usability of the unit of product for its intended purpose, or is a departure from established standards having little bearing on the effective use or operation of the unit.

2.2 METHOD OF CLASSIFYING DEFECTIVES.
A defective is a unit of product which contains one or more defects. Defectives will usually be classified as follows:

2.2.1 CRITICAL DEFECTIVE.
A critical defective contains one or more critical defects and may also contain major and or minor defects. NOTE: For a special provision relating to critical defectives, see 6.3.

2.2.2 MAJOR DEFECTIVE.
A major defective contains one or more major defects, and may also contain minor defects but contains no critical defect.

2.2.3 MINOR DEFECTIVE.
A minor defective contains one or more minor defects but contains no critical or major defect.

3. PERCENT DEFECTIVE AND DEFECTS PER HUNDRED UNITS

3.1 EXPRESSION OF NONCONFORMANCE.
The extent of nonconformance of product shall be expressed either in terms of percent defective or in terms of defects per hundred units.

3.2 PERCENT DEFECTIVE.
The percent defective of any given quantity of units of product is one hunderd times the number of defective units of product contained therein divided by the total number of units of product, i.e.:

$$\text{Percent defective} = \frac{\text{Number of defectives}}{\text{Number of units inspected}} \times 100$$

3.3 DEFECTS PER HUNDRED UNITS.
The number of defects per hundred units of any given quantity of units of product is one hundred times the number of defects contained therein (one or more defects being possible in any unit of product) divided by the total number of units of product, i.e.:

$$\frac{\text{Defects per}}{\text{hundred units}} = \frac{\text{Number of defects}}{\text{Number of units inspected}} \times 100$$

4. ACCEPTABLE QUALITY LEVEL (AQL)

4.1 USE. The AQL, together with the Sample Size Code Letter, is used for indexing the sampling plans provided herein.

4.2 DEFINITION. The AQL is the maximum percent defective (or the maximum number of defects per hundred units) that, for purposes of sampling inspection, can be considered satisfactory as a process average (see 11.2).

4.3 NOTE ON THE MEANING OF AQL. When a consumer designates some specific value of AQL for a certain defect or group of defects, he indicates to the supplier that his (the consumer's) acceptance sampling plan will accept the great majority of the lots or batches that the supplier submits, provided the process average level of percent defective (or defects per hundred units) in these lots or batches be no greater than the designated value of AQL. Thus, the AQL is a designated value of percent defective (or defects per hundred units) that the consumer indicates will be accepted most of the time by the acceptance sampling procedure to be used. The sampling plans provided herein are so arranged that the probability of acceptance at the designated AQL value depends upon the sample size, being generally higher for large samples than for small ones, for a given AQL. The AQL alone does not describe the protection to the consumer for individual lots or batches but more directly relates to what might be expected from a series of lots or batches, provided the steps indicated in this publication are taken. It is necessary to refer to the operating characteristic curve of the plan, to determine what protection the consumer will have.

4.4 LIMITATION. The designation of an AQL shall not imply that the supplier has the right to supply knowingly any defective unit of product.

4.5 SPECIFYING AQLs. The AQL to be used will be designated in the contract or by the responsible authority. Different AQLs may be designated for groups of defects considered collectively, or for individual defects. An AQL for a group of defects may be designated in addition to AQLs for individual defects, or subgroups, within that group. AQL values of 10.0 or less may be expressed either in percent defective or in defects per hundred units; those over 10.0 shall be expressed in defects per hundred units only.

4.6 PREFERRED AQLs. The values of AQLs given in these tables are known as preferred AQLs. If, for any product, an AQL be designated other than a preferred AQL, these tables are not applicable.

5. SUBMISSION OF PRODUCT

5.1 LOT OR BATCH. The term lot or batch shall mean "inspection lot" or "inspection batch," i.e., a collection of units of product from which a sample is to be drawn and inspected to determine conformance with the acceptability criteria, and may differ from a collection of units designated as a lot or batch for other purposes (e.g., production, shipment, etc.).

5.2 FORMATION OF LOTS OR BATCHES. The product shall be assembled into identifiable lots, sublots, batches, or in such other manner as may be prescribed (see 5.4). Each lot or batch shall, as far as is practicable,

5. SUBMISSION OF PRODUCT (Continued)

consist of units of product of a single type, grade, class, size, and composition, manufactured under essentially the same conditions, and at essentially the same time.

5.3 LOT OR BATCH SIZE. The lot or batch size is the number of units of product in a lot or batch.

5.4 PRESENTATION OF LOTS OR BATCHES. The formation of the lots or batches, lot or batch size, and the manner in which each lot or batch is to be presented and identified by the supplier shall be designated or approved by the responsible authority. As necessary, the supplier shall provide adequate and suitable storage space for each lot or batch, equipment needed for proper identification and presentation, and personnel for all handling of product required for drawing of samples.

6. ACCEPTANCE AND REJECTION

6.1 ACCEPTABILITY OF LOTS OR BATCHES. Acceptability of a lot or batch will be determined by the use of a sampling plan or plans associated with the designated AQL or AQLs.

6.2 DEFECTIVE UNITS. The right is reserved to reject any unit of product found defective during inspection whether that unit of product forms part of a sample or not, and whether the lot or batch as a whole is accepted or rejected. Rejected units may be repaired or corrected and resubmitted for inspection with the approval of, and in the manner specified by, the responsible authority.

6.3 SPECIAL RESERVATION FOR CRITICAL DEFECTS. The supplier may be required at the discretion of the responsible authority to inspect every unit of the lot or batch for critical defects. The right is reserved to inspect every unit submitted by the supplier for critical defects, and to reject the lot or batch immediately, when a critical defect is found. The right is reserved also to sample, for critical defects, every lot or batch submitted by the supplier and to reject any lot or batch if a sample drawn therefrom is found to contain one or more critical defects.

6.4 RESUBMITTED LOTS OR BATCHES. Lots or batches found unacceptable shall be resubmitted for reinspection only after all units are re-examined or retested and all defective units are removed or defects corrected. The responsible authority shall determine whether normal or tightened inspection shall be used, and whether reinspection shall include all types or classes of defects or only the particular types or classes of defects which caused initial rejection.

7. DRAWING OF SAMPLES

7.1 SAMPLE. A sample consists of one or more units of product drawn from a lot or batch, the units of the sample being selected at random without regard to their quality. The number of units of product in the sample is the sample size.

7.2 REPRESENTATIVE SAMPLING. When appropriate, the number of units in the sample shall be selected in proportion to the size of sublots or subbatches, or parts of the lot or batch, identified by some rational criterion.

7. DRAWING OF SAMPLES (Continued)

When representative sampling is used, the units from each part of the lot or batch shall be selected at random.

7.3 TIME OF SAMPLING. Samples may be drawn after all the units comprising the lot or batch have been assembled, or samples may be drawn during assembly of the lot or batch.

7.4 DOUBLE OR MULTIPLE SAMPLING. When double or multiple sampling is to be used, each sample shall be selected over the entire lot or batch.

8. NORMAL, TIGHTENED AND REDUCED INSPECTION

8.1 INITIATION OF INSPECTION. Normal inspection will be used at the start of inspection unless otherwise directed by the responsible authority.

8.2 CONTINUATION OF INSPECTION. Normal, tightened or reduced inspection shall continue unchanged for each class of defects or defectives on successive lots or batches except where the switching procedures given below require change. The switching procedures shall be applied to each class of defects or defectives independently.

8.3 SWITCHING PROCEDURES.

8.3.1 NORMAL TO TIGHTENED. When normal inspection is in effect, tightened inspection shall be instituted when 2 out of 5 consecutive lots or batches have been rejected on original inspection (i.e., ignoring resubmitted lots or batches for this procedure).

8.3.2 TIGHTENED TO NORMAL. When tightened inspection is in effect, normal inspection shall be instituted when 5 consecutive lots or batches have been considered acceptable on original inspection.

8.3.3 NORMAL TO REDUCED. When normal inspection is in effect, reduced inspection shall be instituted providing that all of the following conditions are satisfied:

a. The preceding 10 lots or batches (or more, as indicated by the note to Table VIII) have been on normal inspection and none has been rejected on original inspection; and

b. The total number of defectives (or defects) in the samples from the preceding 10 lots or batches (or such other number as was used for condition "a" above) is equal to or less than the applicable number given in Table VIII. If double or multiple sampling is in use, all samples inspected should be included, not "first" samples only; and

c. Production is at a steady rate; and

d. Reduced inspection is considered desirable by the responsible authority.

8.3.4 REDUCED TO NORMAL. When reduced inspection is in effect, normal inspection shall be instituted if any of the following occur on original inspection:

a. A lot or batch is rejected; or

b. A lot or batch is considered acceptable under the procedures of 10.1.4; or

c. Production becomes irregular or delayed; or

d. Other conditions warrant that normal inspection shall be instituted.

8.4 DISCONTINUATION OF INSPECTION. In the event that 10 consecutive lots or batches remain on tightened inspection (or such other number as may be designated by the responsible authority), inspection under the provisions of this document should be discontinued pending action to improve the quality of submitted material.

9. SAMPLING PLANS

9.1 SAMPLING PLAN. A sampling plan indicates the number of units of product from each lot or batch which are to be inspected (sample size or series of sample sizes) and the criteria for determining the acceptability of the lot or batch (acceptance and rejection numbers).

9.2 INSPECTION LEVEL. The inspection level determines the relationship between the lot or batch size and the sample size. The inspection level to be used for any particular requirement will be prescribed by the responsible authority. Three inspection levels: I, II, and III, are given in Table I for general use. Unless otherwise specified, Inspection Level II will be used. However, Inspection Level I may be specified when less discrimination is needed, or Level III may be specified for greater discrimination. Four additional special levels: S–1, S–2, S–3 and S–4, are given in the same table and may be used where relatively small sample sizes are necessary and large sampling risks can or must be tolerated.

NOTE: In the designation of inspection levels S–1 to S–4, care must be exercised to avoid AQLs inconsistent with these inspection levels.

9.3 CODE LETTERS. Sample sizes are designated by code letters. Table I shall be used to find the applicable code letter for the particular lot or batch size and the prescribed inspection level.

9.4 OBTAINING SAMPLING PLAN. The AQL and the code letter shall be used to obtain the sampling plan from Tables II, III or IV. When no sampling plan is available for a given combination of AQL and code letter, the tables direct the user to a different letter. The sample size to be used is given by the new code letter not by the original letter. If this procedure leads to different sample sizes for different classes of defects, the code letter corresponding to the largest sample size derived may be used for all classes of defects when designated or approved by the responsible authority. As an alternative to a single sampling plan with an acceptance number of 0, the plan with an acceptance number of 1 with its correspondingly larger sample size for a designated AQL (where available), may be used when designated or approved by the responsible authority.

9.5 TYPES OF SAMPLING PLANS. Three types of sampling plans: Single, Double and Multiple, are given in Tables II, III and IV, respectively. When several types of plans are available for a given AQL and code letter, any one may be used. A decision as to type of plan, either single, double, or multiple, when available for a given AQL and code letter, will usually be based upon the comparison between the administrative difficulty and the average sample sizes of the available plans. The average sample size of multiple plans is less than for double (except in the case corresponding to single acceptance number 1) and both of these are always less than a single sample size. Usually the administrative difficulty for single sampling and the cost per unit of the sample are less than for double or multiple.

10.   DETERMINATION OF ACCEPTABILITY

10.1   PERCENT DEFECTIVE INSPECTION. To determine acceptability of a lot or batch under percent defective inspection, the applicable sampling plan shall be used in accordance with 10.1.1, 10.1.2, 10.1.3, and 10.1.4.

10.1.1   SINGLE SAMPLING PLAN. The number of sample units inspected shall be equal to the sample size given by the plan. If the number of defectives found in the sample is equal to or less than the acceptance number, the lot or batch shall be considered acceptable. If the number of defectives is equal to or greater than the rejection number, the lot or batch shall be rejected.

10.1.2   DOUBLE SAMPLING PLAN. The number of sample units inspected shall be equal to the first sample size given by the plan. If the number of defectives found in the first sample is equal to or less than the first acceptance number, the lot or batch shall be considered acceptable. If the number of defectives found in the first sample is equal to or greater than the first rejection number, the lot or batch shall be rejected. If the number of defectives found in the first sample is between the first acceptance and rejection numbers, a second sample of the size given by the plan shall be inspected. The number of defectives found in the first and second samples shall be accumulated. If the cumulative number of defectives is equal to or less than the second acceptance number, the lot or batch shall be considered acceptable. If the cumulative number of defectives is equal to or greater than the second rejection number, the lot or batch shall be rejected.

10.1.3   MULTIPLE SAMPLE PLAN. Under multiple sampling, the procedure shall be similar to that specified in 10.1.2, except that the number of successive samples required to reach a decision may be more than two.

10.1.4   SPECIAL PROCEDURE FOR REDUCED INSPECTION. Under reduced inspection, the sampling procedure may terminate without either acceptance or rejection criteria having been met. In these circumstances, the lot or batch will be considered acceptable, but normal inspection will be reinstated starting with the next lot or batch (see 8.3.4 (b)).

10.2   DEFECTS PER HUNDRED UNITS INSPECTION. To determine the acceptability of a lot or batch under Defects per Hundred Units inspection, the procedure specified for Percent Defective inspection above shall be used, except that the word "defects" shall be substituted for "defectives."

11.   SUPPLEMENTARY INFORMATION

11.1   OPERATING CHARACTERISTIC CURVES. The operating characteristic curves for normal inspection, shown in Table X (pages 30–62), indicate the percentage of lots or batches which may be expected to be accepted under the various sampling plans for a given process quality. The curves shown are for single sampling; curves for double and multiple sampling are matched as closely as practicable. The O. C. curves shown for AQLs greater than 10.0 are based on the Poisson distribution and are applicable for defects per hundred units inspection; those for AQLs of 10.0 or less and sample sizes of 80 or less are based on the binomial distribution and are applicable for percent defec-

11.   SUPPLEMENTARY INFORMATION (Continued)

tive inspection; those for AQLs of 10.0 or less and sample sizes larger than 80 are based on the Poisson distribution and are applicable either for defects per hundred units inspection, or for percent defective inspection (the Poisson distribution being an adequate approximation to the binomial distribution under these conditions). Tabulated values, corresponding to selected values of probabilities of acceptance (P_a, in percent) are given for each of the curves shown, and, in addition, for tightened inspection, and for defects per hundred units for AQLs of 10.0 or less and sample sizes of 80 or less.

11.2   PROCESS AVERAGE. The process average is the average percent defective or average number of defects per hundred units (whichever is applicable) of product submitted by the supplier for original inspection. Original inspection is the first inspection of a particular quantity of product as distinguished from the inspection of product which has been resubmitted after prior rejection.

11.3   AVERAGE OUTGOING QUALITY (AOQ). The AOQ is the average quality of outgoing product including all accepted lots or batches, plus all rejected lots or batches after the rejected lots or batches have been effectively 100 percent inspected and all defectives replaced by nondefectives.

11.4   AVERAGE OUTGOING QUALITY LIMIT (AOQL). The AOQL is the maximum of the AOQs for all possible incoming qualities for a given acceptance sampling plan. AOQL values are given in Table V–A for each of the single sampling plans for normal inspection and in Table V–B for each of the single sampling plans for tightened inspection.

11.5   AVERAGE SAMPLE SIZE CURVES. Average sample size curves for double and multiple sampling are in Table IX. These show the average sample sizes which may be expected to occur under the various sampling plans for a given process quality. The curves assume no curtailment of inspection and are approximate to the extent that they are based upon the Poisson distribution, and that the sample sizes for double and multiple sampling are assumed to be 0.631n and 0.25n respectively, where n is the equivalent single sample size.

11.6   LIMITING QUALITY PROTECTION. The sampling plans and associated procedures given in this publication were designed for use where the units of product are produced in a continuing series of lots or batches over a period of time. However, if the lot or batch is of an isolated nature, it is desirable to limit the selection of sampling plans to those, associated with a designated AQL value, that provide not less than a specified limiting quality protection. Sampling plans for this purpose can be selected by choosing a Limiting Quality (LQ) and a consumer's risk to be associated with it. Tables VI and VII give values of LQ for the commonly used consumer's risks of 10 percent and 5 percent respectively. If a different value of consumer's risk is required, the O.C. curves and their tabulated values may be used. The concept of LQ may also be useful in specifying the AQL and Inspection Levels for a series of lots or batches, thus fixing minimum sample size where there is some reason for avoiding (with more than a given consumer's risk) more than a limiting proportion of defectives (or defects) in any single lot or batch.

Table 7-1 *Sample-Size Code Letters (MIL-STD-105D).*

Lot or batch size	Special inspection levels				General inspection levels		
	S-1	S-2	S-3	S-4	I	II	III
2 to 8	A	A	A	A	A	A	B
9 to 15	A	A	A	A	A	B	C
16 to 25	A	A	B	B	B	C	D
26 to 50	A	B	B	C	C	D	E
51 to 90	B	B	C	C	C	E	F
91 to 150	B	B	C	D	D	F	G
151 to 280	B	C	D	E	E	G	H
281 to 500	B	C	D	E	F	H	J
501 to 1200	C	C	E	F	G	J	K
1201 to 3200	C	D	E	G	H	K	L
3201 to 10000	C	D	F	G	J	L	M
10001 to 35000	C	D	F	H	K	M	N
35001 to 150000	D	E	G	J	L	N	P
150001 to 500000	D	E	G	J	M	P	Q
500001 and over	D	E	H	K	N	Q	R

Note.

Small sample inspection levels of MIL-STD-105C

Convert to these special inspection levels

L–1 and L–2 ----------------------- S–1
L–3 and L–4 ----------------------- S–2
L–5 and L–6 ----------------------- S–3
L–7 and L–8 ----------------------- S–4

Table 7-2 Single-Sampling Plans for Normal Inspection (MIL-STD-105D).

Acceptable Quality Levels (normal inspection) — each cell shows **Ac Re** (Ac = Acceptance number, Re = Rejection number). ↓ = use first sampling plan below arrow; ↑ = use first sampling plan above arrow.

Sample size code letter	Sample size	0.010	0.015	0.025	0.040	0.065	0.10	0.15	0.25	0.40	0.65	1.0	1.5	2.5	4.0	6.5	10	15	25	40	65	100	150	250	400	650	1000
A	2	↓	↓	↓	↓	↓	↓	↓	↓	↓	↓	↓	↓	↓	↓	↓	↓	0 1	1 2	2 3	3 4	5 6	7 8	10 11	14 15	21 22	30 31
B	3	↓	↓	↓	↓	↓	↓	↓	↓	↓	↓	↓	↓	↓	↓	↓	0 1	1 2	2 3	3 4	5 6	7 8	10 11	14 15	21 22	30 31	44 45
C	5	↓	↓	↓	↓	↓	↓	↓	↓	↓	↓	↓	↓	↓	↓	0 1	1 2	2 3	3 4	5 6	7 8	10 11	14 15	21 22	30 31	44 45	↑
D	8	↓	↓	↓	↓	↓	↓	↓	↓	↓	↓	↓	↓	↓	0 1	1 2	2 3	3 4	5 6	7 8	10 11	14 15	21 22	30 31	44 45	↑	↑
E	13	↓	↓	↓	↓	↓	↓	↓	↓	↓	↓	↓	↓	0 1	1 2	2 3	3 4	5 6	7 8	10 11	14 15	21 22	30 31	44 45	↑	↑	↑
F	20	↓	↓	↓	↓	↓	↓	↓	↓	↓	↓	↓	0 1	1 2	2 3	3 4	5 6	7 8	10 11	14 15	21 22	30 31	44 45	↑	↑	↑	↑
G	32	↓	↓	↓	↓	↓	↓	↓	↓	↓	↓	0 1	1 2	2 3	3 4	5 6	7 8	10 11	14 15	21 22	30 31	44 45	↑	↑	↑	↑	↑
H	50	↓	↓	↓	↓	↓	↓	↓	↓	↓	0 1	1 2	2 3	3 4	5 6	7 8	10 11	14 15	21 22	30 31	44 45	↑	↑	↑	↑	↑	↑
J	80	↓	↓	↓	↓	↓	↓	↓	↓	0 1	1 2	2 3	3 4	5 6	7 8	10 11	14 15	21 22	30 31	44 45	↑	↑	↑	↑	↑	↑	↑
K	125	↓	↓	↓	↓	↓	↓	↓	0 1	1 2	2 3	3 4	5 6	7 8	10 11	14 15	21 22	30 31	44 45	↑	↑	↑	↑	↑	↑	↑	↑
L	200	↓	↓	↓	↓	↓	↓	0 1	1 2	2 3	3 4	5 6	7 8	10 11	14 15	21 22	30 31	44 45	↑	↑	↑	↑	↑	↑	↑	↑	↑
M	315	↓	↓	↓	↓	↓	0 1	1 2	2 3	3 4	5 6	7 8	10 11	14 15	21 22	30 31	44 45	↑	↑	↑	↑	↑	↑	↑	↑	↑	↑
N	500	↓	↓	↓	↓	0 1	1 2	2 3	3 4	5 6	7 8	10 11	14 15	21 22	30 31	44 45	↑	↑	↑	↑	↑	↑	↑	↑	↑	↑	↑
P	800	↓	↓	↓	0 1	1 2	2 3	3 4	5 6	7 8	10 11	14 15	21 22	30 31	44 45	↑	↑	↑	↑	↑	↑	↑	↑	↑	↑	↑	↑
Q	1250	↓	↓	0 1	1 2	2 3	3 4	5 6	7 8	10 11	14 15	21 22	30 31	44 45	↑	↑	↑	↑	↑	↑	↑	↑	↑	↑	↑	↑	↑
R	2000	↓	0 1	1 2	2 3	3 4	5 6	7 8	10 11	14 15	21 22	30 31	44 45	↑	↑	↑	↑	↑	↑	↑	↑	↑	↑	↑	↑	↑	↑

↓ = Use first sampling plan below arrow. If sample size equals, or exceeds, lot or batch size, do 100 percent inspection.

↑ = Use first sampling plan above arrow.

Ac = Acceptance number.

Re = Rejection number.

154

Table 7-3 Single-Sampling Plans for Tightened Inspection (MIL-STD-105D).

Acceptable Quality Levels (tightened inspection) — each cell shows **Ac Re** (Acceptance number / Rejection number).

Sample size code letter	Sample size	0.010	0.015	0.025	0.040	0.065	0.10	0.15	0.25	0.40	0.65	1.0	1.5	2.5	4.0	6.5	10	15	25	40	65	100	150	250	400	650	1000
A	2	↓	↓	↓	↓	↓	↓	↓	↓	↓	↓	↓	↓	↓	↓	↓	↓	↓	0 1	1 2	2 3	3 4	5 6	8 9	12 13	18 19	27 28
B	3	↓	↓	↓	↓	↓	↓	↓	↓	↓	↓	↓	↓	↓	↓	↓	↓	0 1	1 2	2 3	3 4	5 6	8 9	12 13	18 19	27 28	41 42
C	5	↓	↓	↓	↓	↓	↓	↓	↓	↓	↓	↓	↓	↓	↓	↓	0 1	1 2	2 3	3 4	5 6	8 9	12 13	18 19	27 28	41 42	↑
D	8	↓	↓	↓	↓	↓	↓	↓	↓	↓	↓	↓	↓	↓	↓	0 1	1 2	2 3	3 4	5 6	8 9	12 13	18 19	27 28	41 42	↑	↑
E	13	↓	↓	↓	↓	↓	↓	↓	↓	↓	↓	↓	↓	↓	0 1	1 2	2 3	3 4	5 6	8 9	12 13	18 19	27 28	41 42	↑	↑	↑
F	20	↓	↓	↓	↓	↓	↓	↓	↓	↓	↓	↓	↓	0 1	1 2	2 3	3 4	5 6	8 9	12 13	18 19	27 28	41 42	↑	↑	↑	↑
G	32	↓	↓	↓	↓	↓	↓	↓	↓	↓	↓	↓	0 1	1 2	2 3	3 4	5 6	8 9	12 13	18 19	27 28	41 42	↑	↑	↑	↑	↑
H	50	↓	↓	↓	↓	↓	↓	↓	↓	↓	↓	0 1	1 2	2 3	3 4	5 6	8 9	12 13	18 19	27 28	41 42	↑	↑	↑	↑	↑	↑
J	80	↓	↓	↓	↓	↓	↓	↓	↓	↓	0 1	1 2	2 3	3 4	5 6	8 9	12 13	18 19	27 28	41 42	↑	↑	↑	↑	↑	↑	↑
K	125	↓	↓	↓	↓	↓	↓	↓	↓	0 1	1 2	2 3	3 4	5 6	8 9	12 13	18 19	27 28	41 42	↑	↑	↑	↑	↑	↑	↑	↑
L	200	↓	↓	↓	↓	↓	↓	↓	0 1	1 2	2 3	3 4	5 6	8 9	12 13	18 19	27 28	41 42	↑	↑	↑	↑	↑	↑	↑	↑	↑
M	315	↓	↓	↓	↓	↓	↓	0 1	1 2	2 3	3 4	5 6	8 9	12 13	18 19	27 28	41 42	↑	↑	↑	↑	↑	↑	↑	↑	↑	↑
N	500	↓	↓	↓	↓	↓	0 1	1 2	2 3	3 4	5 6	8 9	12 13	18 19	27 28	41 42	↑	↑	↑	↑	↑	↑	↑	↑	↑	↑	↑
P	800	↓	↓	↓	↓	0 1	1 2	2 3	3 4	5 6	8 9	12 13	18 19	27 28	41 42	↑	↑	↑	↑	↑	↑	↑	↑	↑	↑	↑	↑
Q	1250	↓	↓	↓	0 1	1 2	2 3	3 4	5 6	8 9	12 13	18 19	27 28	41 42	↑	↑	↑	↑	↑	↑	↑	↑	↑	↑	↑	↑	↑
R	2000	↓	↓	0 1	1 2	2 3	3 4	5 6	8 9	12 13	18 19	27 28	41 42	↑	↑	↑	↑	↑	↑	↑	↑	↑	↑	↑	↑	↑	↑
S	3150	↓	0 1	1 2	2 3	3 4	5 6	8 9	12 13	18 19	27 28	41 42	↑	↑	↑	↑	↑	↑	↑	↑	↑	↑	↑	↑	↑	↑	↑

⇩ = Use first sampling plan below arrow. If sample size equals, or exceeds, lot or batch size, do 100 percent inspection.

⇧ = Use first sampling plan above arrow.

Ac = Acceptance number.

Re = Rejection number.

Table 7-4 Single-Sampling Plans for Reduced Inspection (MIL-STD-105D).

Acceptable Quality Levels (reduced inspection)†

(Each AQL cell below gives "Ac Re"; ↓ = use first sampling plan below arrow; ↑ = use first sampling plan above arrow.)

Sample size code letter	Sample size	0.010	0.015	0.025	0.040	0.065	0.10	0.15	0.25	0.40	0.65	1.0	1.5	2.5	4.0	6.5	10	15	25	40	65	100	150	250	400	650	1000
A	2	↓	↓	↓	↓	↓	↓	↓	↓	↓	↓	↓	↓	↓	↓	0 1	↓	↑	1 2	2 3	3 4	5 6	7 8	10 11	14 15	21 22	30 31
B	2	↓	↓	↓	↓	↓	↓	↓	↓	↓	↓	↓	↓	↓	0 1	↓	↓	0 2	1 3	2 4	3 5	5 6	7 8	10 11	14 15	21 22	30 31
C	2	↓	↓	↓	↓	↓	↓	↓	↓	↓	↓	↓	↓	0 1	↓	↓	0 2	1 3	1 4	2 5	3 6	5 8	7 10	10 13	14 17	21 24	↑
D	3	↓	↓	↓	↓	↓	↓	↓	↓	↓	↓	↓	0 1	↓	↓	0 2	1 3	1 4	2 5	3 6	5 8	7 10	10 13	14 17	21 24	↑	↑
E	5	↓	↓	↓	↓	↓	↓	↓	↓	↓	↓	0 1	↓	↓	0 2	1 3	1 4	2 5	3 6	5 8	7 10	10 13	14 17	21 24	↑	↑	↑
F	8	↓	↓	↓	↓	↓	↓	↓	↓	↓	0 1	↓	↓	0 2	1 3	1 4	2 5	3 6	5 8	7 10	10 13	↑	↑	↑	↑	↑	↑
G	13	↓	↓	↓	↓	↓	↓	↓	↓	0 1	↓	↓	0 2	1 3	1 4	2 5	3 6	5 8	7 10	10 13	↑	↑	↑	↑	↑	↑	↑
H	20	↓	↓	↓	↓	↓	↓	↓	0 1	↓	↓	0 2	1 3	1 4	2 5	3 6	5 8	7 10	10 13	↑	↑	↑	↑	↑	↑	↑	↑
J	32	↓	↓	↓	↓	↓	↓	0 1	↓	↓	0 2	1 3	1 4	2 5	3 6	5 8	7 10	10 13	↑	↑	↑	↑	↑	↑	↑	↑	↑
K	50	↓	↓	↓	↓	↓	0 1	↓	↓	0 2	1 3	1 4	2 5	3 6	5 8	7 10	10 13	↑	↑	↑	↑	↑	↑	↑	↑	↑	↑
L	80	↓	↓	↓	↓	0 1	↓	↓	0 2	1 3	1 4	2 5	3 6	5 8	7 10	10 13	↑	↑	↑	↑	↑	↑	↑	↑	↑	↑	↑
M	125	↓	↓	↓	0 1	↓	↓	0 2	1 3	1 4	2 5	3 6	5 8	7 10	10 13	↑	↑	↑	↑	↑	↑	↑	↑	↑	↑	↑	↑
N	200	↓	↓	0 1	↓	↓	0 2	1 3	1 4	2 5	3 6	5 8	7 10	10 13	↑	↑	↑	↑	↑	↑	↑	↑	↑	↑	↑	↑	↑
P	315	↓	0 1	↓	↓	0 2	1 3	1 4	2 5	3 6	5 8	7 10	10 13	↑	↑	↑	↑	↑	↑	↑	↑	↑	↑	↑	↑	↑	↑
Q	500	0 1	↓	↓	0 2	1 3	1 4	2 5	3 6	5 8	7 10	10 13	↑	↑	↑	↑	↑	↑	↑	↑	↑	↑	↑	↑	↑	↑	↑
R	800	↑	↑	0 2	1 3	1 4	2 5	3 6	5 8	7 10	10 13	↑	↑	↑	↑	↑	↑	↑	↑	↑	↑	↑	↑	↑	↑	↑	↑

↓ = Use first sampling plan below arrow. If sample size equals or exceeds lot or batch size, do 100 percent inspection.

↑ = Use first sampling plan above arrow.

Ac = Acceptance number.

Re = Rejection number.

† = If the acceptance number has been exceeded, but the rejection number has not been reached, accept the lot, but reinstate normal inspection (see 10.1.4).

Table 7-5 Double-Sampling Plans for Normal Inspection (MIL-STD-105D).

Acceptable Quality Levels (normal inspection). Each AQL cell shows "Ac Re" (acceptance number, rejection number).

Code	Sample	Sample size	Cum. sample size	0.010	0.015	0.025	0.040	0.065	0.10	0.15	0.25	0.40	0.65	1.0	1.5	2.5	4.0	6.5	10	15	25	40	65	100	150	250	400	650	1000
A				↓	↓	↓	↓	↓	↓	↓	↓	↓	↓	↓	↓	↓	↓	↓	↓	•	•	•	•	•	•	•	•	•	•
B	First	2	2	↓	↓	↓	↓	↓	↓	↓	↓	↓	↓	↓	↓	↓	↓	↓	•	0 2	0 3	1 4	2 5	3 7	5 9	7 11	11 16	17 22	25 31
	Second	2	4																	1 2	3 4	4 5	6 7	8 9	12 13	18 19	26 27	37 38	56 57
C	First	3	3	↓	↓	↓	↓	↓	↓	↓	↓	↓	↓	↓	↓	↓	↓	•	0 2	0 3	1 4	2 5	3 7	5 9	7 11	11 16	17 22	25 31	↑
	Second	3	6																1 2	3 4	4 5	6 7	8 9	12 13	18 19	26 27	37 38	56 57	
D	First	5	5	↓	↓	↓	↓	↓	↓	↓	↓	↓	↓	↓	↓	↓	•	0 2	0 3	1 4	2 5	3 7	5 9	7 11	11 16	17 22	25 31	↑	↑
	Second	5	10															1 2	3 4	4 5	6 7	8 9	12 13	18 19	26 27	37 38	56 57		
E	First	8	8	↓	↓	↓	↓	↓	↓	↓	↓	↓	↓	↓	↓	•	0 2	0 3	1 4	2 5	3 7	5 9	7 11	11 16	17 22	25 31	↑	↑	↑
	Second	8	16														1 2	3 4	4 5	6 7	8 9	12 13	18 19	26 27	37 38	56 57			
F	First	13	13	↓	↓	↓	↓	↓	↓	↓	↓	↓	↓	↓	•	0 2	0 3	1 4	2 5	3 7	5 9	7 11	11 16	17 22	25 31	↑	↑	↑	↑
	Second	13	26													1 2	3 4	4 5	6 7	8 9	12 13	18 19	26 27	37 38	56 57				
G	First	20	20	↓	↓	↓	↓	↓	↓	↓	↓	↓	↓	•	0 2	0 3	1 4	2 5	3 7	5 9	7 11	11 16	17 22	25 31	↑	↑	↑	↑	↑
	Second	20	40												1 2	3 4	4 5	6 7	8 9	12 13	18 19	26 27	37 38	56 57					
H	First	32	32	↓	↓	↓	↓	↓	↓	↓	↓	↓	•	0 2	0 3	1 4	2 5	3 7	5 9	7 11	11 16	17 22	25 31	↑	↑	↑	↑	↑	↑
	Second	32	64											1 2	3 4	4 5	6 7	8 9	12 13	18 19	26 27	37 38	56 57						
J	First	50	50	↓	↓	↓	↓	↓	↓	↓	↓	•	0 2	0 3	1 4	2 5	3 7	5 9	7 11	11 16	17 22	25 31	↑	↑	↑	↑	↑	↑	↑
	Second	50	100										1 2	3 4	4 5	6 7	8 9	12 13	18 19	26 27	37 38	56 57							
K	First	80	80	↓	↓	↓	↓	↓	↓	↓	•	0 2	0 3	1 4	2 5	3 7	5 9	7 11	11 16	17 22	25 31	↑	↑	↑	↑	↑	↑	↑	↑
	Second	80	160									1 2	3 4	4 5	6 7	8 9	12 13	18 19	26 27	37 38	56 57								
L	First	125	125	↓	↓	↓	↓	↓	↓	•	0 2	0 3	1 4	2 5	3 7	5 9	7 11	11 16	17 22	25 31	↑	↑	↑	↑	↑	↑	↑	↑	↑
	Second	125	250								1 2	3 4	4 5	6 7	8 9	12 13	18 19	26 27	37 38	56 57									
M	First	200	200	↓	↓	↓	↓	↓	•	0 2	0 3	1 4	2 5	3 7	5 9	7 11	11 16	17 22	25 31	↑	↑	↑	↑	↑	↑	↑	↑	↑	↑
	Second	200	400							1 2	3 4	4 5	6 7	8 9	12 13	18 19	26 27	37 38	56 57										
N	First	315	315	↓	↓	↓	↓	•	0 2	0 3	1 4	2 5	3 7	5 9	7 11	11 16	17 22	25 31	↑	↑	↑	↑	↑	↑	↑	↑	↑	↑	↑
	Second	315	630						1 2	3 4	4 5	6 7	8 9	12 13	18 19	26 27	37 38	56 57											
P	First	500	500	↓	↓	↓	•	0 2	0 3	1 4	2 5	3 7	5 9	7 11	11 16	17 22	25 31	↑	↑	↑	↑	↑	↑	↑	↑	↑	↑	↑	↑
	Second	500	1000					1 2	3 4	4 5	6 7	8 9	12 13	18 19	26 27	37 38	56 57												
Q	First	800	800	↓	↓	•	0 2	0 3	1 4	2 5	3 7	5 9	7 11	11 16	17 22	25 31	↑	↑	↑	↑	↑	↑	↑	↑	↑	↑	↑	↑	↑
	Second	800	1600				1 2	3 4	4 5	6 7	8 9	12 13	18 19	26 27	37 38	56 57													
R	First	1250	1250	↓	•	0 2	0 3	1 4	2 5	3 7	5 9	7 11	11 16	17 22	25 31	↑	↑	↑	↑	↑	↑	↑	↑	↑	↑	↑	↑	↑	↑
	Second	1250	2500			1 2	3 4	4 5	6 7	8 9	12 13	18 19	26 27	37 38	56 57														

↓ = Use first sampling plan below arrow. If sample size equals or exceeds lot or batch size, do 100 percent inspection.

↑ = Use first sampling plan above arrow.

Ac = Acceptance number.

Re = Rejection number.

• = Use corresponding single sampling plan (or alternatively, use double sampling plan below, where available).

Table 7-6 Double-Sampling Plans for Tightened Inspection (MIL-STD-105D).

Acceptable Quality Levels (tightened inspection)

Legend for acceptance (Ac) / rejection (Re) number cells, given as `First: Ac Re / Second: Ac Re`. Empty cells indicate an arrow (use the first sampling plan above or below the arrow, per the notes) or a single-sampling reference (*).

Code	Sample	Sample size	Cum. sample size	0.010	0.015	0.025	0.040	0.065	0.10	0.15	0.25	0.40	0.65	1.0	1.5	2.5	4.0	6.5	10	15	25	40	65	100	150	250	400	650	1000
A																												
B	First	2	2																		0 2	0 3	1 4	2 5	3 7	6 10	9 14	15 20	23 24
B	Second	2	4																		1 2	3 4	4 5	6 7	11 12	15 16	23 24	34 35	52 53
C	First	3	3																	0 2	0 3	1 4	2 5	3 7	6 10	9 14	15 20	23 24	
C	Second	3	6																	1 2	3 4	4 5	6 7	11 12	15 16	23 24	34 35	52 53	
D	First	5	5																0 2	0 3	1 4	2 5	3 7	6 10	9 14	15 20	23 24		
D	Second	5	10																1 2	3 4	4 5	6 7	11 12	15 16	23 24	34 35	52 53		
E	First	8	8															0 2	0 3	1 4	2 5	3 7	6 10	9 14	15 20	23 24			
E	Second	8	16															1 2	3 4	4 5	6 7	11 12	15 16	23 24	34 35	52 53			
F	First	13	13														0 2	0 3	1 4	2 5	3 7	6 10	9 14	15 20	23 24				
F	Second	13	26														1 2	3 4	4 5	6 7	11 12	15 16	23 24	34 35	52 53				
G	First	20	20													0 2	0 3	1 4	2 5	3 7	6 10	9 14							
G	Second	20	40													1 2	3 4	4 5	6 7	11 12	15 16	23 24							
H	First	32	32												0 2	0 3	1 4	2 5	3 7	6 10	9 14								
H	Second	32	64												1 2	3 4	4 5	6 7	11 12	15 16	23 24								
J	First	50	50											0 2	0 3	1 4	2 5	3 7	6 10	9 14									
J	Second	50	100											1 2	3 4	4 5	6 7	11 12	15 16	23 24									
K	First	80	80										0 2	0 3	1 4	2 5	3 7	6 10	9 14										
K	Second	80	160										1 2	3 4	4 5	6 7	11 12	15 16	23 24										
L	First	125	125									0 2	0 3	1 4	2 5	3 7	6 10	9 14											
L	Second	125	250									1 2	3 4	4 5	6 7	11 12	15 16	23 24											
M	First	200	200								0 2	0 3	1 4	2 5	3 7	6 10	9 14												
M	Second	200	400								1 2	3 4	4 5	6 7	11 12	15 16	23 24												
N	First	315	315							0 2	0 3	1 4	2 5	3 7	6 10	9 14													
N	Second	315	630							1 2	3 4	4 5	6 7	11 12	15 16	23 24													
P	First	500	500						0 2	0 3	1 4	2 5	3 7	6 10	9 14														
P	Second	500	1000						1 2	3 4	4 5	6 7	11 12	15 16	23 24														
Q	First	800	800					0 2	0 3	1 4	2 5	3 7	6 10	9 14															
Q	Second	800	1600					1 2	3 4	4 5	6 7	11 12	15 16	23 24															
R	First	1250	1250				0 2	0 3	1 4	2 5	3 7	6 10	9 14																
R	Second	1250	2500				1 2	3 4	4 5	6 7	11 12	15 16	23 24																
S	First	2000	2000			0 2	0 3	1 4	2 5	3 7	6 10	9 14																	
S	Second	2000	4000			1 2	3 4	4 5	6 7	11 12	15 16	23 24																	

↓ = Use first sampling plan below arrow. If sample size equals or exceeds lot or batch size, do 100 percent inspection.

↑ = Use first sampling plan above arrow.

Ac = Acceptance number.

Re = Rejection number.

* = Use corresponding single sampling plan (or alternatively, use double sampling plan below, where available).

158

Table 7-7 Double-Sampling Plans for Reduced Inspection (MIL-STD-105D).

Acceptable Quality Levels (reduced inspection) †

Note: Each Acceptable Quality Level (AQL) column contains an acceptance number (Ac) and a rejection number (Re). Values below are shown as "Ac Re". ↓ = use first sampling plan below arrow; ↑ = use first sampling plan above arrow; • = use corresponding single sampling plan.

Code	Sample	Sample size	Cum. sample size	0.010	0.015	0.025	0.040	0.065	0.10	0.15	0.25	0.40	0.65	1.0	1.5	2.5	4.0	6.5	10	15	25	40	65	100	150	250	400	650	1000
A				↑	↑	↑	↑	↑	↑	↑	↑	↑	↑	↑	↑	↑	↑	↑	↑	↑	↑	↑	↑	↑	↑	↑	•	•	•
B				↑	↑	↑	↑	↑	↑	↑	↑	↑	↑	↑	↑	↑	↑	↑	↑	↑	↑	↑	↑	↑	↑	↑	•	•	•
C				↑	↑	↑	↑	↑	↑	↑	↑	↑	↑	↑	↑	↑	↑	↑	↑	↑	↑	↑	↑	↑	↑	↑	•	•	•
D	First	2	2	↓	↓	↓	↓	↓	↓	↓	↓	↓	↓	↓	↓	0 2	0 2	0 3	0 3	0 4	0 4	1 5	2 6	3 8	5 10	7 12	11 17	↑	↑
D	Second	2	4	↓	↓	↓	↓	↓	↓	↓	↓	↓	↓	↓	↓	0 2	0 3	0 4	1 4	1 5	3 6	4 7	6 9	8 12	12 16	18 22	26 30	↑	↑
E	First	3	3	↓	↓	↓	↓	↓	↓	↓	↓	↓	↓	↓	0 2	0 2	0 3	0 3	0 4	0 4	1 5	2 6	3 8	5 10	7 12	11 17	↑	↑	↑
E	Second	3	6	↓	↓	↓	↓	↓	↓	↓	↓	↓	↓	↓	0 2	0 3	0 4	1 4	1 5	3 6	4 7	6 9	8 12	12 16	18 22	26 30	↑	↑	↑
F	First	5	5	↓	↓	↓	↓	↓	↓	↓	↓	↓	↓	0 2	0 2	0 3	0 3	0 4	0 4	1 5	2 6	3 8	5 10	7 12	11 17	↑	↑	↑	↑
F	Second	5	10	↓	↓	↓	↓	↓	↓	↓	↓	↓	↓	0 2	0 3	0 4	1 4	1 5	3 6	4 7	6 9	8 12	12 16	18 22	26 30	↑	↑	↑	↑
G	First	8	8	↓	↓	↓	↓	↓	↓	↓	↓	↓	0 2	0 2	0 3	0 3	0 4	0 4	1 5	2 6	3 8	5 10	7 12	11 17	↑	↑	↑	↑	↑
G	Second	8	16	↓	↓	↓	↓	↓	↓	↓	↓	↓	0 2	0 3	0 4	1 4	1 5	3 6	4 7	6 9	8 12	12 16	18 22	26 30	↑	↑	↑	↑	↑
H	First	13	13	↓	↓	↓	↓	↓	↓	↓	↓	0 2	0 2	0 3	0 3	0 4	0 4	1 5	2 6	3 8	5 10	7 12	11 17	↑	↑	↑	↑	↑	↑
H	Second	13	26	↓	↓	↓	↓	↓	↓	↓	↓	0 2	0 3	0 4	1 4	1 5	3 6	4 7	6 9	8 12	12 16	18 22	26 30	↑	↑	↑	↑	↑	↑
J	First	20	20	↓	↓	↓	↓	↓	↓	↓	0 2	0 2	0 3	0 3	0 4	0 4	1 5	2 6	3 8	5 10	7 12	11 17	↑	↑	↑	↑	↑	↑	↑
J	Second	20	40	↓	↓	↓	↓	↓	↓	↓	0 2	0 3	0 4	1 4	1 5	3 6	4 7	6 9	8 12	12 16	18 22	26 30	↑	↑	↑	↑	↑	↑	↑
K	First	32	32	↓	↓	↓	↓	↓	↓	0 2	0 2	0 3	0 3	0 4	0 4	1 5	2 6	3 8	5 10	7 12	11 17	↑	↑	↑	↑	↑	↑	↑	↑
K	Second	32	64	↓	↓	↓	↓	↓	↓	0 2	0 3	0 4	1 4	1 5	3 6	4 7	6 9	8 12	12 16	18 22	26 30	↑	↑	↑	↑	↑	↑	↑	↑
L	First	50	50	↓	↓	↓	↓	↓	0 2	0 2	0 3	0 3	0 4	0 4	1 5	2 6	3 8	5 10	7 12	11 17	↑	↑	↑	↑	↑	↑	↑	↑	↑
L	Second	50	100	↓	↓	↓	↓	↓	0 2	0 3	0 4	1 4	1 5	3 6	4 7	6 9	8 12	12 16	18 22	26 30	↑	↑	↑	↑	↑	↑	↑	↑	↑
M	First	80	80	↓	↓	↓	↓	0 2	0 2	0 3	0 3	0 4	0 4	1 5	2 6	3 8	5 10	7 12	11 17	↑	↑	↑	↑	↑	↑	↑	↑	↑	↑
M	Second	80	160	↓	↓	↓	↓	0 2	0 3	0 4	1 4	1 5	3 6	4 7	6 9	8 12	12 16	18 22	26 30	↑	↑	↑	↑	↑	↑	↑	↑	↑	↑
N	First	125	125	↓	↓	↓	0 2	0 2	0 3	0 3	0 4	0 4	1 5	2 6	3 8	5 10	7 12	11 17	↑	↑	↑	↑	↑	↑	↑	↑	↑	↑	↑
N	Second	125	250	↓	↓	↓	0 2	0 3	0 4	1 4	1 5	3 6	4 7	6 9	8 12	12 16	18 22	26 30	↑	↑	↑	↑	↑	↑	↑	↑	↑	↑	↑
P	First	200	200	↓	↓	0 2	0 2	0 3	0 3	0 4	0 4	1 5	2 6	3 8	5 10	7 12	11 17	↑	↑	↑	↑	↑	↑	↑	↑	↑	↑	↑	↑
P	Second	200	400	↓	↓	0 2	0 3	0 4	1 4	1 5	3 6	4 7	6 9	8 12	12 16	18 22	26 30	↑	↑	↑	↑	↑	↑	↑	↑	↑	↑	↑	↑
Q	First	315	315	↓	0 2	0 2	0 3	0 3	0 4	0 4	1 5	2 6	3 8	5 10	7 12	11 17	↑	↑	↑	↑	↑	↑	↑	↑	↑	↑	↑	↑	↑
Q	Second	315	630	↓	0 2	0 3	0 4	1 4	1 5	3 6	4 7	6 9	8 12	12 16	18 22	26 30	↑	↑	↑	↑	↑	↑	↑	↑	↑	↑	↑	↑	↑
R	First	500	500	0 2	0 2	0 3	0 3	0 4	0 4	1 5	2 6	3 8	5 10	7 12	11 17	↑	↑	↑	↑	↑	↑	↑	↑	↑	↑	↑	↑	↑	↑
R	Second	500	1000	0 2	0 3	0 4	1 4	1 5	3 6	4 7	6 9	8 12	12 16	18 22	26 30	↑	↑	↑	↑	↑	↑	↑	↑	↑	↑	↑	↑	↑	↑

↓ = Use first sampling plan below arrow. If sample size equals or exceeds lot or batch size, do 100 percent inspection.

↑ = Use first sampling plan above arrow.

Ac = Acceptance number.

Re = Rejection number.

• = Use corresponding single sampling plan (or alternatively, use double sampling plan below, when available).

† = If, after the second sample, the acceptance number has been exceeded, but the rejection number has not been reached, accept the lot, but reinstate normal inspection (see 10.14).

Table 7-8 Multiple-Sampling Plans for Normal Inspection (MIL-STD-105D).

Acceptance Quality Levels (normal inspection)

Legend

↓	Use first sampling plan below arrow (refer to continuation of table on following page, when necessary). If sample size equals or exceeds lot or batch size, do 100 percent inspection.
↑	Use first sampling plan above arrow.
Ac	Acceptance number
Re	Rejection number
•	Use corresponding single sampling plan (or alternatively, use multiple sampling plan below, where available)
‡	Use corresponding double sampling plan (or alternatively, use multiple sampling plan below, where available)
*	Acceptance not permitted at this sample size.

Sample size code letters A, B, C: (arrow — use corresponding plan; ‡ and • as marked) — no multiple-sampling plan; use plan below.

Code Letter D (sample size 2 each)

Cumulative sample sizes: 2, 4, 6, 8, 10, 12, 14

AQLs below 10: ↓ (use plan below). AQLs 650, 1000: ↑ (use plan above).

Sample	10 (Ac Re)	15	25	40	65	100	150	250	400
First	* 2	* 3	* 4	* 5	0 6	1 7	2 9	4 12	6 16
Second	* 2	0 3	0 5	1 6	3 8	4 10	7 14	11 19	17 27
Third	0 2	0 4	1 6	2 8	6 10	8 13	13 19	19 27	29 39
Fourth	0 3	1 5	2 7	3 9	8 13	12 17	19 25	27 34	40 49
Fifth	1 3	2 6	3 8	5 11	11 15	17 20	25 29	36 40	53 58
Sixth	1 3	3 6	4 9	7 12	14 17	21 23	31 33	45 47	66 68
Seventh	2 3	4 7	6 10	9 14	18 19	25 26	37 38	53 54	77 78

Code Letter E (sample size 3 each)

Cumulative sample sizes: 3, 6, 9, 12, 15, 18, 21

Sample	6.5 (Ac Re)	10	15	25	40	65	100	150	250
First	* 2	* 3	* 4	* 5	0 6	1 7	2 9	4 12	6 16
Second	* 2	0 3	0 5	1 6	3 8	4 10	7 14	11 19	17 27
Third	0 2	0 4	1 6	2 8	6 10	8 13	13 19	19 27	29 39
Fourth	0 3	1 5	2 7	3 9	8 13	12 17	19 25	27 34	40 49
Fifth	1 3	2 6	3 8	5 11	11 15	17 20	25 29	36 40	53 58
Sixth	1 3	3 6	4 9	7 12	14 17	21 23	31 33	45 47	66 68
Seventh	2 3	4 7	6 10	9 14	18 19	25 26	37 38	53 54	77 78

Code Letter F (sample size 5 each)

Cumulative sample sizes: 5, 10, 15, 20, 25, 30, 35

Sample	4.0 (Ac Re)	6.5	10	15	25	40	65	100	150
First	* 2	* 3	* 4	* 5	0 6	1 7	2 9	4 12	6 16
Second	* 2	0 3	0 5	1 6	3 8	4 10	7 14	11 19	17 27
Third	0 2	0 4	1 6	2 8	6 10	8 13	13 19	19 27	29 39
Fourth	0 3	1 5	2 7	3 9	8 13	12 17	19 25	27 34	40 49
Fifth	1 3	2 6	3 8	5 11	11 15	17 20	25 29	36 40	53 58
Sixth	1 3	3 6	4 9	7 12	14 17	21 23	31 33	45 47	66 68
Seventh	2 3	4 7	6 10	9 14	18 19	25 26	37 38	53 54	77 78

Code Letter G (sample size 8 each)

Cumulative sample sizes: 8, 16, 24, 32, 40, 48, 56

Sample	2.5 (Ac Re)	4.0	6.5	10	15	25	40	65	100
First	* 2	* 3	* 4	* 5	0 6	1 7	2 9	4 12	6 16
Second	* 2	0 3	0 5	1 6	3 8	4 10	7 14	11 19	17 27
Third	0 2	0 4	1 6	2 8	6 10	8 13	13 19	19 27	29 39
Fourth	0 3	1 5	2 7	3 9	8 13	12 17	19 25	27 34	40 49
Fifth	1 3	2 6	3 8	5 11	11 15	17 20	25 29	36 40	53 58
Sixth	1 3	3 6	4 9	7 12	14 17	21 23	31 33	45 47	66 68
Seventh	2 3	4 7	6 10	9 14	18 19	25 26	37 38	53 54	77 78

Code Letter H (sample size 13 each)

Cumulative sample sizes: 13, 26, 39, 52, 65, 78, 91

Sample	1.5 (Ac Re)	2.5	4.0	6.5	10	15	25	40	65
First	* 2	* 3	* 4	* 5	0 6	1 7	2 9	4 12	6 16
Second	* 2	0 3	0 5	1 6	3 8	4 10	7 14	11 19	17 27
Third	0 2	0 4	1 6	2 8	6 10	8 13	13 19	19 27	29 39
Fourth	0 3	1 5	2 7	3 9	8 13	12 17	19 25	27 34	40 49
Fifth	1 3	2 6	3 8	5 11	11 15	17 20	25 29	36 40	53 58
Sixth	1 3	3 6	4 9	7 12	14 17	21 23	31 33	45 47	66 68
Seventh	2 3	4 7	6 10	9 14	18 19	25 26	37 38	53 54	77 78

Code Letter J (sample size 20 each)

Cumulative sample sizes: 20, 40, 60, 80, 100, 120, 140

Sample	1.0 (Ac Re)	1.5	2.5	4.0	6.5	10	15	25	40
First	* 2	* 3	* 4	* 5	0 6	1 7	2 9	4 12	6 16
Second	* 2	0 3	0 5	1 6	3 8	4 10	7 14	11 19	17 27
Third	0 2	0 4	1 6	2 8	6 10	8 13	13 19	19 27	29 39
Fourth	0 3	1 5	2 7	3 9	8 13	12 17	19 25	27 34	40 49
Fifth	1 3	2 6	3 8	5 11	11 15	17 20	25 29	36 40	53 58
Sixth	1 3	3 6	4 9	7 12	14 17	21 23	31 33	45 47	66 68
Seventh	2 3	4 7	6 10	9 14	18 19	25 26	37 38	53 54	77 78

For AQLs to the left of the tabulated values use the first sampling plan below the arrow; for AQLs to the right, use the first sampling plan above the arrow.

Table 7-9 Average Outgoing Quality Limit Factors for Normal Inspection (Single Sampling) (MIL-STD-105D).

Acceptable Quality Level

Code Letter	Sample Size	0.010	0.015	0.025	0.040	0.065	0.10	0.15	0.25	0.40	0.65	1.0	1.5	2.5	4.0	6.5	10	15	25	40	65	100	150	250	400	650	1000
A	2																17	28	42	69	97	160	220	330	470	730	1100
B	3															18	17	27	46	65	110	150	220	310	490	720	1100
C	5														12	11	15	24	39	63	90	130	190	290	430	660	
D	8													7.4	6.5	11	16	24	40	56	82	120	180	270	410		
E	13												4.6	4.2	6.9	9.7	14	22	34	50	72	110	170	250			
F	20											2.8	2.6	4.3	6.1	9.9	13	21	33	47	73						
G	32										1.8	1.7	2.7	3.9	6.3	9.0	12	19	29	46							
H	50									1.2	1.1	1.7	2.4	4.0	5.6	8.2	12	18	29								
J	80								0.74	0.67	1.1	1.6	2.5	3.6	5.2	7.5											
K	125							0.46	0.42	0.69	0.97	1.6	2.2	3.3	4.7	7.3											
L	200						0.29	0.27	0.44	0.62	1.00	1.4	2.1	3.0	4.7												
M	315					0.18	0.17	0.27	0.39	0.63	0.90	1.3	1.9	2.9													
N	500				0.12	0.11	0.17	0.24	0.40	0.56	0.82	1.2	1.8														
P	800			0.074	0.067	0.11	0.16	0.25	0.36	0.52	0.75	1.2															
Q	1250		0.046	0.042	0.069	0.097	0.16	0.22	0.33	0.47	0.73																
R	2000	0.029																									

Note: For the exact AOQL, the above values must be multiplied by $\left(1 - \dfrac{\text{Sample size}}{\text{Lot or Batch size}}\right)$

Table 7-10 *Limiting Quality (percent defective) for Which P_a = 10% (for Normal Inspection. Single Sampling) (MIL-STD-105D).*

Acceptable Quality Level

Code letter	Sample size	0.010	0.015	0.025	0.040	0.065	0.10	0.15	0.25	0.40	0.65	1.0	1.5	2.5	4.0	6.5	10
A	2																
B	3													37	54	68	
C	5																58
D	8											16	25			41	54
E	13										11				27	36	44
F	20													18	25	30	42
G	32									6.9			12	16	20	27	34
H	50								4.5			7.6	10	13	18	22	29
J	80							2.8			4.8	6.5	8.2	11	14	19	24
K	125						1.8			3.1	4.3	5.4	7.4	9.4	12	16	23
L	200					1.2		1.2	2.0	2.7	3.3	4.6	5.9	7.7	10	14	
M	315				0.73				1.7	2.1	2.9	3.7	4.9	6.4	9.0		
N	500			0.46			0.78	1.1	1.3	1.9	2.4	3.1	4.0	5.6			
P	800		0.29			0.49	0.67	0.84	1.2	1.5	1.9	2.5	3.5				
Q	1250	0.18			0.31	0.43	0.53	0.74	0.94	1.2	1.6	2.3					
R	2000			0.20	0.27	0.33	0.46	0.59	0.77	1.0	1.4						

Table 7-11 *Limiting Quality (defects per hundred units) for Which $P_a = 10\%$ (for Normal Inspection, Single Sampling) (MIL-STD-105D).*

Code letter	Sample size	0.010	0.015	0.025	0.040	0.065	0.10	0.15	0.25	0.40	0.65	1.0	1.5	2.5	4.0	6.5	10	15	25	40	65	100	150	250	400	650	1000
																											Acceptable Quality Level
A	2															120			200	270	330	460	590	770	1000	1400	1000
B	3														77			130	180	220	310	390	510	670	940	1300	1900
C	5													46			78	110	130	190	240	310	400	560	770	1100	1800
D	8												29			49	67	84	120	150	190	250	350	480	670		
E	13											18			30	41	51	71	91	120	160	220	300	410			
F	20										12			20	27	33	46	59	77	100	140						
G	32									7.2			12	17	21	29	37	48	63	88							
H	50								4.6			7.8	11	13	19	24	31	40	56								
J	80							2.9			4.9	6.7	8.4	12	15	19	25	35									
K	125						1.8			3.1	4.3	5.4	7.4	9.4	12	16	23										
L	200					1.2			2.0	2.7	3.3	4.6	5.9	7.7	10	14											
M	315				0.73			1.2	1.7	2.1	2.9	3.7	4.9	6.4	9.0												
N	500			0.46			0.78	1.1	1.3	1.9	2.4	3.1	4.0	5.6													
P	800		0.29			0.49	0.67	0.84	1.2	1.5	1.9	2.5	3.5														
Q	1250	0.18			0.31	0.43	0.53	0.74	0.94	1.2	1.6	2.3															
R	2000			0.20	0.27	0.33	0.46	0.59	0.77	1.0	1.4																

Table 7-12 Limit Numbers for Reduced Inspection (MIL-STD-105D).

Number of sample units from last 10 lots or batches	Acceptable Quality Level																									
	0.010	0.015	0.025	0.040	0.065	0.10	0.15	0.25	0.40	0.65	1.0	1.5	2.5	4.0	6.5	10	15	25	40	65	100	150	250	400	650	1000
20 - 29	*	*	*	*	*	*	*	*	*	*	*	*	*	*	*	0	0	2	4	8	14	22	40	68	115	181
30 - 49	*	*	*	*	*	*	*	*	*	*	*	*	*	*	0	0	1	3	7	13	22	36	63	105	178	277
50 - 79	*	*	*	*	*	*	*	*	*	*	*	*	*	0	0	2	3	7	14	25	40	63	110	181	301	
80 - 129	*	*	*	*	*	*	*	*	*	*	*	*	0	0	2	4	7	14	24	42	68	105	181	297		
130 - 199	*	*	*	*	*	*	*	*	*	*	*	0	0	2	4	7	13	25	42	72	115	177	301	490		
200 - 319	*	*	*	*	*	*	*	*	*	*	0	0	2	4	8	14	22	40	68	115	181	277	471			
320 - 499	*	*	*	*	*	*	*	*	*	0	0	1	4	8	14	24	39	68	113	189						
500 - 799	*	*	*	*	*	*	*	*	0	0	2	3	7	14	25	40	63	110	181							
800 - 1249	*	*	*	*	*	*	*	0	0	2	4	7	14	24	42	68	105	181								
1250 - 1999	*	*	*	*	*	*	0	0	2	4	7	13	24	40	69	110	169									
2000 - 3149	*	*	*	*	*	0	0	2	4	8	14	22	40	68	115	181										
3150 - 4999	*	*	*	*	0	0	1	4	8	14	24	38	67	111	186											
5000 - 7999	*	*	*	0	0	2	3	7	14	25	40	63	110	181												
8000 - 12499	*	*	0	0	2	4	7	14	24	42	68	105	181													
12500 - 19999	*	0	0	2	4	7	13	24	40	69	110	169														
20000 - 31499	0	0	2	4	8	14	22	40	68	115	181															
31500 - 49999	0	1	4	8	14	24	38	67	111	186																
50000 & Over	2	3	7	14	25	40	63	110	181	301																

* Denotes that the number of sample units from the last ten lots or batches is not sufficient for reduced inspection for this AQL. In this instance more than ten lots or batches may be used for the calculation, provided that the lots or batches used are the most recent ones in sequence, that they have all been on normal inspection, and that none has been rejected while on original inspection.

Table 7-13 *Average Sample-Size Curves for Double and Multiple Sampling (Normal and Tightened Inspection) (MIL-STD-105D).*

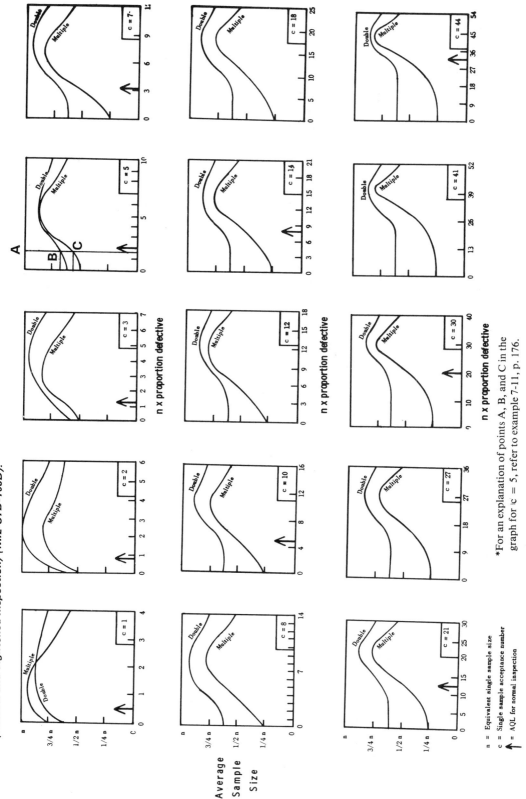

n = Equivalent single sample size

c = Single sample acceptance number

↱ = AQL for normal inspection

*For an explanation of points A, B, and C in the graph for ic = 5, refer to example 7-11, p. 176.

Table 7-14 Tables for Sample-Size Code Letter: J (MIL-STD-105D).

(a) Operating Characteristic Curves for Single Sampling Plans

(Curves for Double and Multiple Sampling Are Matched as Closely as Practicable)

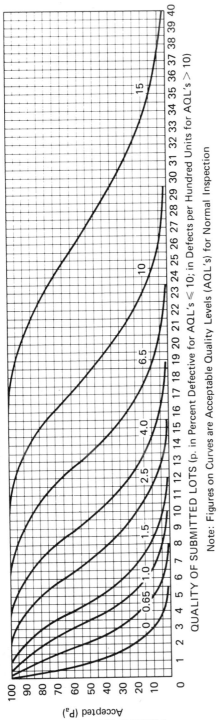

QUALITY OF SUBMITTED LOTS (p. in Percent Defective for AQL's ≤ 10; in Defects per Hundred Units for AQL's > 10)

Note: Figures on Curves are Acceptable Quality Levels (AQL's) for Normal Inspection

(b) Tabulated Values for Operating Characteristic Curves for Single Sampling Plans

Acceptable Quality Levels (normal inspection)

p (in Percent Defective)

P_a	0.15	0.65	1.0	1.5	2.5	4.0	✕	6.5	✕	10
99.0	0.013	0.188	0.550	1.05	2.30	3.72	4.50	6.13	7.88	9.75
95.0	0.064	0.444	1.03	1.73	3.32	5.06	5.98	7.91	9.89	11.9
90.0	0.132	0.666	1.38	2.20	3.98	5.91	6.91	8.95	11.0	13.2
75.0	0.359	1.202	2.16	3.18	5.30	7.50	8.62	10.9	13.2	15.5
50.0	0.863	2.09	3.33	4.57	7.06	9.55	10.8	13.3	15.8	18.3
25.0	1.72	3.33	4.84	6.31	9.14	11.9	13.3	16.0	18.6	21.3
10.0	2.84	4.78	6.52	8.16	11.3	14.2	15.7	18.6	21.4	24.2
5.0	3.68	5.80	7.66	9.39	12.7	15.8	17.3	20.3	23.2	26.0
1.0	5.59	8.00	10.1	12.0	15.6	18.9	20.5	23.6	26.5	29.5
(tightened)		0.25	1.0	1.5	2.5	4.0		6.5		10

p (in Defects per Hundred Units)

P_a	0.15	0.65	1.0	1.5	2.5	4.0	✕	6.5	✕	10	✕	15
99.0	0.013	0.186	0.545	1.03	2.23	3.63	4.38	5.96	7.62	9.35	12.9	15.7
95.0	0.064	0.444	1.02	1.71	3.27	4.98	5.87	7.71	9.61	11.6	15.6	18.6
90.0	0.131	0.665	1.38	2.18	3.94	5.82	6.79	8.78	10.8	12.9	17.1	20.3
75.0	0.360	1.20	2.16	3.17	5.27	7.45	8.55	10.8	13.0	15.3	19.9	23.4
50.0	0.866	2.10	3.34	4.59	7.09	9.59	10.8	13.3	15.8	18.3	23.3	27.1
25.0	1.73	3.37	4.90	6.39	9.28	12.1	13.5	16.3	19.0	21.8	27.2	31.2
10.0	2.88	4.86	6.65	8.35	11.6	14.7	16.2	19.3	22.2	25.2	30.9	35.2
5.0	3.75	5.93	7.87	9.69	13.1	16.4	18.0	21.2	24.3	27.4	33.4	37.8
1.0	5.76	8.30	10.5	12.6	16.4	20.0	21.8	25.2	28.5	31.8	38.2	42.9
(tightened)		0.25	1.0	1.5	2.5	4.0		6.5		10		15

Acceptable Quality Levels (tightened inspection)

Note: All Values Given in Above Table Based on Poisson Distribution as an Approximation to the Binomial Binomial distribution used for percent defective computations; Poisson for defects per hundred units''.

Table 7-14 (c)

Acceptable Quality Levels (normal inspection)

(Each AQL cell shows Ac / Re, where Ac = acceptance number and Re = rejection number.)

Type of sampling plan	Cumulative sample size	<0.15	0.15	0.25	0.40	0.65	1.0	1.5	2.5	4.0	6.5	10	15	>15	Cumulative sample size
Single	80	▽	0 / 1	Use Letter H	Use Letter	1 / 2	2 / 3	3 / 4	5 / 6	7 / 8	10 / 11	14 / 15	21 / 22	△	80
Double	50	▽	*	Use Letter L	Use Letter K	0 / 2	0 / 3	1 / 4	2 / 5	3 / 7	5 / 9	7 / 11	11 / 16	△	50
	100					1 / 2	3 / 4	4 / 5	6 / 7	8 / 9	12 / 13	18 / 19	26 / 27		100
Multiple	20	▽	*			# / 2	# / 2	# / 3	# / 4	0 / 4	0 / 5	1 / 7	2 / 9	△	20
	40					# / 2	0 / 3	0 / 3	1 / 5	1 / 6	3 / 8	4 / 10	7 / 14		40
	60					0 / 2	0 / 3	1 / 4	2 / 6	3 / 8	6 / 10	8 / 13	13 / 19		60
	80					0 / 3	1 / 4	2 / 5	3 / 7	5 / 10	8 / 13	12 / 17	19 / 25		80
	100					1 / 3	2 / 5	3 / 6	5 / 8	7 / 11	11 / 15	17 / 20	25 / 29		100
	120					1 / 3	3 / 5	4 / 6	7 / 9	10 / 12	14 / 17	21 / 24	31 / 33		120
	140					2 / 3	4 / 5	6 / 7	9 / 10	13 / 14	18 / 19	25 / 26	37 / 38		140

Acceptable Quality Levels (tightened inspection) — label row aligned (shifted) beneath the normal columns:

Less than 0.25	0.25	0.40	0.65	1.0	1.5	2.5	4.0	6.5	10	15	Higher than 15

△ = Use next preceding sample size code letter for which acceptance and rejection numbers are available.

▽ = Use next subsequent sample size code letter for which acceptance and rejection numbers are available.

Ac = Acceptance number

Re = Rejection number

* = Use single sampling plan above (or alternatively use letter M)

= Acceptance not permitted at this sample size.

Table 7-15 Tables for Sample-Size Code Letter: K (MIL-STD-105D).

(a) OPERATING CHARACTERISTIC CURVES FOR SINGLE SAMPLING PLANS

(Curves for double and multiple sampling are matched as closely as practicable)

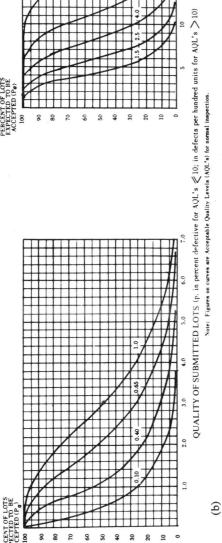

Note: Figures on curves are Acceptable Quality Levels (AQL's) for normal inspection.

QUALITY OF SUBMITTED LOTS (p, in percent defective for AQL's ≪10; in defects per hundred units for AQL's >10)

(b) TABULATED VALUES FOR OPERATING CHARACTERISTIC CURVES FOR SINGLE SAMPLING PLANS

P_a	Acceptable Quality Levels (normal inspection)								
	0.10	0.40	0.65	1.0	1.5	2.5	4.0	6.5	10
	p (in percent defective or defects per hundred units)								
99.0	0.0081	0.119	0.349	0.658	1.43	2.33	3.82	5.98	10.1
95.0	0.0410	0.284	0.654	1.09	2.09	3.19	4.94	7.40	11.9
90.0	0.0840	0.426	0.882	1.40	2.52	3.73	5.62	8.24	13.0
75.0	0.230	0.769	1.382	2.03	3.38	4.77	6.90	9.79	14.9
50.0	0.554	1.34	2.14	2.94	4.54	6.14	8.53	11.7	17.3
25.0	1.11	2.15	3.14	4.09	5.94	7.75	10.4	13.9	20.0
10.0	1.84	3.11	4.26	5.35	7.42	9.42	12.3	16.1	22.5
5.0	2.40	3.80	5.04	6.20	8.41	10.5	13.6	17.5	24.2
1.0	3.68	5.31	6.73	8.04	10.5	12.8	16.1	20.4	27.5
	0.15	0.65	1.0	1.5	2.5	4.0	6.5	10	X
	Acceptable Quality Levels (tightened inspection)								

Note: All Values Given in Above Table Based on Poisson Distribution as an Approximation to the Binomial

Table 7-15 (c) Sampling Plans for Sample Size Code Letter: K

Acceptable Quality Levels (normal inspection)

Type of sampling plan	Cumulative sample size	Less than 0.10 (Ac/Re)	0.10	0.15	0.25	0.40	0.65	1.0	1.5	2.5	4.0	6.5	✕	10	Higher than 10
Single	125	▽	0 1	Use Letter J	Use Letter L	1 2	2 3	3 4	5 6	7 8	10 11	14 15	18 19	21 22	△
Double	80	▽	*	Use Letter J	Use Letter M	0 2	0 3	1 4	2 5	3 7	5 9	7 11	9 14	11 16	△
Double	160					1 2	3 4	4 5	6 7	8 9	12 13	18 19	23 24	26 27	

Multiple (sample size 32 each; cumulative sizes 32–224)

Type of sampling plan	Cumulative sample size	Less than 0.10	0.10	0.15	0.25	0.40	0.65	1.0	1.5	2.5	4.0	6.5	✕	10	Higher than 10
Multiple	32	▽	*	J	L	# 2	# 2	# 3	# 4	0 4	0 5	1 6	1 8	2 9	△
Multiple	64					# 2	0 3	0 3	1 5	1 6	3 8	4 9	6 12	7 14	
Multiple	96					0 2	0 3	1 4	2 6	3 8	6 10	8 13	11 17	13 19	
Multiple	128					0 3	1 4	2 5	3 7	5 10	8 13	12 17	16 22	19 25	
Multiple	160					1 3	2 4	3 6	5 8	7 11	11 15	17 20	22 25	25 29	
Multiple	192					1 3	3 5	4 6	7 9	10 12	14 17	21 23	27 29	31 33	
Multiple	224					2 3	4 5	6 7	9 10	13 14	18 19	25 26	32 33	37 38	

Acceptable Quality Levels (tightened inspection) — label row aligned beneath the columns above:

Less than 0.15	✕	0.15	0.25	0.40	0.65	1.0	1.5	2.5	4.0	✕	6.5	✕	10	Higher than 10

The two rightmost sub-column headings for each AQL are **Ac** (left) and **Re** (right).

Legend

△ = Use next preceding sample size code letter for which acceptance and rejection numbers are available.

▽ = Use next subsequent sample size code letter for which acceptance and rejection numbers are available.

Ac = Acceptance number

Re = Rejection number

* = Use single sampling plan above (or alternatively use letter N).

\# = Acceptance not permitted at this sample size.

169

7-3 EXAMPLES—MIL-STD-105D

The following examples illustrate the use of MIL-STD-105D.

EXAMPLE 7-1

Given Production lot of 750 units,

$$AQL = 2.5\%$$

single-sampling plan, normal inspection.

Questions

(a) What is the sample size n?
(b) What is the maximum number of defective units that may be tolerated in the sample without causing rejection of the production lot?
(c) How many defective units in the sample require rejection of the production lot?

Solution

(a) Choose general inspection level II.
(b) Determine the sample-size code letter from Table 7-1 of MIL-STD-105D (p. 153). The production lot size of 750 units falls in the range 501 to 1200. For this range of lot sizes and under general inspection level II, find sample-size code letter J.
(c) Determine the sample size n from Table 7-2 (p. 154) for sample-size code letter J. The sample size is $n = 80$.
(d) Determine the acceptance number A_c and the rejection number R_e from Table 7-2 for the sample-size code letter J and for an acceptable quality level AQL = 2.5%:

$$A_c = 5 \qquad R_e = 6$$

Therefore, the maximum number of tolerable defective units in the sample of 80 is 5. Should six or more defective units be found in the sample of 80, this would require rejection of the production lot of 750 units.

EXAMPLE 7-2

Given Same specifications as for Example 7-1, except tightened inspection is used.

Questions Same as for Example 7-1.

Solution

(a) Choose general inspection level II.
(b) The sample-size code letter and sample size remain unchanged (code letter J, sample size 80).
(c) The accept and reject numbers are, from Table 7-3 (p. 155),

$$A_c = 3.0 \qquad R_e = 4.0$$

Hence, if the sample lot of $n = 80$ contains up to 3 defective units, the production lot of 750 units is acceptable. Should 4 or more defective units be found, the production lot is rejected.

EXAMPLE 7-3

Given   Production lot of 3000 units,

$$AQL = 1.0\%$$

single-sampling plan, normal inspection.

Questions

- (a) What is the probability that this sampling plan will reject the production lot if it contains
  1. 2% defectives.
  2. 3% defectives.
- (b) What is the minimum number of defectives which, when found in the sample, will result in rejection of the production lot?

Solution   The sample-size code letter for a production lot of 3000 units and for general inspection level II, is determined from Table 7-1 (p. 153); it is K (125 units). Turn to Table 7-15 (p. 168), part (a). The graph on the left side of the page is based on lot quality expressed in percent defectives (the graph on the right side is for lot quality expressed in defects per hundred units; it is used when the AQL exceeds 10%). The family of OC curves depicted by the graph covers a range of AQL values (0.1%, 0.4%, 0.65%, and 1%).

To find the probability of acceptance of a production lot containing 2% defectives, the vertical line through the 2% point is followed to the intersection of the $AQL = 1\%$ OC curve. The associated percent of lots expected to be accepted is read off the vertical scale. In this case, $P_a = 75\%$. When the production lot is 3% defective, $P_a = 48\%$. Hence, 2% and 3% defective production lots have a chance of 75% and 48%, respectively, of being accepted by this plan.

Table 7-2 (p. 154) is used to determine the minimum number of defectives in the sample lot, R_e, which will cause rejection of the production lot. For sample size, K and $AQL = 1\%$, $R_e = 4$. Therefore, if 4 defectives are found in the 125-unit sample, the production lot of 3000 units is rejected.

EXAMPLE 7-4

Given   Same specification as for Example 7-3.

Question   Determine the LTPD value for this plan.

Solution   The LTPD value is that product quality which has a 10% chance of acceptance. Table 7-15 (p. 168) applies. The intersection of the 10% horizontal probability line with the $AQL = 1\%$ OC curve is determined and the corresponding product quality determined on the horizontal axis; $LTPD = 5.4\%$.

EXAMPLE 7-5

Given   Production lot of 400 units,

$$\text{AQL} = 4\%$$

single-sampling plan, normal inspection.

Question   What is the limiting quality (in percent defective) for which the probability of acceptance is 10% (LTPD)?

Solution

(a) Choose general inspection level II.
(b) Determine the sample-size code letter from Table 7-1 (p. 153) for a lot size of 400; it is *H*.
(c) Since we are interested in an LTPD, turn to Table 7-10 (p. 162), which covers limiting quality sampling for which $P_a = 10\%$. The limiting quality level for code letter *H* and an AQL = 4.0% is found to be 18% defectives.

Hence, if the production lot runs 4% defective (16 defective units in the lot of 400), there is a 95% chance of acceptance (AQL); if quality deteriorates to 18% defectives (72 defective units in the lot of 400 units), there is only a 10% chance of acceptance (LTPD). Acceptance, under the terms of this plan, means finding no more than 5 defective units in a sample of 50 units drawn from the lot of 400 units.

EXAMPLE 7-6

Given   Production lot of 150 units,

$$\text{AQL} = 0.25\%$$

single-sampling plan, tightened inspection.

Question   How many defective units must be found in the required sample to cause rejection of the production lot?

Solution

(a) Choose general inspection level II.
(b) Determine the sample-size code letter from Table 7-1 (p. 153); it is *F*.
(c) Enter Table 7-3 (p. 155) with code letter *F* (sample size 20 units). There is no accept/reject criterion at the intersection of the horizontal code letter *F* line and the AQL = 0.25% column. There is, however, an arrow pointing downward. The applicable accept/reject criterion is found at the tip of the arrow; it is $A_c = 0$, $R_e = 1.0$. As will be noted, this changes the corresponding required sample size from 20 to 80 (new sample size code letter *J*).

Hence, since the reject criterion is 1.0, the lot of 150 units will be rejected if one or more defective units are found in a sample of 80 units.

EXAMPLE 7-7

Given   Production lot of 2500 units,

$$\text{AQL} = 1\%$$

double-sampling plan, normal inspection.

Questions

(a) What is the size of the first sample?

(b) Determine whether to accept or reject the production lot if the following results are obtained in testing the initial sample:
   1. zero defectives.
   2. one defective.
   3. two defectives.
   4. three defectives.
   5. four defectives.
   6. five defectives.

(c) How many defectives have to be found in the first sample to require drawing a second sample? What is the size of the second sample?

(d) What is the accept/reject criterion for the second sample?

(e) Should the production lot be accepted if the first sample has 2 defectives and the second sample 5 defectives?

Solution   Determine the sample-size code letter from Table 7-1 (p. 153); for general inspection level II and a lot size of 2500 units, the code letter is K. Turn to Table 7-5 (p. 157) for double-sampling plans for normal inspection. Enter the table at sample-size code letter K. Columns 2, 3, and 4 of the table furnish the sample sizes for the first and second sample (in case a second sample is required); the first sample is 80 units, the second sample is also 80 units, giving a cumulative sample size of 160 units.

For an AQL $= 1.0\%$ and code letter K, the accept/reject criterion for the first sample is $A_c = 1.0$, $R_e = 4.0$. Similarly, the accept/reject criterion for the second sample is $A_c = 4.0$, $R_e = 5.0$. Therefore, the following decisions regarding the assumed 0 to 5 defectives in the initial sample apply:

Number of Defectives in Initial Sample	Decision Regarding Acceptability of Production Lot of 2500 Units
0	Accept
1	Accept
2	Pick second sample lot
3	Pick second sample lot
4	Reject
5	Reject

It is seen that if the number of defectives found in the initial sample equals a number that falls between the accept number and reject number (2.0 and 3.0 in this case), a second sample of 80 units has to be drawn. If the total number of defectives found in both samples does not exceed 4, the production lot is still acceptable. Five or more defectives found in the two samples will cause rejection of the production lot.

Finally, regarding the last question—if the first sample has 2 defectives and the second sample 5 defectives, the production lot is to be rejected.

EXAMPLE 7-8

Given   Production lot of 1000 units,

$$AQL = 1.0\%$$

limiting quality plan, consumer's risk is 10%, normal inspection, single sampling.

Questions

    (a) What is the size of the sample?
    (b) What is the limiting quality in percent defectives which corresponds to the consumer's risk of 10%?
    (c) What are the accept and reject numbers for this plan?
    (d) What is the probability of lot acceptance if the submitted lots contain 1%, 3%, and $4\frac{1}{2}\%$ defectives, respectively?
    (e) Discuss the results.

Solution   Determine the sample-size code letter from Table 7-1 (p. 153). For general inspection level II and a lot size of 1000 units, the code letter is *J*. From Table 7-10 (p. 162), determine the sample size corresponding to the code letter *J* in column 2. The size of the sample is 80. From the same table, for an AQL = 1% and code letter *J*, the limiting quality for a probability of acceptance of 10% is 6.5%.

From Table 7-2 (p. 154), determine the accept and reject numbers for sample-size code letter *J* and AQL = 1.0%,

$$A_c = 2.0 \qquad R_e = 3$$

Turn to Table 7-14 (p. 166) and determine the probability of lot acceptance for the given values of lot quality from the family of OC curves.

Quality of Submitted Lots	*Probability of Lot Acceptance*
1%	95% (producer's risk)
3%	60%
$4\frac{1}{2}\%$	30%

Hence, submitted lots that have a quality level of 1% defectives have a 95% chance of being accepted (AQL = 1%). This means that, 95 times out of 100, samples of size 80, taken from production lots which are 1% defective, which contain not more than 2 defective units will be accepted. Submitted lots which have the limiting quality level of 6.5% defectives have a 10% chance of being accepted (LTPD = 10%). The consumer is thus assured that if a lot of 1000 units, for example, contains 65 defective units, this sampling plan will pass such a lot only 10 of 100 times. The producer is assured that if he submits lots which are 1% defective (10 defective units in a lot of 1000), he has, on the average, a 95% chance of having such lots accepted.

EXAMPLE 7-9

Given   A manufacturer receives an invitation to bid on a government contract for special brass inserts for aircraft use. The contract is for 114,000 inserts and specifies a 1.0% AQL sampling plan. Production and delivery is in weekly batches of 1200 inserts. Based on MIL-STD-105D requirements, the weekly inspection sample size is $n = 80$.

Question   Disregarding quantity discounts, what should the manufacturer's unit price be if past experience on similar orders showed average production to be 1% defective and if it costs him \$2.00 to produce an insert and 35 cents to inspect it (overhead costs are included)? To cover his profit the manufacturer adds 10% to his total cost figure.

Solution

(a) First, the manufacturer has to determine how many inserts he will have to produce to assure a yield of 114,000 units acceptable by the inspection plan. Since his own product quality matches the required AQL, he is justified to expect 95% of his lots to be accepted and 5% to be rejected. Hence,

$$\frac{114,000}{0.95} = 120,000$$

Therefore, he has to produce 120,000 inserts.

(b) The quantity of inserts requiring inspection is 80 per production lot of 1200. There will be $120,000/1200 = 100$ production lots; hence, a total of $80 \times 100 = 8000$ inserts will be inspected.

(c) The total cost of production plus markup is therefore:

cost to produce 120,000 inserts, at \$2.00/unit	\$240,000
cost of inspecting 8000 inserts, at \$0.35/unit	2,800
Total	\$242,800
10% profit	24,280
	\$267,080

(d) The per-unit insert price (not considering quantity discounts) is therefore

$$\frac{267,080}{114,000} = \$2.34/\text{insert}$$

EXAMPLE 7-10

Given   A production lot consists of 1000 units. A single-sampling plan with AQL = 1.5% is in effect. In accordance with MIL-STD-105D, an inspection sample of 80 units is required. The acceptance number is $A_c = 3$.

Question   Determine the AOQL value.

Solution From Table 7-9 (p. 161) the factor for AQL = 1.5% and a sample size of 80 is 2.4. Per equation (57), the AOQL is

$$\text{AOQL} = 2.4\left(1 - \frac{80}{1000}\right) = \underline{2.21\%}$$

EXAMPLE 7-11

Given A single-sampling plan is in effect. The sample-size code letter is *K* and the AQL = 1.5%.

Question Determine the expected ASN for single sampling, double sampling, and multiple sampling.

Solution For a sample-size code letter *K* and an AQL = 1.5%, the single-sampling plan (see p. 169) requires a sample size of 125 and the acceptance number $A_c = 5$ ($c = 5$ or fewer defectives).

First, we determine the expected number of defectives: ($n \times$ proportion defective) = $125 \times 0.015 = 1.875$, where 0.015 represents the AQL = 1.5%. In the top row of Table 7-13 (p. 165) we locate the family of curves for $c = 5$. A vertical line drawn through $n \times$ proportion defective = 1.875 intersects the single-sampling line at *A* and the double- and multiple-sampling curves at *B* and *C*, respectively.

Point *A* represents 100%, or 125 units; point *B* is read off as approximately 68% of 125, or 85 units; and point *C* is approximately 56% of 125, or 70 units. Hence, the expected ASN for single sampling, double sampling, and multiple sampling will be 125, 85, and 70, respectively.

7-4 TEXT OF MIL-STD-414 (SECTION A)*

GENERAL DESCRIPTION OF SAMPLING PLANS

A1. SCOPE

A1.1 Purpose. This Standard establishes sampling plans and procedures for inspection by variables for use in Government procurement, supply and storage, and maintenance inspection operations. When applicable this Standard shall be referenced in the specification, contract, or inspection instructions, and the provisions set forth herein shall govern.

A1.2 Inspection. Inspection is the process of measuring, examining, testing, gaging, or otherwise comparing the "unit of product" (See A1.4) with the applicable requirements.

A1.3 Inspection by Variables. Inspection by variables is inspection wherein a specified quality characteristic (See A1.5) on a unit of product is measured on a continuous scale, such as pounds, inches, feet per second, etc., and a measurement is recorded.

A1.4 Unit of Product. The unit of product is the entity of product inspected in order to determine its measurable quality characteristic. This may be a single article, a pair, a set, a component of an end product, or the end product itself. The unit of product may or may not be the same as the unit of purchase, supply, production, or shipment.

A1.5 Quality Characteristic. The quality characteristic for variables inspection is that characteristic of a unit of product that is actually measured, to determine conformance with a given requirement.

A1.6 Specification Limits. The specification limit(s) is the requirement that a quality characteristic should meet. This requirement may be expressed as an upper specification limit; or a lower specification

* For MIL-STD-414 Tables A1, A2, A3 see Tables 7-16, 7-17, 7-18 in this book; (Table A3 is not reproduced in its entirety).

limit, called herein a single specification limit; or both upper and lower specification limits, called herein a double specification limit.

A1.7 <u>Sampling Plans</u>. A sampling plan is a procedure which specifies the number of units of product from a lot which are to be inspected, and the criterion for acceptability of the lot. Sampling plans designated in this Standard are applicable to the inspection of a single quality characteristic of a unit of product. These plans may be used whether procurement inspection is performed at the plant of a prime contractor, subcontractor or vendor, or at destination, and also may be used when appropriate in supply and storage, and maintenance inspection operations.

A2. CLASSIFICATION OF DEFECTS

A2.1 <u>Method of Classifying Defects</u>. A classification of defects is the enumeration of defects of the unit of product classified according to their importance. A defect is a deviation of the unit of product from requirements of the specifications, drawings, purchase descriptions, and any changes thereto in the contract or order. Defects normally belong to one of the following classes; however, defects may be placed in other classes.

A2.1.1 <u>Critical Defects</u>. A critical defect is one that judgment and experience indicate could result in hazardous or unsafe conditions for individuals using or maintaining the product; or, for major end items units of product, such as ships, aircraft, or tanks, a defect that could prevent performance of their tactical function.

A2.1.2 <u>Major Defects</u>. A major defect is a defect, other than critical, that could result in failure, or materially reduce the usability of the unit of product for its intended purpose.

A2.1.3 <u>Minor Defects</u>. A minor defect is one that does not materially reduce the usability of the unit of product for its intended purpose, or is a departure from established standards having no significant bearing on the effective use or operation of the unit.

A3. PERCENT DEFECTIVE

A3.1 <u>Expression of Nonconformance</u>. The extent of nonconformance of product shall be expressed in terms of percent defective.

A3.2 <u>Percent Defective</u>. The percent defective for a quality characteristic of a given lot of product is the number of units of product defective for that characteristic divided by the total number of units of product and multiplied by one hundred. Expressed as an equation: Percent defective =

$$\frac{\text{Number of defectives x 100}}{\text{Number of units}}$$

A4. ACCEPTABLE QUALITY LEVEL

A4.1 <u>Acceptable Quality Level</u>. The acceptable quality level (AQL) is a nominal value expressed in terms of percent defective specified for a single quality characteristic. Certain numerical values of AQL ranging from .04 to 15.00 percent are shown in Table A-1. When a range of AQL values is specified, it shall be treated as if it were equal to the value of AQL for which sampling plans are furnished and which is included within the AQL range. When the specified AQL is a particular value other than those for which sampling plans are furnished, the AQL, which is to be used in applying the provisions of this Standard, shall be as shown in Table A-1.

A4.2 <u>Specifying AQL's</u>. The particular AQL value to be used for a single quality characteristic of a given product must be specified. In the case of a double specification limit, either an AQL value is specified for the total percent defective outside of both upper and lower specification limits, or two AQL values are specified, one for the upper limit and another for the lower limit.

A5. SUBMITTAL OF PRODUCT

A5.1 <u>Lot</u>. The term "lot" shall mean "inspection lot," i.e., a collection of units of product from which a sample is drawn and inspected to determine compliance with the acceptability criterion.

A5.1.1 <u>Formation of Lots</u>. Each lot shall, as far as is practicable, consist of units of product of a single type, grade, class, size, or composition manufactured under essentially the same conditions.

A5.2 <u>Lot Size</u>. The lot size is the number of units of product in a lot, and may differ from the quantity designated in the contract or order as a lot for production, shipment, or other purposes.

A6. LOT ACCEPTABILITY

A6.1 <u>Acceptability Criterion</u>. The acceptability of a lot of material submitted for inspection shall be determined by use of one of the sampling plans associated with a specified value of the AQL(s). This Standard provides sampling plans based on known and unknown variability. In the latter case two alternative methods are provided, one based on the estimate of lot standard deviation and the other on the average range of the sample. These are referred to as the standard deviation method and the range method. For the case of a single specification limit, the acceptability criterion is given in two forms. These are identified as Form 1 and Form 2.

A6.2 Choice of Sampling Plans. Sampling plans and procedures are provided in Section B if variability is unknown and the standard deviation method is used, in Section C if variability is unknown and the range method is used, and in Section D if variability is known. Unless otherwise specified, unknown variability, standard deviation method sampling plans, and the acceptability criterion of Form 2 (for the single specification limit case) shall be used.

A7. SAMPLE SELECTION

A7.1 Determination of Sample Size. The sample size is the number of units of product drawn from a lot. Relative sample sizes are designated by code letters. The sample size code letter depends on the inspection level and the lot size. There are five inspection levels: I, II, III, IV, and V. Unless otherwise specified inspection level IV shall be used. The sample size code letter applicable to the specified inspection level and for lots of given size shall be obtained from Table A-2.

NOTICE—Special Reservation for Critical Characteristics. The Government reserves the right to inspect every unit submitted by the supplier for critical characteristics, and to reject the remainder of the lot immediately after a defect is found. The Government also reserves the right to sample for critical defects every lot submitted by the supplier and to reject any lot if a sample drawn therefrom is found to contain one or more critical defects.

A7.2 Drawing of Samples. A sample is one or more units of product drawn from a lot. Units of the sample shall be selected without regard to their quality.

A8. ESTIMATION OF PROCESS AVERAGE AND SEVERITY OF INSPECTION

Procedures for estimating the process average and criteria for tightened and reduced inspection based on the inspection results of preceding lots are provided in Part III of Sections B, C, and D.

A9. SPECIAL PROCEDURE FOR APPLICATION OF MIXED VARIABLES-ATTRIBUTES SAMPLING PLANS

A9.1 Applicability. A mixed variables and attributes sampling plan may be used under either of the two following conditions: (NOTE: No Operating Characteristic Curves are provided for the mixed variables-attributes sampling plans herein and that those in Table A-3 are not applicable.)

Condition A. Ample evidence exists that the product submitted for inspection is selected by the supplier to meet the specification limit(s) by a screening process from a larger quantity of product which is not being produced within the specification limit(s).

Condition B. Other conditions exist that warrant the use of a variables-attributes sampling plan.

A9.2 Definitions.

A9.2.1 Inspection by Attributes. Inspection by attributes is inspection wherein the unit of product is classified simply as defective or nondefective with respect to a given requirement or set of requirements.

A9.2.2 Mixed Variables-Attributes Inspection. Mixed variables-attributes inspection is inspection of a sample by attributes, in addition to inspection by variables already made of a previous sample, before a decision as to acceptability or rejectability of a lot can be made.

A9.3 Selection of Sampling Plans. The mixed variables-attributes sampling plan shall be selected in accordance with the following:

A9.3.1 Select the variables sampling plan in accordance with Section B, C, or D.

A9.3.2 Select the attributes sampling plan from MIL-STD-105, paragraph 10, using a single sampling plan and tightened inspection. The same AQL value(s) shall be used for the attributes sampling plan as used for the variables plan of paragraph A9.3.1. (Additional sample items may be drawn, as necessary, to satisfy the requirements for sample size of the attributes sampling plan.

Table 7-16 *AQL Conversion Table (MIL-STD-414).*

For specified AQL values falling within these ranges	Use this AQL value
—— to 0.049	0.04
0.050 to 0.069	0.065
0.070 to 0.109	0.10
0.110 to 0.164	0.15
0.165 to 0.279	0.25
0.280 to 0.439	0.40
0.440 to 0.699	0.65
0.700 to 1.09	1.0
1.10 to 1.64	1.5
1.65 to 2.79	2.5
2.80 to 4.39	4.0
4.40 to 6.99	6.5
7.00 to 10.9	10.0
11.00 to 16.4	15.0

Count as a defective each sample item falling outside of specification limit(s).)

A9.4 Determination of Acceptability. A lot meets the acceptability criterion if one of the following conditions is satisfied:

Condition A. The lot complies with the appropriate variables acceptability criterion of Section B, C, or D.

Condition B. The lot complies with the acceptability criterion of paragraph 11.1.2 of MIL-STD-105.

A9.4.1 If Condition A is not satisfied, proceed in accordance with the attributes sampling plan to meet Condition B.

A9.4.2 If Condition B is not satisfied, the lot does not meet the acceptability criterion.

A9.5 Severity of Inspection. The procedures for severity of inspection referred to in paragraph A8 are not applicable for mixed variables-attributes inspection.

Table 7-17 *Sample-Size Code Letters[1] (MIL-STD-414).*

Lot Size		Inspection Levels				
		I	II	III	IV	V
3 to	8	B	B	B	B	C
9 to	15	B	B	B	B	D
16 to	25	B	B	B	C	E
26 to	40	B	B	B	D	F
41 to	65	B	B	C	E	G
66 to	110	B	B	D	F	H
111 to	180	B	C	E	G	I
181 to	300	B	D	F	H	J
301 to	500	C	E	G	I	K
501 to	800	D	F	H	J	L
801 to	1,300	E	G	I	K	L
1,301 to	3,200	F	H	J	L	M
3,201 to	8,000	G	I	L	M	N
8,001 to	22,000	H	J	M	N	O
22,001 to	110,000	I	K	N	O	P
110,001 to	550,000	I	K	O	P	Q
550,001 and over		I	K	P	Q	Q

[1] Sample size code letters given in body of table are applicable when the indicated inspection levels are to be used.

Table 7-18 *Operating Characteristic Curves for Sampling Plans Based on Standard Deviation Method, Sample-Size Code Letter D (MIL-STD-414).*

(Curves for Sampling Plans Based on Range Method and Known Variability Are Essentially Equivalent)

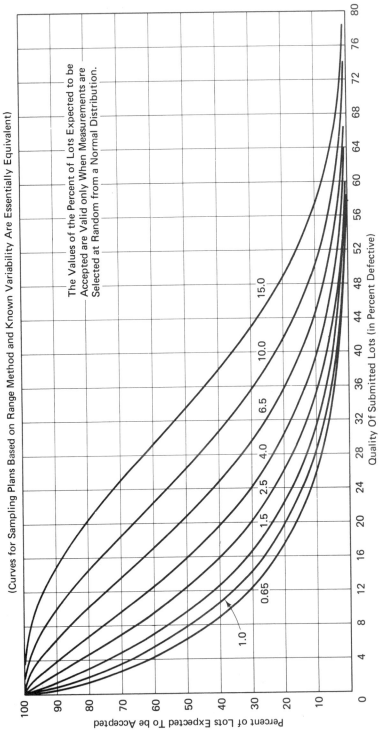

The Values of the Percent of Lots Expected to be Accepted are Valid only When Measurements are Selected at Random from a Normal Distribution.

Percent of Lots Expected To be Accepted

Quality Of Submitted Lots (in Percent Defective)

Note: Figures on Curves are Acceptable Quality Levels for Normal Inspection.

Table 7-19 *(MIL-STD-414) Master Table for Normal and Tightened Inspection for Plans Based on Variability Unknown, Standard Deviation Method (Single-Specification Limit, Form 1).*

Sample size code letter	Sample size	Acceptable Quality Levels (normal inspection)													
		.04	.065	.10	.15	.25	.40	.65	1.00	1.50	2.50	4.00	6.50	10.00	15.00
		k	k	k	k	k	k	k	k	k	k	k	k	k	k
B	3	↓	↓	↓	↓	↓	↓	↓	↓	↓	1.12	.958	.765	.566	.341
C	4	↓	↓	↓	↓	↓	↓	↓	1.45	1.34	1.17	1.01	.814	.617	.393
D	5	↓	↓	↓	↓	↓	↓	1.65	1.53	1.40	1.24	1.07	.874	.675	.455
E	7	↓	↓	↓	↓	2.00	1.88	1.75	1.62	1.50	1.33	1.15	.955	.755	.536
F	10	↓	↓	↓	2.24	2.11	1.98	1.84	1.72	1.58	1.41	1.23	1.03	.828	.611
G	15	2.64	2.53	2.42	2.32	2.20	2.06	1.91	1.79	1.65	1.47	1.30	1.09	.886	.664
H	20	2.69	2.58	2.47	2.36	2.24	2.11	1.96	1.82	1.69	1.51	1.33	1.12	.917	.695
I	25	2.72	2.61	2.50	2.40	2.26	2.14	1.98	1.85	1.72	1.53	1.35	1.14	.936	.712
J	30	2.73	2.61	2.51	2.41	2.28	2.15	2.00	1.86	1.73	1.55	1.36	1.15	.946	.723
K	35	2.77	2.65	2.54	2.45	2.31	2.18	2.03	1.89	1.76	1.57	1.39	1.18	.969	.745
L	40	2.77	2.66	2.55	2.44	2.31	2.18	2.03	1.89	1.76	1.58	1.39	1.18	.971	.746
M	50	2.83	2.71	2.60	2.50	2.35	2.22	2.08	1.93	1.80	1.61	1.42	1.21	1.00	.774
N	75	2.90	2.77	2.66	2.55	2.41	2.27	2.12	1.98	1.84	1.65	1.46	1.24	1.03	.804
O	100	2.92	2.80	2.69	2.58	2.43	2.29	2.14	2.00	1.86	1.67	1.48	1.26	1.05	.819
P	150	2.96	2.84	2.73	2.61	2.47	2.33	2.18	2.03	1.89	1.70	1.51	1.29	1.07	.841
Q	200	2.97	2.85	2.73	2.62	2.47	2.33	2.18	2.04	1.89	1.70	1.51	1.29	1.07	.845
		.065	.10	.15	.25	.40	.65	1.00	1.50	2.50	4.00	6.50	10.00	15.00	
		Acceptable Quality Levels (tightened inspection)													

All AQL values are in percent defective.

↓ Use first sampling plan below arrow, that is, both sample size as well as k value. When sample size equals or exceeds lot size, every item in the lot must be inspected.

181

Table 7-20 (MIL-STD-414) Master Table for Normal and Tightened Inspection for Plans Based on Variability Unknown, Standard Deviation Method (Double-Specification Limit and Form 2—Single-Specification Limit).

Sample size code letter	Sample size	Acceptable Quality Levels (normal inspection)													
		.04	.065	.10	.15	.25	.40	.65	1.00	1.50	2.50	4.00	6.50	10.00	15.00
		M	M	M	M	M	M	M	M	M	M	M	M	M	M
B	3	↓	↓	↓	↓	↓	→	→	↓	↓	7.59	18.86	26.94	33.69	40.47
C	4	↓	↓	↓	↓	↓	→	→	1.53	5.50	10.92	16.45	22.86	29.45	36.90
D	5	↓	↓	↓	↓	→	→	1.33	3.32	5.83	9.80	14.39	20.19	26.56	33.99
E	7	↓	↓	↓	↓	0.422	1.06	2.14	3.55	5.35	8.40	12.20	17.35	23.29	30.50
F	10	↓	↓	↓	0.349	0.716	1.30	2.17	3.26	4.77	7.29	10.54	15.17	20.74	27.57
G	15	0.099	0.186	0.312	0.503	0.818	1.31	2.11	3.05	4.31	6.56	9.46	13.71	18.94	25.61
H	20	0.135	0.228	0.365	0.544	0.846	1.29	2.05	2.95	4.09	6.17	8.92	12.99	18.03	24.53
I	25	0.155	0.250	0.380	0.551	0.877	1.29	2.00	2.86	3.97	5.97	8.63	12.57	17.51	23.97
J	30	0.179	0.280	0.413	0.581	0.879	1.29	1.98	2.83	3.91	5.86	8.47	12.36	17.24	23.58
K	35	0.170	0.264	0.388	0.535	0.847	1.23	1.87	2.68	3.70	5.57	8.10	11.87	16.65	22.91
L	40	0.179	0.275	0.401	0.566	0.873	1.26	1.88	2.71	3.72	5.58	8.09	11.85	16.61	22.86
M	50	0.163	0.250	0.363	0.503	0.789	1.17	1.71	2.49	3.45	5.20	7.61	11.23	15.87	22.00
N	75	0.147	0.228	0.330	0.467	0.720	1.07	1.60	2.29	3.20	4.87	7.15	10.63	15.13	21.11
O	100	0.145	0.220	0.317	0.447	0.689	1.02	1.53	2.20	3.07	4.69	6.91	10.32	14.75	20.66
P	150	0.134	0.203	0.293	0.413	0.638	0.949	1.43	2.05	2.89	4.43	6.57	9.88	14.20	20.02
Q	200	0.135	0.204	0.294	0.414	0.637	0.945	1.42	2.04	2.87	4.40	6.53	9.81	14.12	19.92
		.065	.10	.15	.25	.40	.65	1.00	1.50	2.50	4.00	6.50	10.00	15.00	
		Acceptability Quality Levels (tightened inspection)													

All AQL and table values are in percent defective.

↓ Use first sampling plan below arrow, that is, both sample size as well as M value. When sample size equals or exceeds lot size, every item in the lot must be inspected.

Table 7-21 *(MIL-STD-414) Table for Estimating the Lot Percent Defective Using Standard Deviation Method.** *

Q_U or Q_L	Sample Size															
	3	4	5	7	10	15	20	25	30	35	40	50	75	100	150	200
0	50.00	50.00	50.00	50.00	50.00	50.00	50.00	50.00	50.00	50.00	50.00	50.00	50.00	50.00	50.00	50.00
.1	47.24	46.67	46.44	46.26	46.16	46.10	46.08	46.06	46.05	46.05	46.04	46.04	46.03	46.03	46.02	46.02
.2	44.46	43.33	42.90	42.54	42.35	42.24	42.19	42.16	42.15	42.13	42.13	42.11	42.10	42.09	42.06	42.06
.3	41.63	40.00	39.37	38.87	38.60	38.44	38.37	38.33	38.31	38.29	38.28	38.27	38.25	38.24	38.22	38.22
.31	41.35	39.67	39.02	38.50	38.23	38.06	37.99	37.95	37.93	37.91	37.90	37.89	37.87	37.86	37.84	37.84
.32	41.06	39.33	38.67	38.14	37.86	37.69	37.62	37.58	37.55	37.54	37.52	37.51	37.49	37.48	37.46	37.46
.33	40.77	39.00	38.32	37.78	37.49	37.31	37.24	37.20	37.18	37.16	37.15	37.13	37.11	37.10	37.09	37.06
.34	40.49	38.67	37.97	37.42	37.12	36.94	36.87	36.83	36.80	36.78	36.77	36.75	36.73	36.72	36.71	36.71
.35	40.20	38.33	37.62	37.06	36.75	36.57	36.49	36.45	36.43	36.41	36.40	36.38	36.36	36.35	36.33	36.33
.36	39.91	38.00	37.28	36.69	36.38	36.20	36.12	36.06	36.05	36.04	36.02	36.01	35.98	35.97	35.96	35.96
.37	39.62	37.67	36.93	36.33	36.02	35.83	35.75	35.71	35.68	35.66	35.65	35.63	35.61	35.60	35.59	35.58
.38	39.33	37.33	36.58	35.98	35.65	35.46	35.38	35.34	35.31	35.29	35.28	35.26	35.24	35.23	35.22	35.21
.39	39.03	37.00	36.23	35.62	35.29	35.10	35.01	34.97	34.94	34.93	34.91	34.89	34.87	34.86	34.85	34.84
.40	38.74	36.67	35.88	35.26	34.93	34.73	34.65	34.60	34.58	34.56	34.54	34.53	34.50	34.49	34.48	34.47
.41	38.45	36.33	35.54	34.90	34.57	34.37	34.28	34.24	34.21	34.19	34.18	34.16	34.13	34.12	34.11	34.10
.42	38.15	36.00	35.19	34.55	34.21	34.00	33.92	33.87	33.85	33.83	33.81	33.79	33.77	33.76	33.74	33.74
.43	37.85	35.67	34.85	34.19	33.85	33.64	33.56	33.51	33.48	33.46	33.45	33.43	33.40	33.39	33.38	33.37
.44	37.56	35.33	34.50	33.84	33.49	33.28	33.20	33.15	33.12	33.10	33.09	33.07	33.04	33.03	33.02	33.01
.45	37.26	35.00	34.16	33.49	33.13	32.92	32.84	32.79	32.76	32.74	32.73	32.71	32.68	32.67	32.66	32.65
.46	36.96	34.67	33.81	33.13	32.78	32.57	32.48	32.43	32.40	32.38	32.37	32.35	32.32	32.31	32.30	32.29
.47	36.66	34.33	33.47	32.78	32.42	32.21	32.12	32.07	32.04	32.02	32.01	31.99	31.96	31.95	31.94	31.93
.48	36.35	34.00	33.12	32.43	32.07	31.85	31.77	31.72	31.69	31.67	31.65	31.63	31.61	31.60	31.58	31.58
.49	36.05	33.67	32.78	32.08	31.72	31.50	31.41	31.36	31.33	31.31	31.30	31.28	31.25	31.24	31.23	31.22
.50	35.75	33.33	32.44	31.74	31.37	31.15	31.06	31.01	30.98	30.96	30.95	30.93	30.90	30.89	30.87	30.87
.51	35.44	33.00	32.10	31.39	31.02	30.80	30.71	30.66	30.63	30.61	30.60	30.57	30.55	30.54	30.52	30.52
.52	35.13	32.67	31.76	31.04	30.67	30.45	30.36	30.31	30.28	30.26	30.25	30.23	30.20	30.19	30.17	30.17
.53	34.82	32.33	31.42	30.70	30.32	30.10	30.01	29.96	29.93	29.91	29.90	29.88	29.85	29.84	29.83	29.82
.54	34.51	32.00	31.08	30.36	29.98	29.76	29.67	29.62	29.59	29.57	29.55	29.53	29.51	29.49	29.48	29.48
.55	34.20	31.67	30.74	30.01	29.64	29.41	29.32	29.27	29.24	29.22	29.21	29.19	29.16	29.15	29.14	29.13
.56	33.88	31.33	30.40	29.67	29.29	29.07	28.96	28.93	28.90	28.88	28.87	28.85	28.82	28.81	28.79	28.79
.57	33.57	31.00	30.06	29.33	28.95	28.73	28.64	28.59	28.56	28.54	28.53	28.51	28.48	28.47	28.45	28.45
.58	33.25	30.67	29.73	28.99	28.61	28.39	28.30	28.25	28.22	28.20	28.19	28.17	28.14	28.13	28.12	28.11
.59	32.93	30.33	29.39	28.66	28.28	28.05	27.96	27.92	27.89	27.87	27.85	27.83	27.81	27.79	27.78	27.77
.60	32.61	30.00	29.05	28.32	27.94	27.72	27.63	27.58	27.55	27.53	27.52	27.50	27.47	27.46	27.45	27.44
.61	32.28	29.67	28.72	27.98	27.60	27.39	27.30	27.25	27.22	27.20	27.18	27.16	27.14	27.13	27.11	27.11
.62	31.96	29.33	28.39	27.65	27.27	27.05	26.96	26.92	26.89	26.87	26.85	26.83	26.81	26.80	26.78	26.78
.63	31.63	29.00	28.05	27.32	26.94	26.72	26.63	26.59	26.56	26.54	26.52	26.50	26.48	26.47	26.45	26.45
.64	31.30	28.67	27.72	26.99	26.61	26.39	26.31	26.26	26.23	26.21	26.20	26.18	26.15	26.14	26.13	26.12
.65	30.97	28.33	27.39	26.66	26.28	26.07	25.98	25.93	25.90	25.88	25.87	25.85	25.83	25.82	25.80	25.80
.66	30.63	28.00	27.06	26.33	25.96	25.74	25.66	25.61	25.58	25.56	25.55	25.53	25.51	25.49	25.48	25.48
.67	30.30	27.67	26.73	26.00	25.63	25.42	25.33	25.29	25.26	25.24	25.23	25.21	25.19	25.17	25.16	25.16
.68	29.96	27.33	26.40	25.68	25.31	25.10	25.01	24.97	24.94	24.92	24.91	24.89	24.87	24.86	24.84	24.84
.69	29.61	27.00	26.07	25.35	24.99	24.78	24.70	24.65	24.62	24.60	24.59	24.57	24.55	24.54	24.53	24.52

[1] Values tabulated are read in percent.

Table 7-21 (Continued) **Table for Estimating the Lot Percent Defective Using Standard Deviation Method.**

Q_U or Q_L	Sample Size															
	3	4	5	7	10	15	20	25	30	35	40	50	75	100	150	200
.70	29.27	26.67	25.74	25.03	24.67	24.46	24.38	24.33	24.31	24.29	24.28	24.26	24.24	24.23	24.21	24.21
.71	28.92	26.33	25.41	24.71	24.35	24.15	24.06	24.02	23.99	23.98	23.96	23.95	23.92	23.91	23.90	23.90
.72	28.57	26.00	25.09	24.39	24.03	23.83	23.75	23.71	23.68	23.67	23.65	23.64	23.61	23.60	23.59	23.59
.73	28.22	25.67	24.76	24.07	23.72	23.52	23.44	23.40	23.37	23.36	23.34	23.33	23.31	23.30	23.29	23.28
.74	27.86	25.33	24.44	23.75	23.41	23.21	23.13	23.09	23.07	23.05	23.04	23.02	23.00	22.99	22.98	22.98
.75	27.50	25.00	24.11	23.44	23.10	22.90	22.83	22.79	22.76	22.75	22.73	22.72	22.70	22.69	22.68	22.67
.76	27.13	24.67	23.79	23.12	22.79	22.60	22.52	22.48	22.46	22.44	22.43	22.42	22.40	22.39	22.38	22.37
.77	26.77	24.33	23.47	22.81	22.48	22.30	22.22	22.18	22.16	22.14	22.13	22.12	22.10	22.09	22.08	22.08
.78	26.39	24.00	23.15	22.50	22.18	21.99	21.92	21.89	21.86	21.85	21.84	21.82	21.80	21.79	21.78	21.78
.79	26.02	23.67	22.83	22.19	21.87	21.70	21.63	21.59	21.57	21.55	21.54	21.53	21.51	21.50	21.49	21.49
.80	25.64	23.33	22.51	21.88	21.57	21.40	21.33	21.29	21.27	21.26	21.25	21.23	21.22	21.21	21.20	21.20
.81	25.25	23.00	22.19	21.58	21.27	21.10	21.04	21.00	20.98	20.97	20.96	2.094	20.93	20.92	20.91	20.91
.82	24.86	22.67	21.87	21.27	20.98	20.81	20.75	20.71	20.69	20.68	20.67	20.65	20.64	20.63	20.62	20.62
.83	24.47	22.33	21.56	20.97	20.68	20.52	20.46	20.42	20.40	20.39	20.38	20.37	20.35	20.35	20.34	20.34
.84	24.07	22.00	21.24	20.67	20.39	20.23	20.17	20.14	20.12	20.11	20.10	20.09	20.07	20.06	20.06	20.05
.85	23.67	21.67	20.93	20.37	20.10	19.94	19.89	19.86	19.84	19.82	19.82	19.80	19.79	19.78	19.78	19.77
.86	23.26	21.33	20.62	20.07	19.81	19.66	19.60	19.57	19.56	19.54	19.54	19.53	19.51	19.51	19.50	19.50
.87	22.84	21.00	20.31	19.78	19.52	19.38	19.32	19.30	19.28	19.27	19.26	19.25	19.24	19.23	19.22	19.22
.88	22.42	20.67	20.00	19.48	19.23	19.10	19.04	19.02	19.00	18.99	18.98	18.98	18.96	18.96	18.95	18.95
.89	21.99	20.33	19.69	19.19	18.95	18.82	18.77	18.74	18.73	18.72	18.71	18.70	18.69	18.69	18.68	18.68
.90	21.55	20.00	19.38	18.90	18.67	18.54	18.50	18.47	18.46	18.45	18.44	18.43	18.42	18.42	18.41	18.41
.91	21.11	19.67	19.07	18.61	18.39	18.27	18.22	18.20	18.19	18.18	18.17	18.17	18.16	18.15	18.15	18.15
.92	20.66	19.33	18.77	18.33	18.11	18.00	17.96	17.94	17.92	17.92	17.91	17.90	17.89	17.89	17.88	17.88
.93	20.20	19.00	18.46	18.04	17.84	17.73	17.69	17.67	17.66	17.65	17.65	17.64	17.63	17.63	17.62	17.62
.94	19.74	18.67	18.16	17.76	17.57	17.46	17.43	17.41	17.40	17.39	17.39	17.38	17.37	17.37	17.36	17.36
.95	19.25	18.33	17.86	17.48	17.29	17.20	17.17	17.15	17.14	17.13	17.13	17.12	17.12	17.11	17.11	17.11
.96	18.76	18.00	17.56	17.20	17.03	16.94	16.91	16.89	16.88	16.88	16.87	16.87	16.86	16.86	16.86	16.85
.97	18.25	17.67	17.25	16.92	16.76	16.68	16.65	16.63	16.63	16.62	16.62	16.61	16.61	16.61	16.60	16.60
.98	17.74	17.33	16.96	16.65	16.49	16.42	16.39	16.38	16.37	16.37	16.37	16.36	16.36	16.36	16.36	16.36
.99	17.21	17.00	16.66	16.37	16.23	16.16	16.14	16.13	16.12	16.12	16.12	16.12	16.11	16.11	16.11	16.11
1.00	16.67	16.67	16.36	16.10	15.97	15.91	15.89	15.88	15.88	15.87	15.87	15.87	15.87	15.87	15.87	15.87
1.01	16.11	16.33	16.07	15.83	15.72	15.66	15.64	15.63	15.63	15.63	15.63	15.63	15.62	15.62	15.62	15.62
1.02	15.53	16.00	15.78	15.56	15.46	15.41	15.40	15.39	15.39	15.39	15.39	15.38	15.38	15.38	15.38	15.38
1.03	14.93	15.67	15.48	15.30	15.21	15.17	15.15	15.15	15.15	15.15	15.15	15.15	15.15	15.15	15.15	15.15
1.04	14.31	15.33	15.19	15.03	14.96	14.92	14.91	14.91	14.91	14.91	14.91	14.91	14.91	14.91	14.91	14.91
1.05	13.66	15.00	14.91	14.77	14.71	14.68	14.67	14.67	14.67	14.67	14.68	14.68	14.68	14.68	14.68	14.68
1.06	12.98	14.67	14.62	14.51	14.46	14.44	14.44	14.44	14.44	14.44	14.44	14.45	14.45	14.45	14.45	14.45
1.07	12.27	14.33	14.33	14.26	14.22	14.20	14.20	14.21	14.21	14.21	14.21	14.22	14.22	14.22	14.22	14.23
1.08	11.51	14.00	14.05	14.00	13.97	13.97	13.97	13.98	13.98	13.98	13.99	13.99	13.99	14.00	14.00	14.00
1.09	10.71	13.67	13.76	13.75	13.73	13.74	13.74	13.75	13.75	13.67	13.76	13.77	13.77	13.77	13.78	13.78

Table 7-21 (*Continued*) *Table for Estimating the Lot Percent Defective Using Standard Deviation Method.*

Q_U or Q_L	Sample Size															
	3	4	5	7	10	15	20	25	30	35	40	50	75	100	150	200
1.10	9.84	13.33	13.48	13.49	13.50	13.51	13.52	13.52	13.53	13.54	13.54	13.54	13.55	13.55	13.56	13.56
1.11	8.89	13.00	13.20	13.25	13.26	13.28	13.29	13.30	13.31	13.31	13.32	13.32	13.33	13.34	13.34	13.34
1.12	7.82	12.67	12.93	13.00	13.03	13.05	13.07	13.08	13.09	13.10	13.10	13.11	13.12	13.12	13.12	13.13
1.13	6.60	12.33	12.65	12.75	12.80	12.83	12.85	12.86	12.87	12.88	12.89	12.89	12.90	12.91	12.91	12.92
1.14	5.08	12.00	12.37	12.51	12.57	12.61	12.63	12.65	12.66	12.67	12.67	12.68	12.69	12.70	12.70	12.70
1.15	0.29	11.67	12.10	12.27	12.34	12.39	12.42	12.44	12.45	12.46	12.46	12.47	12.48	12.49	12.49	12.50
1.16	0.00	11.33	11.83	12.03	12.12	12.18	12.21	12.22	12.24	12.25	12.25	12.26	12.28	12.28	12.29	12.29
1.17	0.00	11.00	11.56	11.79	11.90	11.96	12.00	12.02	12.03	12.04	12.05	12.06	12.07	12.08	12.08	12.09
1.18	0.00	10.67	11.29	11.56	11.68	11.75	11.79	11.81	11.82	11.84	11.84	11.85	11.87	11.88	11.88	11.89
1.19	0.00	10.33	11.02	11.33	11.46	11.54	11.58	11.61	11.62	11.63	11.64	11.65	11.67	11.68	11.69	11.69
1.20	0.00	10.00	10.76	11.10	11.24	11.34	11.38	11.41	11.42	11.43	11.44	11.46	11.47	11.48	11.49	11.49
1.21	0.00	9.67	10.50	10.87	11.03	11.13	11.18	11.21	11.22	11.24	11.25	11.26	11.28	11.29	11.30	11.30
1.22	0.00	9.33	10.23	10.65	10.82	10.93	10.98	11.01	11.03	11.04	11.05	11.07	11.09	11.09	11.10	11.11
1.23	0.00	9.00	9.97	10.42	10.61	10.73	10.78	10.81	10.84	10.85	10.86	10.88	10.90	10.91	10.91	10.92
1.24	0.00	8.67	9.72	10.20	10.41	10.53	10.59	10.62	10.64	10.66	10.67	10.69	10.71	10.72	10.73	10.73
1.25	0.00	8.33	9.46	9.98	10.21	10.34	10.40	10.43	10.46	10.47	10.48	10.50	10.52	10.53	10.54	10.55
1.26	0.00	8.00	9.21	9.77	10.00	10.15	10.21	10.25	10.27	10.29	10.30	10.32	10.34	10.35	10.36	10.37
1.27	0.00	7.67	8.96	9.55	9.81	9.96	10.02	10.06	10.09	10.10	10.12	10.13	10.16	10.17	10.18	10.19
1.28	0.00	7.33	8.71	9.34	9.61	9.77	9.84	9.88	9.90	9.92	9.94	9.95	9.98	9.99	10.00	10.01
1.29	0.00	7.00	8.46	9.13	9.42	9.58	9.65	9.70	9.72	9.74	9.76	9.78	9.80	9.82	9.83	8.83
1.30	0.00	6.67	8.21	8.93	9.22	9.40	9.48	9.52	9.55	9.57	9.58	9.60	9.63	9.64	9.65	9.66
1.31	0.00	6.33	7.97	8.72	9.03	9.22	9.30	9.34	9.37	9.39	9.41	9.43	9.46	9.47	9.48	9.49
1.32	0.00	6.00	7.73	8.52	8.85	9.04	9.12	9.17	9.20	9.22	9.24	9.26	9.29	9.30	9.31	9.32
1.33	0.00	5.67	7.49	8.32	8.66	8.86	8.95	9.00	9.03	9.05	9.07	9.09	9.12	9.13	9.15	9.15
1.34	0.00	5.33	7.25	8.12	8.48	8.69	8.78	8.83	8.86	8.88	8.90	8.92	8.95	8.97	8.98	8.99
1.35	0.00	5.00	7.02	7.92	8.30	8.52	8.61	8.66	8.69	8.72	8.74	8.76	8.79	8.81	8.82	8.83
1.36	0.00	4.67	6.79	7.73	8.12	8.35	8.44	8.50	8.53	8.55	8.57	8.60	8.63	8.65	8.66	8.67
1.37	0.00	4.33	6.56	7.54	7.95	8.18	8.28	8.33	8.37	8.39	8.41	8.44	8.47	8.49	8.50	8.51
1.38	0.00	4.00	6.33	7.35	7.77	8.01	8.12	8.17	8.21	8.24	8.25	8.28	8.31	8.33	8.35	8.35
1.39	0.00	3.67	6.10	7.17	7.60	7.85	7.96	8.01	8.05	8.08	8.10	8.12	8.16	8.18	8.19	8.20
1.40	0.00	3.33	5.88	6.98	7.44	7.69	7.80	7.86	7.90	7.92	7.94	7.97	8.01	8.02	8.04	8.05
1.41	0.00	3.00	5.66	6.80	7.27	7.53	7.64	7.70	7.74	7.77	7.79	7.82	7.86	7.87	7.89	7.90
1.42	0.00	2.67	5.44	6.62	7.10	7.37	7.49	7.55	7.59	7.62	7.64	7.67	7.71	7.73	7.74	7.75
1.43	0.00	2.33	5.23	6.45	6.94	7.22	7.34	7.40	7.44	7.47	7.50	7.52	7.56	7.58	7.60	7.61
1.44	0.00	2.00	5.01	6.27	6.78	7.07	7.19	7.26	7.30	7.33	7.35	7.38	7.42	7.44	7.46	7.47
1.45	0.00	1.67	4.81	6.10	6.63	6.92	7.04	7.11	7.15	7.18	7.21	7.24	7.28	7.30	7.31	7.33
1.46	0.00	1.33	4.60	5.93	6.47	6.77	6.90	6.97	7.01	7.04	7.07	7.10	7.14	7.16	7.18	7.19
1.47	0.00	1.00	4.39	5.77	6.32	6.63	6.75	6.83	6.87	6.90	6.93	6.96	7.00	7.02	7.04	7.05
1.48	0.00	.67	4.19	5.60	6.17	6.48	6.61	6.69	6.73	6.77	6.79	6.82	6.86	6.88	6.90	6.91
1.49	0.00	.33	3.99	5.44	6.02	6.34	6.48	6.55	6.60	6.63	6.65	6.69	6.73	6.75	6.77	6.78

Table 7-21 (*Continued*) **Table for Estimating the Lot Percent Defective Using Standard Deviation Method.**

Q_U or Q_L	Sample Size															
	3	4	5	7	10	15	20	25	30	35	40	50	75	100	150	200
1.50	0.00	0.00	3.80	5.28	5.87	6.20	6.34	6.41	6.46	6.50	6.52	6.55	6.60	6.62	6.64	6.65
1.51	0.00	0.00	3.61	5.13	5.73	6.06	6.20	6.28	6.33	6.36	6.39	6.42	6.47	6.49	6.51	6.52
1.52	0.00	0.00	3.42	4.97	5.59	5.93	6.07	6.15	6.20	6.23	6.26	6.29	6.34	6.36	6.38	6.39
1.53	0.00	0.00	3.23	4.82	5.45	5.80	5.94	6.02	6.07	6.11	6.13	6.17	6.21	6.24	6.26	6.27
1.54	0.00	0.00	3.05	4.67	5.31	5.67	5.81	5.89	5.95	5.98	6.01	6.04	6.09	6.11	6.13	6.15
1.55	0.00	0.00	2.87	4.52	5.18	5.54	5.69	5.77	5.82	5.86	5.88	5.92	5.97	5.99	6.01	6.02
1.56	0.00	0.00	2.69	4.38	5.05	5.41	5.56	5.65	5.70	5.74	5.76	5.80	5.85	5.87	5.89	5.90
1.57	0.00	0.00	2.52	4.24	4.92	5.29	5.44	5.53	5.58	5.62	5.64	5.68	5.73	5.75	5.78	5.79
1.58	0.00	0.00	2.35	4.10	4.79	5.16	5.32	5.41	5.46	5.50	5.53	5.56	5.61	5.64	5.66	5.67
1.59	0.00	0.00	2.19	3.96	4.66	5.04	5.20	5.29	5.34	5.38	5.41	5.45	5.50	5.52	5.54	5.56
1.60	0.00	0.00	2.03	3.83	4.54	4.92	5.09	5.17	5.23	5.27	5.30	5.33	5.38	5.41	5.43	5.44
1.61	0.00	0.00	1.87	3.69	4.41	4.81	4.97	5.06	5.12	5.16	5.18	5.22	5.27	5.30	5.32	5.33
1.62	0.00	0.00	1.72	3.57	4.30	4.69	4.86	4.95	5.01	5.04	5.07	5.11	5.16	5.19	5.21	5.23
1.63	0.00	0.00	1.57	3.44	4.18	4.58	4.75	4.84	4.90	4.94	4.97	5.01	5.06	5.08	5.11	5.12
1.64	0.00	0.00	1.42	3.31	4.06	4.47	4.64	4.73	4.79	4.83	4.86	4.90	4.95	4.98	5.00	5.01
1.65	0.00	0.00	1.28	3.19	3.95	4.36	4.53	4.62	4.68	4.72	4.75	4.79	4.85	4.87	4.90	4.91
1.66	0.00	0.00	1.15	3.07	3.84	4.25	4.43	4.52	4.58	4.62	4.65	4.69	4.74	4.77	4.80	4.81
1.67	0.00	0.00	1.02	2.95	3.73	4.15	4.32	4.42	4.48	4.52	4.55	4.59	4.64	4.67	4.70	4.71
1.68	0.00	0.00	0.89	2.84	3.62	4.05	4.22	4.32	4.38	4.42	4.45	4.49	4.55	4.57	4.60	4.61
1.69	0.00	0.00	0.77	2.73	3.52	3.94	4.12	4.22	4.28	4.32	4.35	4.39	4.45	4.47	4.50	4.51
1.70	0.00	0.00	0.66	2.62	3.41	3.84	4.02	4.12	4.18	4.22	4.25	4.30	4.35	4.38	4.41	4.42
1.71	0.00	0.00	0.55	2.51	3.31	3.75	3.93	4.02	4.09	4.13	4.16	4.20	4.26	4.29	4.31	4.32
1.72	0.00	0.00	0.45	2.41	3.21	3.65	3.83	3.93	3.99	4.04	4.07	4.11	4.17	4.19	4.22	4.23
1.73	0.00	0.00	0.36	2.30	3.11	3.56	3.74	3.84	3.90	3.94	3.98	4.02	4.08	4.10	4.13	4.14
1.74	0.00	0.00	0.27	2.20	3.02	3.46	3.65	3.75	3.81	3.85	3.89	3.93	3.99	4.01	4.04	4.05
1.75	0.00	0.00	0.19	2.11	2.93	3.37	3.56	3.66	3.72	3.77	3.80	3.84	3.90	3.93	3.95	3.97
1.76	0.00	0.00	0.12	2.01	2.83	3.28	3.47	3.57	3.63	3.68	3.71	3.76	3.81	3.84	3.87	3.88
1.77	0.00	0.00	0.06	1.92	2.74	3.20	3.38	3.48	3.55	3.59	3.63	3.67	3.73	3.76	3.78	3.80
1.78	0.00	0.00	0.02	1.83	2.66	3.11	3.30	3.40	3.47	3.51	3.54	3.59	3.64	3.67	3.70	3.71
1.79	0.00	0.00	0.00	1.74	2.57	3.03	3.21	3.32	3.38	3.43	3.46	3.51	3.56	3.59	3.63	3.63
1.80	0.00	0.00	0.00	1.65	2.49	2.94	3.13	3.24	3.30	3.35	3.38	3.43	3.48	3.51	3.54	3.55
1.81	0.00	0.00	0.00	1.57	2.40	2.86	3.05	3.16	3.22	3.27	3.30	3.35	3.40	3.43	3.46	3.47
1.82	0.00	00.0	0.00	1.49	2.32	2.79	2.98	3.08	3.15	3.19	3.22	3.27	3.33	3.36	3.38	3.40
1.83	0.00	0.00	0.00	1.41	2.25	2.71	2.90	3.00	3.07	3.11	3.15	3.19	3.25	3.28	3.31	3.32
1.84	0.00	0.00	0.00	1.34	2.17	2.63	2.82	2.93	2.99	3.04	3.07	3.12	3.18	3.21	3.23	3.25
1.85	0.00	0.00	0.00	1.26	2.09	2.56	2.75	2.85	2.92	2.97	3.00	3.05	3.10	3.13	3.16	3.17
1.86	0.00	0.00	0.00	1.19	2.02	2.48	2.68	2.78	2.85	2.89	2.93	2.97	3.03	3.06	3.09	3.10
1.87	0.00	0.00	0.00	1.12	1.95	2.41	2.61	2.71	2.78	2.82	2.86	2.90	2.96	2.99	3.02	3.03
1.88	0.00	0.00	0.00	1.06	1.88	2.34	2.54	2.64	2.71	2.75	2.79	2.83	2.89	2.92	2.95	2.96
1.89	0.00	0.00	0.00	0.99	1.81	2.28	2.47	2.57	2.64	2.69	2.72	2.77	2.83	2.85	2.88	2.90

Q_U or Q_L	Sample Size															
	3	4	5	7	10	15	20	25	30	35	40	50	75	100	150	200
1.90	0.00	0.00	0.00	0.93	1.75	2.21	2.40	2.51	2.57	2.62	2.65	2.70	2.76	2.79	2.82	2.83
1.91	0.00	0.00	0.00	0.87	1.68	2.14	2.34	2.44	2.51	2.56	2.59	2.63	2.69	2.72	2.75	2.77
1.92	0.00	0.00	0.00	0.81	1.62	2.08	2.27	2.38	2.45	2.49	2.52	2.57	2.63	2.66	2.69	2.70
1.93	0.00	0.00	0.00	0.76	1.56	2.02	2.21	2.32	2.38	2.43	2.46	2.51	2.57	2.60	2.62	2.64
1.94	0.00	0.00	0.00	0.70	1.50	1.96	2.15	2.25	2.32	2.37	2.40	2.45	2.51	2.54	2.56	2.58
1.95	0.00	0.00	0.00	0.65	1.44	1.90	2.09	2.19	2.26	2.31	2.34	2.39	2.45	2.48	2.50	2.52
1.96	0.00	0.00	0.00	0.60	1.38	1.84	2.03	2.14	2.20	2.25	2.28	2.33	2.39	2.42	2.44	2.46
1.97	0.00	0.00	0.00	0.56	1.33	1.78	1.97	2.08	2.14	2.19	2.22	2.27	2.33	2.36	2.39	2.40
1.98	0.00	0.00	0.00	0.51	1.27	1.73	1.92	2.02	2.09	2.13	2.17	2.21	2.27	2.30	2.33	2.34
1.99	0.00	0.00	0.00	0.47	1.22	1.67	1.86	1.97	2.03	2.08	2.11	2.16	2.22	2.25	2.27	2.29
2.00	0.00	0.00	0.00	0.43	1.17	1.62	1.81	1.91	1.98	2.03	2.06	2.10	2.16	2.19	2.22	2.23
2.01	0.00	0.00	0.00	0.39	1.12	1.57	1.76	1.86	1.93	1.97	2.01	2.05	2.11	2.14	2.17	2.18
2.02	0.00	0.00	0.00	0.36	1.07	1.52	1.71	1.81	1.87	1.92	1.95	2.00	2.06	2.09	2.11	2.13
2.03	0.00	0.00	0.00	0.32	1.03	1.47	1.66	1.76	1.82	1.87	1.90	1.95	2.01	2.04	2.06	2.08
2.04	0.00	0.00	0.00	0.29	0.98	1.42	1.61	1.71	1.77	1.82	1.85	1.90	1.96	1.99	2.01	2.03
2.05	0.00	0.00	0.00	0.26	0.94	1.37	1.56	1.66	1.73	1.77	1.80	1.85	1.91	1.94	1.96	1.98
2.06	0.00	0.00	0.00	0.23	0.90	1.33	1.51	1.61	1.68	1.72	1.76	1.80	1.86	1.89	1.92	1.93
2.07	0.00	0.00	0.00	0.21	0.86	1.28	1.47	1.57	1.63	1.68	1.71	1.76	1.81	1.84	1.87	1.88
2.08	0.00	0.00	0.00	0.18	0.82	1.24	1.42	1.52	1.59	1.63	1.66	1.71	1.77	1.79	1.82	1.84
2.09	0.00	0.00	0.00	0.16	0.78	1.20	1.38	1.48	1.54	1.59	1.62	1.66	1.72	1.75	1.78	1.79
2.10	0.00	0.00	0.00	0.14	0.74	1.16	1.34	1.44	1.50	1.54	1.58	1.62	1.68	1.71	1.73	1.75
2.11	0.00	0.00	0.00	0.12	0.71	1.12	1.30	1.39	1.46	1.50	1.53	1.58	1.63	1.66	1.69	1.70
2.12	0.00	0.00	0.00	0.10	0.67	1.08	1.26	1.35	1.42	1.46	1.49	1.54	1.59	1.62	1.65	1.66
2.13	0.00	0.00	0.00	0.08	0.64	1.04	1.22	1.31	1.38	1.42	1.45	1.50	1.55	1.58	1.61	1.62
2.14	0.00	0.00	0.00	0.07	0.61	1.00	1.18	1.28	1.34	1.38	1.41	1.46	1.51	1.54	1.57	1.58
2.15	0.00	0.00	0.00	0.06	0.58	0.97	1.14	1.24	1.30	1.34	1.37	1.42	1.47	1.50	1.53	1.54
2.16	0.00	0.00	0.00	0.05	0.55	0.93	1.10	1.20	1.26	1.30	1.34	1.38	1.43	1.46	1.49	1.50
2.17	0.00	0.00	0.00	0.04	0.52	0.90	1.07	1.16	1.22	1.27	1.30	1.34	1.40	1.42	1.45	1.46
2.18	0.00	0.00	0.00	0.03	0.49	0 87	1.03	1.13	1.19	1.23	1.26	1.30	1.36	1.39	1.41	1.42
2.19	0.00	0.00	0.00	0.02	0.46	0.83	1.00	1.09	1.15	1.20	1.23	1.27	1.32	1.35	1.38	1.39
2.20	0.000	0.000	0.000	0.015	0.437	0.803	0.968	1.061	1.120	1.161	1.192	1.233	1.287	1.314	1.340	1.352
2.21	0.000	0.000	0.000	0.010	0.413	0.772	0.936	1.028	1.067	1.128	1.158	1.199	1.253	1.279	1.305	1.318
2.22	0.000	0.000	0.000	0.006	0.389	0.743	0.905	0.996	1.054	1.095	1.125	1.166	1.219	1.245	1.271	1.283
2.23	0.000	0.000	0.000	0.003	0.366	0.715	0.875	0.965	1.023	1.063	1.093	1.134	1.186	1.212	1.238	1.250
2.24	0.000	0.000	0.000	0.002	0.345	0.687	0.845	0.935	0.992	1.032	1.061	1.102	1.154	1.180	1.205	1.218
2.25	0.000	0.000	0.000	0.001	0.324	0.660	0.816	0.905	0.962	1.002	1.031	1.071	1.123	1.148	1.173	1.186
2.26	0.000	0.000	0.000	0.000	0.304	0.634	0.789	0.876	0.933	0.972	1.001	1.041	1.092	1.117	1.142	1.155
2.27	0.000	0.000	0.000	0.000	0.285	0.609	0.762	0.848	0.904	0.943	0.972	1.011	1.062	1.087	1.112	1.124
2.28	0.000	0.000	0.000	0.000	0.267	0.585	0.735	0.821	0.876	0.915	0.943	0.982	1.033	1.058	1.082	1.094
2.29	0.000	0.000	0.000	0.000	0.250	0.561	0.710	0.794	0.849	0.887	0.915	0.954	1.004	1.029	1.053	1.065

Table 7-21 (Continued) *Table for Estimating the Lot Percent Defective Using Standard Deviation Method.*

Q_U or Q_L	Sample Size															
	3	4	5	7	10	15	20	25	30	35	40	50	75	100	150	200
2.30	0.000	0.000	0.000	0.000	0.233	0.538	0.685	0.769	0.823	0.861	0.888	0.927	0.977	1.001	1.025	1.037
2.31	0.000	0.000	0.000	0.000	0.218	0.516	0.661	0.743	0.797	0.834	0.862	0.900	0.949	0.974	0.997	1.009
2.32	0.000	0.000	0.000	0.000	0.203	0.495	0.637	0.719	0.772	0.809	0.836	0.874	0.923	0.947	0.971	0.982
2.33	0.000	0.000	0.000	0.000	0.189	0.474	0.614	0.695	0.748	0.784	0.811	0.848	0.897	0.921	0.944	0.956
2.34	0.000	0.000	0.000	0.000	0.175	0.454	0.592	0.672	0.724	0.760	0.787	0.824	0.872	0.895	0.915	0.930
2.35	0.000	0.000	0.000	0.000	0.163	0.435	0.571	0.650	0.701	0.736	0.763	0.799	0.847	0.870	0.893	0.905
2.36	0.000	0.000	0.000	0.000	0.151	0.416	0.550	0.628	0.678	0.714	0.740	0.776	0.823	0.846	0.869	0.880
2.37	0.000	0.000	0.000	0.000	0.139	0.398	0.530	0.606	0.656	0.691	0.717	0.753	0.799	0.822	0.845	0.856
2.38	0.000	0.000	0.000	0.000	0.128	0.381	0.510	0.586	0.635	0.670	0.695	0.730	0.777	0.799	0.822	0.833
2.39	0.000	0.000	0.000	0.000	0.118	0.364	0.491	0.566	0.614	0.648	0.674	0.709	0.754	0.777	0.799	0.810
2.40	0.000	0.000	0.000	0.000	0.109	0.348	0.473	0.546	0.594	0.628	0.653	0.687	0.732	0.755	0.777	0.787
2.41	0.000	0.000	0.000	0.000	0.100	0.332	0.455	0.527	0.575	0.608	0.633	0.667	0.711	0.733	0.755	0.766
2.42	0.000	0.000	0.000	0.000	0.091	0.317	0.437	0.509	0.555	0.588	0.613	0.646	0.691	0.712	0.734	0.744
2.43	0.000	0.000	0.000	0.000	0.083	0.302	0.421	0.491	0.537	0.569	0.593	0.627	0.670	0.692	0.713	0.724
2.44	0.000	0.000	0.000	0.000	0.076	0.288	0.404	0.474	0.519	0.551	0.575	0.608	0.651	0.672	0.693	0.703
2.45	0.000	0.000	0.000	0.000	0.069	0.275	0.389	0.457	0.501	0.533	0.556	0.589	0.632	0.653	0.673	0.684
2.46	0.000	0.000	0.000	0.000	0.063	0.262	0.373	0.440	0.484	0.516	0.539	0.571	0.613	0.634	0.654	0.664
2.47	0.000	0.000	0.000	0.000	0.057	0.249	0.359	0.425	0.468	0.499	0.521	0.553	0.595	0.615	0.635	0.646
2.48	0.000	0.000	0.000	0.000	0.051	0.237	0.344	0.409	0.452	0.482	0.505	0.536	0.577	0.597	0.617	0.627
2.49	0.000	0.000	0.000	0.000	0.046	0.226	0.331	0.394	0.436	0.466	0.488	0.519	0.560	0.580	0.600	0.609
2.50	0.000	0.000	0.000	0.000	0.041	0.214	0.317	0.380	0.421	0.451	0.473	0.503	0.543	0.563	0.582	0.592
2.51	0.000	0.000	0.000	0.000	0.037	0.204	0.304	0.366	0.407	0.436	0.457	0.487	0.527	0.546	0.565	0.575
2.52	0.000	0.000	0.000	0.000	0.033	0.193	0.292	0.352	0.392	0.421	0.442	0.472	0.511	0.530	0.549	0.558
2.53	0.000	0.000	0.000	0.000	0.029	0.184	0.280	0.339	0.379	0.407	0.428	0.457	0.495	0.514	0.533	0.542
2.54	0.000	0.000	0.000	0.000	0.026	0.174	0.268	0.326	0.365	0.393	0.413	0.442	0.480	0.499	0.517	0.527
2.55	0.000	0.000	0.000	0.000	0.023	0.165	0.257	0.314	0.352	0.379	0.400	0.428	0.465	0.484	0.502	0.511
2.56	0.000	0.000	0.000	0.000	0.020	0.156	0.246	0.302	0.340	0.366	0.386	0.414	0.451	0.469	0.487	0.496
2.57	0.000	0.000	0.000	0.000	0.017	0.148	0.236	0.291	0.327	0.354	0.373	0.401	0.437	0.455	0.473	0.482
2.58	0.000	0.000	0.000	0.000	0.015	0.140	0.226	0.279	0.316	0.341	0.361	0.388	0.424	0.441	0.459	0.468
2.59	0.000	0.000	0.000	0.000	0.013	0.133	0.216	0.269	0.304	0.330	0.349	0.375	0.410	0.428	0.445	0.454
2.60	0.000	0.000	0.000	0.000	0.011	0.125	0.207	0.258	0.293	0.318	0.337	0.363	0.398	0.415	0.432	0.441
2.61	0.000	0.000	0.000	0.000	0.009	0.118	0.198	0.248	0.282	0.307	0.325	0.351	0.385	0.402	0.419	0.428
2.62	0.000	0.000	0.000	0.000	0.008	0.112	0.189	0.238	0.272	0.296	0.314	0.339	0.373	0.390	0.406	0.415
2.63	0.000	0.000	0.000	0.000	0.007	0.105	0.181	0.229	0.262	0.285	0.303	0.328	0.361	0.378	0.394	0.402
2.64	0.000	0.000	0.000	0.000	0.005	0.099	0.172	0.220	0.252	0.275	0.293	0.317	0.350	0.366	0.382	0.390
2.65	0.000	0.000	0.000	0.000	0.005	0.094	0.165	0.211	0.243	0.265	0.282	0.307	0.339	0.355	0.371	0.379
2.66	0.000	0.000	0.000	0.000	0.004	0.088	0.157	0.202	0.233	0.256	0.273	0.296	0.328	0.344	0.359	0.367
2.67	0.000	0.000	0.000	0.000	0.003	0.083	0.150	0.194	0.224	0.246	0.263	0.286	0.317	0.333	0.348	0.356
2.68	0.000	0.000	0.000	0.000	0.002	0.078	0.143	0.186	0.216	0.237	0.254	0.277	0.307	0.322	0.338	0.345
2.69	0.000	0.000	0.000	0.000	0.002	0.073	0.136	0.179	0.208	0.229	0.245	0.267	0.297	0.312	0.327	0.335

Q_U or Q_L	Sample Size															
	3	4	5	7	10	15	20	25	30	35	40	50	75	100	150	200
2.70	0.000	0.000	0.000	0.000	0.001	0.069	0.130	0.171	0.200	0.220	0.236	0.258	0.288	0.302	0.317	0.325
2.71	0.000	0.000	0.000	0.000	0.001	0.064	0.124	0.164	0.192	0.212	0.227	0.249	0.278	0.293	0.307	0.315
2.72	0.000	0.000	0.000	0.000	0.000	0.060	0.118	0.157	0.184	0.204	0.219	0.241	0.269	0.283	0.298	0.305
2.73	0.000	0.000	0.000	0.000	0.000	0.057	0.112	0.151	0.177	0.197	0.211	0.232	0.260	0.274	0.288	0.296
2.74	0.000	0.000	0.000	0.000	0.000	0.053	0.107	0.144	0.170	0.189	0.204	0.224	0.252	0.266	0.279	0.286
2.75	0.000	0.000	0.000	0.000	0.000	0.049	0.102	0.138	0.163	0.182	0.196	0.216	0.243	0.257	0.271	0.277
2.76	0.000	0.000	0.000	0.000	0.000	0.046	0.097	0.132	0.157	0.175	0.189	0.209	0.235	0.249	0,262	0.269
2.77	0.000	0.000	0.000	0.000	0.000	0.043	0.092	0.126	0.151	0.168	0.182	0.201	0.227	0.241	0.254	0.260
2.78	0.000	0.000	0.000	0.000	0.000	0.040	0.087	0.121	0.145	0.162	0.175	0.194	0.220	0.233	0.246	0.252
2.79	0.000	0.000	0.000	0.000	0.000	0.037	0.083	0.115	0.139	0.156	0.169	0.187	0.212	0.225	0.238	0.244
2.80	0.000	0.000	0.000	0.000	0.000	0.035	0.079	0.110	0.133	0.150	0.162	0.181	0.205	0.218	0.230	0.237
2.81	0.000	0.000	0.000	0.000	0.000	0.032	0.075	0.105	0.128	0.144	0.156	0.174	0.198	0.211	0.223	0.229
2.82	0.000	0.000	0.000	0.000	0.000	0.030	0.071	0.101	0.122	0.138	0.150	0.168	0.192	0.204	0.216	0.222
2.83	0.000	0.000	0.000	0.000	0.000	0.028	0.067	0.096	0.117	0.133	0.145	0.162	0.185	0.197	0.209	0.215
2.84	0.000	0.000	0.000	0.000	0.000	0.026	0.064	0.092	0.112	0.128	0.139	0.156	0.179	0.190	0.202	0.208
2.85	0.000	0.000	0.000	0.000	0.000	0.024	0.060	0.088	0.108	0.122	0.134	0.150	0.173	0.184	0.195	0.201
2.86	0.000	0.000	0.000	0.000	0.000	0.022	0.057	0.084	0.103	0.118	0.129	0.145	0.167	0.178	0.189	0.195
2.87	0.000	0.000	0.000	0.000	0.000	0.020	0.054	0.080	0.099	0.113	0.124	0.139	0.161	0.172	0.183	0.188
2.88	0.000	0.000	0.000	0.000	0.000	0.019	0.051	0.076	0.094	0.108	0.119	0.134	0.155	0.166	0.177	0.182
2.89	0.000	0.000	0.000	0.000	0.000	0.017	0.048	0.073	0.090	0.104	0.114	0.129	0.150	0.160	0.171	0.176
2.90	0.000	0.000	0.000	0.000	0.000	0.016	0.046	0.069	0.087	0.100	0.110	0.125	0.145	0.155	0.165	0.171
2.91	0.000	0.000	0.000	0.000	0.000	0.015	0.043	0.066	0.083	0.096	0.106	0.120	0.140	0.150	0.160	0.165
2.92	0.000	0.000	0.000	0.000	0.000	0.013	0.041	0.063	0.079	0.092	0.101	0.115	0.135	0.145	0.155	0.160
2.93	0.000	0.000	0.000	0.000	0.000	0.012	0.038	0.060	0.076	0.088	0.097	0.111	0.130	0.140	0.149	0.154
2.94	0.000	0.000	0.000	0.000	0.000	0.011	0.036	0.057	0.072	0.084	0.093	0.107	0.125	0.135	0.144	0.149
2.95	0.000	0.000	0.000	0.000	0.000	0.010	0.034	0.054	0.069	0.081	0.090	0.103	0.121	0.130	0.140	0.144
2.96	0.000	0.000	0.000	0.000	0.000	0.009	0.032	0.051	0.066	0.077	0.086	0.099	0.117	0.126	0.135	0.140
2.97	0.000	0.000	0.000	0.000	0.000	0.009	0.030	0.049	0.063	0.074	0.083	0.095	0.112	0.121	0.130	0.135
2.98	0.000	0.000	0.000	0.000	0.000	0.008	0.028	0.046	0.060	0.071	0.079	0.091	0.108	0.117	0.126	0.130
2.99	0.000	0.000	0.000	0.000	0.000	0.007	0.027	0.044	0.057	0.068	0.076	0.088	0.104	0.113	0.122	0.126
3.00	0.000	0.000	0.000	0.000	0.000	0.006	0.025	0.042	0.055	0.065	0.073	0.084	0.101	0.109	0.118	0.122
3.01	0.000	0.000	0.000	0.000	0.000	0.006	0.024	0.040	0.052	0.062	0.070	0.081	0.097	0.105	0.114	0.118
3.02	0.000	0.000	0.000	0.000	0.000	0.005	0.022	0.038	0.050	0.059	0.067	0.078	0.093	0.101	0.110	0.114
3.03	0.000	0.000	0.000	0.000	0.000	0.005	0.021	0.036	0.048	0.057	0.064	0.075	0.090	0.098	0.106	0.110
3.04	0.000	0.000	0.000	0.000	0.000	0.004	0.019	0.034	0.045	0.054	0.061	0.072	0.087	0.094	0.102	0.106
3.05	0.000	0.000	0.000	0.000	0.000	0.004	0.018	0.032	0.043	0.052	0.059	0.069	0.083	0.091	0.099	0.103
3.06	0.000	0.000	0.000	0.000	0.000	0.003	0.017	0.030	0.041	0.050	0.056	0.066	0.080	0.088	0.095	0.099
3.07	0.000	0.000	0.000	0.000	0.000	0.003	0.016	0.029	0.039	0.047	0.054	0.064	0.077	0.085	0.092	0.096
3.08	0.000	0.000	0.000	0.000	0.000	0.003	0.015	0.027	0.037	0.045	0.052	0.061	0.074	0.081	0.089	0.092
3.09	0.000	0.000	0.000	0.000	0.000	0.002	0.014	0.026	0.036	0.043	0.049	0.059	0.072	0.079	0.086	0.089

Q_U or Q_L	Sample Size															
	3	4	5	7	10	15	20	25	30	35	40	50	75	100	150	200
3.10	0.000	0.000	0.000	0.000	0.000	0.002	0.013	0.024	0.034	0.041	0.047	0.056	0.069	0.076	0.083	0.086
3.11	0.000	0.000	0.000	0.000	0.000	0.002	0.012	0.023	0.032	0.039	0.045	0.054	0.066	0.073	0.080	0.083
3.12	0.000	0.000	0.000	0.000	0.000	0.002	0.011	0.022	0.031	0.038	0.043	0.052	0.064	0.070	0.077	0.080
3.13	0.000	0.000	0.000	0.000	0.000	0.002	0.011	0.021	0.029	0.036	0.041	0.050	0.061	0.068	0.074	0.077
3.14	0.000	0.000	0.000	0.000	0.000	0.001	0.010	0.019	0.028	0.034	0.040	0.048	0.059	0.065	0.071	0.075
3.15	0.000	0.000	0.000	0.000	0.000	0.001	0.009	0.018	0.026	0.033	0.038	0.046	0.057	0.063	0.069	0.072
3.16	0.000	0.000	0.000	0.000	0.000	0.001	0.009	0.017	0.025	0.031	0.036	0.044	0.055	0.060	0.066	0.069
3.17	0.000	0.000	0.000	0.000	0.000	0.001	0.008	0.016	0.024	0.030	0.035	0.042	0.053	0.058	0.064	0.067
3.18	0.000	0.000	0.000	0.000	0.000	0.001	0.007	0.015	0.022	0.028	0.033	0.040	0.050	0.056	0.062	0.065
3.19	0.000	0.000	0.000	0.000	0.000	0.001	0.007	0.015	0.021	0.027	0.032	0.038	0.049	0.054	0.059	0.062
3.20	0.000	0.000	0.000	0.000	0.000	0.001	0.006	0.014	0.020	0.026	0.030	0.037	0.047	0.052	0.057	0.060
3.21	0.000	0.000	0.000	0.000	0.000	0.000	0.006	0.013	0.019	0.024	0.029	0.035	0.045	0.050	0.055	0.058
3.22	0.000	0.000	0.000	0.000	0.000	0.000	0.005	0.012	0.018	0.023	0.027	0.034	0.043	0.048	0.053	0.056
3.23	0.000	0.000	0.000	0.000	0.000	0.000	0.005	0.011	0.017	0.022	0.026	0.032	0.041	0.046	0.051	0.054
3.24	0.000	0.000	0.000	0.000	0.000	0.000	0.005	0.011	0.016	0.021	0.025	0.031	0.040	0.044	0.049	0.052
3.25	0.000	0.000	0.000	0.000	0.000	0.000	0.004	0.010	0.015	0.020	0.024	0.030	0.038	0.043	0.048	0.050
3.26	0.000	0.000	0.000	0.000	0.000	0.000	0.004	0.009	0.015	0.019	0.023	0.028	0.037	0.041	0.046	0.048
3.27	0.000	0.000	0.000	0.000	0.000	0.000	0.004	0.009	0.014	0.019	0.022	0.027	0.035	0.040	0.044	0.046
3.28	0.000	0.000	0.000	0.000	0.000	0.000	0.003	0.008	0.013	0.017	0.021	0.026	0.034	0.038	0.042	0.045
3.29	0.000	0.000	0.000	0.000	0.000	0.000	0.003	0.008	0.012	0.016	0.020	0.025	0.032	0.037	0.041	0.043
3.30	0.000	0.000	0.000	0.000	0.000	0.000	0.003	0.007	0.012	0.015	0.019	0.024	0.031	0.035	0.039	0.042
3.31	0.000	0.000	0.000	0.000	0.000	0.000	0.003	0.007	0.011	0.015	0.018	0.023	0.030	0.034	0.038	0.040
3.32	0.000	0.000	0.000	0.000	0.000	0.000	0.002	0.006	0.010	0.014	0.017	0.022	0.029	0.032	0.036	0.039
3.33	0.000	0.000	0.000	0.000	0.000	0.000	0.002	0.006	0.010	0.013	0.016	0.021	0.027	0.031	0.035	0.037
3.34	0.000	0.000	0.000	0.000	0.000	0.000	0.002	0.006	0.009	0.013	0.015	0.020	0.026	0.030	0.034	0.036
3.35	0.000	0.000	0.000	0.000	0.000	0.000	0.002	0.005	0.009	0.012	0.015	0.019	0.025	0.029	0.032	0.034
3.36	0.000	0.000	0.000	0.000	0.000	0.000	0.002	0.005	0.008	0.011	0.014	0.018	0.024	0.028	0.031	0.033
3.37	0.000	0.000	0.000	0.000	0.000	0.000	0.002	0.005	0.008	0.011	0.013	0.017	0.023	0.026	0.030	0.032
3.38	0.000	0.000	0.000	0.000	0.000	0.000	0.001	0.004	0.007	0.010	0.013	0.016	0.022	0.025	0.029	0.031
3.39	0.000	0.000	0.000	0.000	0.000	0.000	0.001	0.004	0.007	0.010	0.012	0.016	0.021	0.024	0.028	0.029
3.40	0.000	0.000	0.000	0.000	0.000	0.000	0.001	0.004	0.007	0.009	0.011	0.015	0.020	0.023	0.027	0.028
3.41	0.000	0.000	0.000	0.000	0.000	0.000	0.001	0.003	0.006	0.009	0.011	0.014	0.020	0.022	0.026	0.027
3.42	0.000	0.000	0.000	0.000	0.000	0.000	0.001	0.003	0.006	0.008	0.010	0.014	0.019	0.022	0.025	0.026
3.43	0.000	0.000	0.000	0.000	0.000	0.000	0.001	0.003	0.005	0.008	0.010	0.013	0.018	0.021	0.024	0.025
3.44	0.000	0.000	0.000	0.000	0.000	0.000	0.001	0.003	0.005	0.007	0.009	0.012	0.017	0.020	0.023	0.024
3.45	0.000	0.000	0.000	0.000	0.000	0.000	0.001	0.003	0.005	0.007	0.009	0.012	0.016	0.019	0.022	0.023
3.46	0.000	0.000	0.000	0.000	0.000	0.000	0.001	0.002	0.005	0.007	0.008	0.011	0.016	0.018	0.021	0.022
3.47	0.000	0.000	0.000	0.000	0.000	0.000	0.001	0.002	0.004	0.006	0.008	0.011	0.015	0.017	0.020	0.022
3.48	0.000	0.000	0.000	0.000	0.000	0.000	0.001	0.002	0.004	0.006	0.007	0.010	0.014	0.017	0.019	0.021
3.49	0.000	0.000	0.000	0.000	0.000	0.000	0.000	0.002	0.004	0.005	0.007	0.010	0.014	0.016	0.019	0.020

Q_U or Q_L	Sample Size															
	3	4	5	7	10	15	20	25	30	35	40	50	75	100	150	200
3.50	0.000	0.000	0.000	0.000	0.000	0.000	0.000	0.002	0.003	0.005	0.007	0.009	0.013	0.015	0.018	0.019
3.51	0.000	0.000	0.000	0.000	0.000	0.000	0.000	0.002	0.003	0.005	0.006	0.009	0.013	0.015	0.017	0.018
3.52	0.000	0.000	0.000	0.000	0.000	0.000	0.000	0.002	0.003	0.005	0.006	0.008	0.012	0.014	0.017	0.018
3.53	0.000	0.000	0.000	0.000	0.000	0.000	0.000	0.001	0.003	0.004	0.006	0.008	0.012	0.014	0.016	0.017
3.54	0.000	0.000	0.000	0.000	0.000	0.000	0.000	0.001	0.003	0.004	0.005	0.008	0.011	0.013	0.015	0.016
3.55	0.000	0.000	0.000	0.000	0.000	0.000	0.000	0.001	0.003	0.004	0.005	0.007	0.011	0.012	0.015	0.016
3.56	0.000	0.000	0.000	0.000	0.000	0.000	0.000	0.001	0.002	0.004	0.005	0.007	0.010	0.012	0.014	0.015
3.57	0.000	0.000	0.000	0.000	0.000	0.000	0.000	0.001	0.002	0.003	0.005	0.006	0.010	0.011	0.013	0.014
3.58	0.000	0.000	0.000	0.000	0.000	0.000	0.000	0.001	0.002	0.003	0.004	0.006	0.009	0.011	0.013	0.014
3.59	0.000	0.000	0.000	0.000	0.000	0.000	0.000	0.001	0.002	0.003	0.004	0.006	0.009	0.010	0.012	0.013
3.60	0.000	0.000	0.000	0.000	0.000	0.000	0.000	0.001	0.002	0.003	0.004	0.006	0.008	0.010	0.012	0.013
3.61	0.000	0.000	0.000	0.000	0.000	0.000	0.000	0.001	0.002	0.003	0.004	0.005	0.008	0.010	0.011	0.012
3.62	0.000	0.000	0.000	0.000	0.000	0.000	0.000	0.001	0.002	0.003	0.003	0.005	0.008	0.009	0.011	0.012
3.63	0.000	0.000	0.000	0.000	0.000	0.000	0.000	0.001	0.001	0.002	0.003	0.005	0.007	0.009	0.010	0.011
3.64	0.000	0.000	0.000	0.000	0.000	0.000	0.000	0.001	0.001	0.002	0.003	0.004	0.007	0.008	0.010	0.011
3.65	0.000	0.000	0.000	0.000	0.000	0.000	0.000	0.001	0.001	0.002	0.003	0.004	0.007	0.008	0.010	0.010
3.66	0.000	0.000	0.000	0.000	0.000	0.000	0.000	0.000	0.001	0.002	0.003	0.004	0.006	0.008	0.009	0.010
3.67	0.000	0.000	0.000	0.000	0.000	0.000	0.000	0.000	0.001	0.002	0.003	0.004	0.006	0.007	0.009	0.010
3.68	0.000	0.000	0.000	0.000	0.000	0.000	0.000	0.000	0.001	0.002	0.002	0.004	0.006	0.007	0.008	0.009
3.69	0.000	0.000	0.000	0.000	0.000	0.000	0.000	0.000	0.001	0.002	0.002	0.003	0.005	0.007	0.008	0.009
3.70	0.000	0.000	0.000	0.000	0.000	0.000	0.000	0.000	0.001	0.002	0.002	0.003	0.005	0.006	0.008	0.008
3.71	0.000	0.000	0.000	0.000	0.000	0.000	0.000	0.000	0.001	0.001	0.002	0.003	0.005	0.006	0.007	0.008
3.72	0.000	0.000	0.000	0.000	0.000	0.000	0.000	0.000	0.001	0.001	0.002	0.003	0.005	0.006	0.007	0.008
3.73	0.000	0.000	0.000	0.000	0.000	0.000	0.000	0.000	0.001	0.001	0.002	0.003	0.005	0.006	0.007	0.007
3.74	0.000	0.000	0.000	0.000	0.000	0.000	0.000	0.000	0.001	0.001	0.002	0.003	0.004	0.005	0.007	0.007
3.75	0.000	0.000	0.000	0.000	0.000	0.000	0.000	0.000	0.001	0.001	0.002	0.002	0.004	0.005	0.006	0.007
3.76	0.000	0.000	0.000	0.000	0.000	0.000	0.000	0.000	0.001	0.001	0.001	0.002	0.004	0.005	0.006	0.007
3.77	0.000	0.000	0.000	0.000	0.000	0.000	0.000	0.000	0.001	0.001	0.001	0.002	0.004	0.005	0.006	0.006
3.78	0.000	0.000	0.000	0.000	0.000	0.000	0.000	0.000	0.000	0.001	0.001	0.002	0.004	0.004	0.005	0.006
3.79	0.000	0.000	0.000	0.000	0.000	0.000	0.000	0.000	0.000	0.001	0.001	0.002	0.003	0.004	0.005	0.006
3.80	0.000	0.000	0.000	0.000	0.000	0.000	0.000	0.000	0.000	0.001	0.001	0.002	0.003	0.004	0.005	0.006
3.81	0.000	0.000	0.000	0.000	0.000	0.000	0.000	0.000	0.000	0.001	0.001	0.002	0.003	0.004	0.005	0.005
3.82	0.000	0.000	0.000	0.000	0.000	0.000	0.000	0.000	0.000	0.001	0.001	0.002	0.003	0.004	0.005	0.005
3.83	0.000	0.000	0.000	0.000	0.000	0.000	0.000	0.000	0.000	0.001	0.001	0.002	0.003	0.004	0.004	0.005
3.84	0.000	0.000	0.000	0.000	0.000	0.000	0.000	0.000	0.000	0.001	0.001	0.001	0.003	0.003	0.004	0.005
3.85	0.000	0.000	0.000	0.000	0.000	0.000	0.000	0.000	0.000	0.001	0.001	0.001	0.002	0.003	0.004	0.004
3.86	0.000	0.000	0.000	0.000	0.000	0.000	0.000	0.000	0.000	0.000	0.001	0.001	0.002	0.003	0.004	0.004
3.87	0.000	0.000	0.000	0.000	0.000	0.000	0.000	0.000	0.000	0.000	0.001	0.001	0.002	0.003	0.004	0.004
3.88	0.000	0.000	0.000	0.000	0.000	0.000	0.000	0.000	0.000	0.000	0.001	0.001	0.002	0.003	0.004	0.004
3.89	0.000	0.000	0.000	0.000	0.000	0.000	0.000	0.000	0.000	0.000	0.001	0.001	0.002	0.003	0.003	0.004
3.90	0.000	0.000	0.000	0.000	0.000	0.000	0.000	0.000	0.000	0.000	0.001	0.001	0.002	0.003	0.003	0.004

Table 7-22 (MIL-STD-414) Master Table for Normal and Tightened Inspection for Plans Based on Variability Unknown, Range Method (Single-Specification Limit, Form 1).

Sample size code letter	Sample size	Acceptable Quality Levels (normal inspection)													
		.04	.065	.10	.15	.25	.40	.65	1.00	1.50	2.50	4.00	6.50	10.00	15.00
		k	k	k	k	k	k	k	k	k	k	k	k	k	k
B	3	↓	↓	↓	↓	↓	↓	↓	▼	▼	.587	.502	.401	.296	.178
C	4	↓	↓	↓	↓	↓	↓	↓	.651	.598	.525	.450	.364	.276	.176
D	5	↓	↓	↓	↓	↓	↓	.663	.614	.565	.498	.431	.352	.272	.184
E	7	↓	↓	↓	↓	.702	.659	.613	.569	.525	.465	.405	.336	.266	.189
F	10	↓	↓	↓	.916	.863	.811	.755	.703	.650	.579	.507	.424	.341	.252
G	15	1.09	1.04	.999	.958	.903	.850	.792	.738	.684	.610	.536	.452	.368	.276
H	25	1.14	1.10	1.05	1.01	.951	.896	.835	.779	.723	.647	.571	.484	.398	.305
I	30	1.15	1.10	1.06	1.02	.959	.904	.843	.787	.730	.654	.577	.490	.403	.310
J	35	1.16	1.11	1.07	1.02	.964	.908	.848	.791	.734	.658	.581	.494	.406	.313
K	40	1.18	1.13	1.08	1.04	.978	.921	.860	.803	.746	.668	.591	.503	.415	.321
L	50	1.19	1.14	1.09	1.05	.988	.931	.893	.812	.754	.676	.598	.510	.421	.327
M	60	1.21	1.16	1.11	1.06	1.00	.948	.885	.826	.768	.689	.610	.521	.432	.336
N	85	1.23	1.17	1.13	1.08	1.02	.962	.899	.839	.780	.701	.621	.530	.441	.345
O	115	1.24	1.19	1.14	1.09	1.03	.975	.911	.851	.791	.711	.631	.539	.449	.353
P	175	1.26	1.21	1.16	1.11	1.05	.994	.929	.868	.807	.726	.644	.552	.460	.363
Q	230	1.27	1.21	1.16	1.12	1.06	.996	.931	.870	.809	.728	.645	.553	.462	.364
		.065	.10	.15	.25	.40	.65	1.00	1.50	2.50	4.00	6.50	10.00	15.00	
		Acceptable Quality Levels (tightened inspection)													

All AQL values are in percent defective.
↓ Use first sampling plan below arrow, that is, both sample size as well as k value. When sample size equals or exceeds lot size, every item in the lot must be inspected.

Table 7-23 *(MIL-STD-414) Master Table for Normal and Tightened Inspection for Plans Based on Variability Unknown. Range Method (Double-Specification Limit and for Form 2—Single-Specification Limit).*

Sample size code letter	Sample size	c factor	Acceptable Quality Levels (normal inspection)													
			.04	.065	.10	.15	.25	.40	.65	1.00	1.50	2.50	4.00	6.50	10.00	15.00
			M	M	M	M	M	M	M	M	M	M	M	M	M	M
B	3	1.910	→	→	→	→	→	→	→	▶	▶	7.59	18.86	26.94	33.69	40.47
C	4	2.234	→	→	→	→	→	→	→	1.53	5.50	10.92	16.45	22.86	29.45	36.90
D	5	2.474	→	→	→	→	→	→	1.42	3.44	5.93	9.90	14.47	20.27	26.59	33.95
E	7	2.830	→	→	→	→	→	.89	1.99	3.46	5.32	8.47	12.35	17.54	23.50	30.66
F	10	2.405	→	→	→	.23	.58	1.14	2.05	3.23	4.77	7.42	10.79	15.49	21.06	27.90
G	15	2.379	.061	.136	.253	.430	.786	1.30	2.10	3.11	4.44	6.76	9.76	14.09	19.30	25.92
H	25	2.358	.125	.214	.336	.506	.827	1.27	1.95	2.82	3.96	5.98	8.65	12.59	17.48	23.79
I	30	2.353	.147	.240	.366	.537	.856	1.29	1.96	2.81	3.92	5.88	8.50	12.36	17.19	23.42
J	35	2.349	.165	.261	.391	.564	.883	1.33	1.98	2.82	3.90	5.85	8.42	12.24	17.03	23.21
K	40	2.346	.160	.252	.375	.539	.842	1.25	1.88	2.69	3.73	5.61	8.11	11.84	16.55	22.38
L	50	2.342	.169	.261	.381	.542	.838	1.25	1.60	2.63	3.64	5.47	7.91	11.57	16.20	22.26
M	60	2.339	.158	.244	.356	.504	.781	1.16	1.74	2.47	3.44	5.17	7.54	11.10	15.64	21.63
N	85	2.335	.156	.242	.350	.493	.755	1.12	1.67	2.37	3.30	4.97	7.27	10.73	15.17	21.05
O	115	2.333	.153	.230	.333	.468	.718	1.06	1.58	2.25	3.14	4.76	6.99	10.37	14.74	20.57
P	175	2.331	.139	.210	.303	.427	.655	.972	1.46	2.08	2.93	4.47	6.60	9.89	14.15	19.88
Q	230	2.330	.142	.215	.308	.432	.661	.976	1.47	2.08	2.92	4.46	6.57	9.84	14.10	19.82
			.065	.10	.15	.25	.40	.65	1.00	1.50	2.50	4.00	6.50	10.00	15.00	
			Acceptable Quality Levels (tightened inspection)													

All AQL and table values are in percent defective.

↓ Use first sampling plan below arrow, that is, both sample size as well as M value. When sample size equals or exceeds lot size, every item in the lot must be inspected.

Table 7-24 *(MIL-STD-414) Table for Estimating the Lot Percent Defective Using Range Method.*[1]

Q_U or Q_L	Sample Size															
	3	4	5	7	10	15	25	30	35	40	50	60	85	115	175	230
0	50.00	50.00	50.00	50.00	50.00	50.00	50.00	50.00	50.00	50.00	50.00	50.00	50.00	50.00	50.00	50.00
.1	47.24	46.67	46.44	46.29	46.20	46.13	46.08	46.07	46.06	46.05	46.05	46.04	46.03	46.03	46.02	46.02
.2	44.46	43.33	42.90	42.60	42.42	42.29	42.19	42.17	42.16	42.15	42.13	42.12	42.10	42.10	42.08	42.08
.3	41.63	40.00	39.37	38.95	38.70	38.51	38.38	38.34	38.32	38.31	38.28	38.27	38.26	38.24	38.23	38.22
.31	41.35	39.67	39.02	38.59	38.33	38.14	38.00	37.96	37.94	37.93	37.90	37.89	37.88	37.86	37.85	37.84
.32	41.06	39.33	38.67	38.23	37.96	37.77	37.63	37.59	37.57	37.55	37.53	37.51	37.50	37.48	37.47	37.46
.33	40.77	39.00	38.32	37.87	37.60	37.39	37.25	37.21	37.19	37.18	37.15	37.14	37.12	37.11	37.09	37.09
.34	40.49	38.67	37.97	37.51	37.23	37.02	36.88	36.84	36.82	36.80	36.77	36.76	36.74	36.73	36.71	36.71
.35	40.20	38.33	37.62	37.15	36.87	36.65	36.50	36.46	36.44	36.43	36.40	36.39	36.37	36.36	36.34	36.33
.36	39.91	38.00	37.28	36.79	36.50	36.29	36.13	36.09	36.07	36.05	36.03	36.01	35.99	35.97	35.96	35.96
.37	39.62	37.67	36.93	36.43	36.14	35.92	35.76	35.72	35.70	35.68	35.65	35.64	35.62	35.61	35.59	35.59
.38	39.33	37.33	36.58	36.07	35.78	35.55	35.39	35.35	35.33	35.31	35.28	35.27	35.25	35.24	35.22	35.22
.39	39.03	37.00	36.23	35.72	35.41	35.19	35.02	34.98	34.96	34.94	34.92	34.90	34.88	34.87	34.85	34.85
.40	38.74	36.67	35.88	35.36	35.05	34.82	34.66	34.62	34.59	34.58	34.55	34.53	34.51	34.49	34.48	34.48
.41	38.45	36.33	35.54	35.01	34.69	34.46	34.29	34.25	34.23	34.21	34.18	34.17	34.14	34.12	34.11	34.11
.42	38.15	36.00	35.19	34.65	34.33	34.10	33.93	33.89	33.86	33.85	33.82	33.80	33.78	33.77	33.75	33.74
.43	37.85	35.67	34.85	34.30	33.98	33.74	33.57	33.53	33.50	33.48	33.45	33.44	33.41	33.39	33.38	33.38
.44	37.56	35.33	34.50	33.95	33.62	33.38	33.21	33.17	33.14	33.12	33.09	33.08	33.05	33.03	33.02	33.02
.45	37.26	35.00	34.16	33.60	33.27	33.02	32.85	32.81	32.78	32.76	32.73	32.72	32.69	32.67	32.66	32.66
.46	36.96	34.67	33.81	33.24	32.91	32.66	32.49	32.45	32.42	32.40	32.37	32.36	32.33	32.31	32.30	32.30
.47	36.66	34.33	33.47	32.89	32.56	32.31	32.13	32.08	32.06	32.04	32.01	32.00	31.97	31.95	31.94	31.94
.48	36.35	34.00	33.12	32.55	32.21	31.96	31.78	31.74	31.71	31.69	31.66	31.64	31.62	31.61	31.59	31.58
.49	36.05	33.67	32.78	32.20	31.86	31.60	31.42	31.38	31.35	31.33	31.30	31.29	31.26	31.24	31.23	31.23
.50	35.75	33.33	32.44	31.85	31.51	31.25	31.07	31.03	31.00	30.98	30.95	30.94	30.91	30.89	30.88	30.87
.51	35.44	33.00	32.10	31.51	31.16	30.90	30.72	30.68	30.65	30.63	30.60	30.59	30.55	30.55	30.53	30.52
.52	35.13	32.67	31.76	31.16	30.81	30.55	30.37	30.33	30.30	30.28	30.25	30.24	30.21	30.19	30.18	30.17
.53	34.82	32.33	31.42	30.82	30.46	30.21	30.02	29.98	29.95	29.93	29.90	29.89	29.86	29.84	29.83	29.83
.54	34.51	32.00	31.08	30.47	30.12	29.86	29.68	29.64	29.61	29.59	29.56	29.54	29.52	29.50	29.48	29.48
.55	34.20	31.67	30.74	30.13	29.78	29.52	29.33	29.29	29.26	29.24	29.21	29.20	29.17	29.15	29.14	29.14
.56	33.88	31.33	30.40	29.79	29.44	29.18	28.99	28.95	28.92	28.90	28.87	28.86	28.83	28.81	28.80	28.79
.57	33.57	31.00	30.06	29.45	29.09	28.83	28.65	28.61	28.58	28.56	28.53	28.52	28.49	28.47	28.46	28.45
.58	33.25	30.67	29.73	29.11	28.76	28.50	28.31	28.27	28.24	28.22	28.19	28.18	28.15	28.13	28.12	28.12
.59	32.93	30.33	29.39	28.77	28.42	28.16	27.97	27.93	27.91	27.89	27.86	27.84	27.82	27.80	27.78	27.78
.60	32.61	30.00	29.05	28.44	28.08	27.82	27.64	27.60	27.57	27.55	27.52	27.51	27.48	27.46	27.45	27.45
.61	32.28	29.67	28.72	28.10	27.75	27.49	27.31	27.27	27.24	27.22	27.19	27.17	27.15	27.14	27.12	27.11
.62	31.96	29.33	28.39	27.77	27.41	27.16	26.97	26.93	26.91	26.89	26.86	26.84	26.82	26.81	26.79	26.78
.63	31.63	29.00	28.05	27.44	27.08	26.82	26.64	26.60	26.58	26.56	26.53	26.51	26.49	26.48	26.46	26.45
.64	31.30	28.67	27.72	27.11	26.75	26.50	26.32	26.28	26.25	26.23	26.20	26.19	26.16	26.14	26.13	26.13
.65	30.97	28.33	27.39	26.78	26.42	26.17	25.99	25.95	25.92	25.90	25.87	25.86	25.84	25.83	25.81	25.80
.66	30.63	28.00	27.06	26.45	26.10	25.84	25.67	25.63	25.60	25.58	25.55	25.54	25.52	25.50	25.48	25.48
.67	30.30	27.67	26.73	26.12	25.77	25.52	25.34	25.30	25.28	25.26	25.23	25.22	25.20	25.18	25.16	25.16
.68	29.96	27.33	26.40	25.79	25.45	25.20	25.02	24.98	24.96	24.94	24.91	24.90	24.88	24.87	24.85	24.84
.69	29.61	27.00	26.07	25.47	25.12	24.88	24.71	24.67	24.64	24.62	24.59	24.58	24.56	24.55	24.53	24.53

[1] Values tabulated are read in percent.

Q_U or Q_L	Sample Size															
	3	4	5	7	10	15	25	30	35	40	50	60	85	115	175	230
.70	29.27	26.67	25.74	25.14	24.80	24.56	24.39	24.35	24.32	24.31	24.28	24.27	24.25	24.24	24.22	24.21
.71	28.92	26.33	25.41	24.82	24.48	24.24	24.07	24.03	24.01	23.99	23.97	23.95	23.93	23.91	23.90	23.90
.72	28.57	26.00	25.09	24.50	24.17	23.93	23.76	23.72	23.70	23.68	23.66	23.64	23.62	23.60	23.59	23.59
.73	28.22	25.67	24.76	24.18	23.85	23.61	23.45	23.41	23.39	23.37	23.35	23.33	23.32	23.30	23.29	23.29
.74	27.86	25.33	24.44	23.86	23.54	23.30	23.14	23.10	23.08	23.07	23.04	23.03	23.01	23.00	22.98	22.98
.75	27.50	25.00	24.11	23.55	23.22	22.99	22.84	22.80	22.78	22.76	22.74	22.72	22.71	22.69	22.68	22.68
.76	27.13	24.67	23.79	23.23	22.91	22.69	22.53	22.49	22.47	22.46	22.43	22.42	22.41	22.39	22.38	22.38
.77	26.77	24.33	23.47	22.92	22.60	22.38	22.23	22.19	22.17	22.16	22.13	22.12	22.11	22.09	22.08	22.08
.78	26.39	24.00	23.15	22.60	22.30	22.08	21.93	21.90	21.88	21.86	21.85	21.83	21.81	21.80	21.78	21.78
.79	26.02	23.67	22.83	22.29	21.99	21.78	21.64	21.60	21.58	21.57	21.54	21.53	21.52	21.50	21.49	21.49
.80	25.64	23.33	22.51	21.98	21.69	21.48	21.34	21.30	21.28	21.27	21.26	21.24	21.22	21.22	21.20	21.20
.81	25.25	23.00	22.19	21.68	21.39	21.18	21.04	21.01	20.99	20.98	20.97	20.95	20.93	20.93	20.91	20.91
.82	24.86	22.67	21.87	21.37	21.09	20.89	20.75	20.72	20.70	20.69	20.68	20.66	20.64	20.64	20.62	20.62
.83	24.47	22.33	21.56	21.06	20.79	20.59	20.46	20.43	20.41	20.40	20.38	20.37	20.36	20.35	20.34	20.34
.84	24.07	22.00	21.24	20.76	20.49	20.30	20.17	20.15	20.13	20.12	20.10	20.09	20.08	20.06	20.06	20.06
.85	23.67	21.67	20.93	20.46	20.20	20.01	19.89	19.87	19.85	19.84	19.82	19.81	19.79	19.79	19.78	19.78
.86	23.26	21.33	20.62	20.16	19.90	19.73	19.60	19.58	19.57	19.56	19.54	19.54	19.52	19.51	19.50	19.50
.87	22.84	21.00	20.31	19.86	19.61	19.44	19.32	19.31	19.29	19.28	19.26	19.25	19.24	19.24	19.22	19.22
.88	22.42	20.67	20.00	19.57	19.33	19.16	19.04	19.03	19.01	19.00	18.98	18.98	18.97	18.96	18.95	18.95
.89	21.99	20.33	19.69	19.27	19.04	18.88	18.77	18.75	18.74	18.73	18.71	18.70	18.69	18.69	18.68	18.68
.90	21.55	20.00	19.38	18.98	18.75	18.60	18.50	18.48	18.47	18.46	18.44	18.43	18.42	18.42	18.41	18.41
.91	21.11	19.67	19.07	18.69	18.47	18.32	18.22	18.21	18.20	18.19	18.17	18.17	18.16	18.15	18.15	18.15
.92	20.66	19.33	18.77	18.40	18.19	18.05	17.96	17.95	17.93	17.92	17.92	17.90	17.89	17.89	17.88	17.88
.93	20.20	19.00	18.46	18.11	17.91	17.78	17.69	17.68	17.67	17.66	17.65	17.65	17.63	17.63	17.62	17.62
.94	19.74	18.67	18.16	17.82	17.64	17.51	17.43	17.42	17.41	17.40	17.39	17.39	17.37	17.37	17.36	17.36
.95	19.25	18.33	17.86	17.54	17.36	17.24	17.17	17.16	17.15	17.14	17.13	17.13	17.12	17.12	17.11	17.11
.96	18.76	18.00	17.56	17.26	17.09	16.98	16.91	16.90	16.89	16.88	16.88	16.87	16.86	16.86	16.86	16.86
.97	18.25	17.67	17.25	16.97	16.82	16.71	16.65	16.64	16.63	16.63	16.62	16.62	16.61	16.61	16.60	16.60
.98	17.74	17.33	16.96	16.70	16.55	16.45	16.39	16.38	16.38	16.37	16.37	16.37	16.36	16.36	16.36	16.36
.99	17.21	17.00	16.66	16.42	16.28	16.19	16.14	16.13	16.13	16.12	16.12	16.12	16.11	16.11	16.11	16.11
1.00	16.67	16.67	16.36	16.14	16.02	15.94	15.89	15.88	15.88	15.88	15.87	15.87	15.87	15.87	15.87	15.87
1.01	16.11	16.33	16.07	15.87	15.76	15.68	15.64	15.63	15.63	15.63	15.63	15.63	15.62	15.62	15.62	15.62
1.02	15.53	16.00	15.78	15.60	15.50	15.43	15.40	15.39	15.39	15.39	15.39	15.39	15.38	15.38	15.38	15.38
1.03	14.93	15.67	15.48	15.33	15.24	15.18	15.15	15.15	15.15	15.15	15.15	15.15	15.15	15.15	15.15	15.15
1.04	14.31	15.33	15.19	15.06	14.98	14.94	14.91	14.91	14.91	14.91	14.91	14.91	14.91	14.91	14.91	14.91
1.05	13.66	15.00	14.91	14.79	14.73	14.69	14.67	14.67	14.67	14.67	14.67	14.68	14.68	14.68	14.68	14.68
1.06	12.98	14.67	14.62	14.53	14.48	14.45	14.44	14.44	14.44	14.44	14.44	14.44	14.44	14.45	14.45	14.45
1.07	12.27	14.33	14.33	14.27	14.23	14.21	14.20	14.21	14.21	14.21	14.21	14.21	14.22	14.22	14.22	14.22
1.08	11.51	14.00	14.05	14.01	13.98	13.97	13.97	13.98	13.98	13.98	13.98	13.99	13.99	13.99	14.00	14.00
1.09	10.71	13.67	13.76	13.75	13.74	13.73	13.74	13.75	13.75	13.75	13.76	13.76	13.77	13.77	13.78	13.78

Q_U or Q_L	Sample Size															
	3	4	5	7	10	15	25	30	35	40	50	60	85	115	175	230
1.10	9.84	13.33	13.48	13.50	13.49	13.50	13.52	13.52	13.52	13.53	13.54	13.54	13.55	13.55	13.56	13.56
1.11	8.89	13.00	13.20	13.24	13.25	13.27	13.29	13.30	13.30	13.31	13.31	13.32	13.32	13.33	13.34	13.34
1.12	7.82	12.67	12.93	12.99	13.02	13.04	13.07	13.08	13.08	13.09	13.10	13.10	13.12	13.12	13.12	13.12
1.13	6.60	12.33	12.65	12.74	12.78	12.81	12.85	12.86	12.86	12.87	12.89	12.89	12.89	12.90	12.91	12.91
1.14	5.08	12.00	12.37	12.49	12.55	12.59	12.63	12.64	12.65	12.66	12.67	12.67	12.69	12.69	12.70	12.70
1.15	0.29	11.67	12.10	12.25	12.31	12.37	12.42	12.43	12.44	12.45	12.46	12.46	12.48	12.48	12.49	12.49
1.16	0.00	11.33	11.83	12.00	12.08	12.15	12.21	12.22	12.23	12.24	12.25	12.25	12.27	12.28	12.29	12.29
1.17	0.00	11.00	11.56	11.76	11.86	11.93	12.00	12.01	12.02	12.03	12.05	12.06	12.07	12.07	12.08	12.08
1.18	0.00	10.67	11.29	11.52	11.63	11.71	11.79	11.80	11.81	11.82	11.84	11.84	11.86	11.88	11.88	11.88
1.19	0.00	10.33	11.02	11.29	11.41	11.50	11.58	11.60	11.61	11.62	11.64	11.65	11.66	11.68	11.69	11.69
1.20	0.00	10.00	10.76	11.05	11.19	11.29	11.38	11.40	11.41	11.42	11.43	11.45	11.47	11.47	11.49	11.49
1.21	0.00	9.67	10.50	10.82	10.97	11.08	11.18	11.20	11.21	11.22	11.25	11.26	11.27	11.29	11.30	11.30
1.22	0.00	9.33	10.23	10.59	10.76	10.88	10.98	11.00	11.02	11.03	11.04	11.06	11.08	11.09	11.10	11.10
1.23	0.00	9.00	9.97	10.36	10.54	10.67	10.78	10.80	10.82	10.84	10.85	10.87	10.89	10.90	10.91	10.91
1.24	0.00	8.67	9.72	10.13	10.33	10.47	10.58	10.61	10.63	10.64	10.67	10.68	10.70	10.71	10.73	10.73
1.25	0.00	8.33	9.46	9.91	10.12	10.27	10.39	10.42	10.44	10.46	10.47	10.49	10.51	10.52	10.54	10.54
1.26	0.00	8.00	9.21	9.69	9.92	10.08	10.20	10.24	10.26	10.27	10.30	10.31	10.33	10.34	10.36	10.36
1.27	0.00	7.67	8.96	9.47	9.71	9.88	10.01	10.05	10.07	10.09	10.11	10.13	10.15	10.17	10.18	10.18
1.28	0.00	7.33	8.71	9.25	9.51	9.69	9.83	9.87	9.89	9.90	9.93	9.95	9.97	9.99	10.00	10.00
1.29	0.00	7.00	8.46	9.04	9.31	9.50	9.64	9.68	9.71	9.72	9.75	9.77	9.79	9.81	9.83	9.83
1.30	0.00	6.67	8.21	8.83	9.11	9.32	9.47	9.51	9.53	9.55	9.58	9.59	9.62	9.64	9.65	9.65
1.31	0.00	6.33	7.97	8.62	8.92	9.13	9.29	9.33	9.35	9.37	9.40	9.42	9.45	9.47	9.48	9.48
1.32	0.00	6.00	7.73	8.41	8.73	8.95	9.11	9.15	9.18	9.20	9.23	9.25	9.28	9.30	9.31	9.31
1.33	0.00	5.67	7.49	8.20	8.54	8.77	8.94	8.98	9.01	9.03	9.06	9.08	9.11	9.13	9.14	9.15
1.34	0.00	5.33	7.25	8.00	8.35	8.59	8.77	8.81	8.84	8.86	8.89	8.91	8.94	8.96	8.98	8.98
1.35	0.00	5.00	7.02	7.80	8.16	8.41	8.60	8.64	8.67	8.69	8.73	8.75	8.78	8.80	8.82	8.82
1.36	0.00	4.67	6.79	7.60	7.98	8.24	8.43	8.48	8.51	8.53	8.56	8.59	8.62	8.64	8.66	8.66
1.37	0.00	4.33	6.56	7.40	7.80	8.07	8.27	8.31	8.34	8.37	8.40	8.43	8.46	8.48	8.50	8.50
1.38	0.00	4.00	6.33	7.21	7.62	7.90	8.11	8.15	8.18	8.21	8.25	8.26	8.30	8.32	8.34	8.35
1.39	0.00	3.67	6.10	7.02	7.45	7.73	7.95	7.99	8.02	8.05	8.09	8.11	8.14	8.17	8.19	8.19
1.40	0.00	3.33	5.88	6.83	7.27	7.57	7.79	7.84	7.88	7.90	7.93	7.96	8.00	8.02	8.03	8.04
1.41	0.00	3.00	5.66	6.65	7.10	7.41	7.63	7.68	7.71	7.74	7.78	7.81	7.85	7.87	7.88	7.89
1.42	0.00	2.67	5.44	6.46	6.93	7.25	7.48	7.53	7.56	7.59	7.63	7.66	7.70	7.72	7.74	7.74
1.43	0.00	2.33	5.23	6.28	6.76	7.09	7.33	7.38	7.41	7.44	7.49	7.51	7.54	7.57	7.59	7.60
1.44	0.00	2.00	5.01	6.10	6.60	6.93	7.18	7.24	7.28	7.30	7.34	7.37	7.41	7.43	7.45	7.46
1.45	0.00	1.67	4.81	5.93	6.44	6.78	7.03	7.09	7.13	7.15	7.20	7.23	7.27	7.29	7.30	7.32
1.46	0.00	1.33	4.60	5.75	6.28	6.63	6.89	6.95	6.99	7.01	7.06	7.09	7.13	7.15	7.17	7.18
1.47	0.00	1.00	4.39	5.58	6.12	6.48	6.74	6.80	6.85	6.87	6.92	6.95	6.99	7.01	7.03	7.04
1.48	0.00	0.67	4.19	5.41	5.96	6.34	6.60	6.66	6.71	6.73	6.78	6.81	6.85	6.87	6.89	6.90
1.49	0.00	0.33	3.99	5.24	5.81	6.19	6.47	6.53	6.57	6.60	6.64	6.67	6.72	6.74	6.76	6.77

Table 7-24 (*Continued*) *Table for Estimating the Lot Percent Defective Using Range Method.*

Q_U or Q_L	\multicolumn{16}{c}{Sample Size}															
	3	4	5	7	10	15	25	30	35	40	50	60	85	115	175	230
1.50	0.00	0.00	3.80	5.08	5.66	6.05	6.33	6.39	6.43	6.46	6.51	6.54	6.58	6.61	6.63	6.64
1.51	0.00	0.00	3.61	4.92	5.51	5.91	6.19	6.25	6.30	6.33	6.38	6.41	6.45	6.48	6.50	6.51
1.52	0.00	0.00	3.42	4.76	5.37	5.77	6.06	6.12	6.17	6.20	6.25	6.28	6.32	6.35	6.37	6.38
1.53	0.00	0.00	3.23	4.60	5.22	5.64	5.93	5.99	6.04	6.07	6.12	6.15	6.20	6.22	6.25	6.26
1.54	0.00	0.00	3.05	4.45	5.08	5.50	5.80	5.86	5.91	5.95	6.00	6.03	6.07	6.10	6.12	6.14
1.55	0.00	0.00	2.87	4.30	4.94	5.37	5.68	5.74	5.79	5.82	5.87	5.90	5.95	5.98	6.00	6.01
1.56	0.00	0.00	2.69	4.15	4.81	5.24	5.55	5.62	5.67	5.70	5.75	5.78	5.83	5.86	5.88	5.89
1.57	0.00	0.00	2.52	4.01	4.67	5.11	5.43	5.50	5.55	5.58	5.63	5.66	5.71	5.74	5.77	5.79
1.58	0.00	0.00	2.35	3.86	4.54	4.99	5.31	5.38	5.43	5.46	5.52	5.55	5.59	5.62	5.65	5.66
1.59	0.00	0.00	2.19	3.72	4.41	4.86	5.19	5.26	5.31	5.34	5.40	5.43	5.48	5.51	5.53	5.55
1.60	0.00	0.00	2.03	3.58	4.28	4.74	5.08	5.14	5.19	5.23	5.29	5.32	5.36	5.39	5.42	5.43
1.61	0.00	0.00	1.87	3.45	4.16	4.62	4.96	5.03	5.08	5.12	5.17	5.20	5.25	5.28	5.31	5.32
1.62	0.00	0.00	1.72	3.31	4.03	4.51	4.85	4.92	4.97	5.01	5.06	5.09	5.14	5.17	5.20	5.22
1.63	0.00	0.00	1.57	3.18	3.91	4.39	4.74	4.81	4.86	4.90	4.96	4.99	5.04	5.07	5.10	5.12
1.64	0.00	0.00	1.42	3.06	3.79	4.28	4.63	4.70	4.75	4.79	4.85	4.88	4.93	4.96	4.99	5.00
1.65	0.00	0.00	1.28	2.93	3.68	4.17	4.52	4.59	4.64	4.68	4.74	4.77	4.83	4.86	4.89	4.91
1.66	0.00	0.00	1.15	2.81	3.56	4.06	4.41	4.49	4.54	4.58	4.64	4.67	4.72	4.75	4.79	4.81
1.67	0.00	0.00	1.02	2.69	3.45	3.95	4.31	4.39	4.44	4.48	4.54	4.57	4.62	4.65	4.69	4.71
1.68	0.00	0.00	0.89	2.57	3.34	3.85	4.21	4.29	4.34	4.38	4.44	4.47	4.53	4.56	4.59	4.61
1.69	0.00	0.00	0.77	2.46	3.23	3.74	4.10	4.19	4.24	4.28	4.34	4.37	4.43	4.46	4.49	4.51
1.70	0.00	0.00	0.66	2.35	3.13	3.64	4.00	4.09	4.14	4.18	4.24	4.28	4.33	4.36	4.40	4.42
1.71	0.00	0.00	0.55	2.24	3.02	3.54	3.92	3.99	4.05	4.09	4.15	4.18	4.24	4.27	4.30	4.31
1.72	0.00	0.00	0.45	2.13	2.92	3.45	3.82	3.90	3.95	3.99	4.06	4.09	4.15	4.18	4.21	4.23
1.73	0.00	0.00	0.36	2.03	2.82	3.35	3.73	3.81	3.86	3.90	3.96	4.00	4.06	4.09	4.12	4.14
1.74	0.00	0.00	0.27	1.93	2.73	3.26	3.63	3.72	3.77	3.81	3.87	3.91	3.97	4.00	4.03	4.05
1.75	0.00	0.00	0.19	1.83	2.63	3.16	3.54	3.63	3.68	3.72	3.79	3.82	3.88	3.91	3.94	3.96
1.76	0.00	0.00	0.12	1.73	2.54	3.07	3.45	3.54	3.59	3.63	3.70	3.74	3.79	3.82	3.86	3.88
1.77	0.00	0.00	0.06	1.64	2.45	2.99	3.37	3.45	3.51	3.55	3.61	3.65	3.71	3.74	3.77	3.79
1.78	0.00	0.00	0.02	1.55	2.36	2.90	3.28	3.37	3.43	3.47	3.53	3.57	3.62	3.65	3.69	3.71
1.79	0.00	0.00	0.00	1.46	2.27	2.81	3.20	3.28	3.34	3.38	3.45	3.49	3.54	3.57	3.61	3.63
1.80	0.00	0.00	0.00	1.38	2.19	2.73	3.11	3.20	3.26	3.30	3.37	3.41	3.46	3.49	3.53	3.55
1.81	0.00	0.00	0.00	1.29	2.10	2.65	3.03	3.12	3.18	3.22	3.29	3.33	3.38	3.41	3.45	3.47
1.82	0.00	0.00	0.00	1.21	2.02	2.57	2.96	3.05	3.11	3.15	3.21	3.25	3.31	3.34	3.37	3.39
1.83	0.00	0.00	0.00	1.14	1.94	2.49	2.88	2.97	3.03	3.07	3.13	3.17	3.23	3.26	3.30	3.32
1.84	0.00	0.00	0.00	1.06	1.87	2.42	2.80	2.89	2.95	2.99	3.06	3.10	3.16	3.19	3.22	3.24
1.85	0.00	0.00	0.00	0.99	1.79	2.34	2.73	2.82	2.88	2.92	2.99	3.03	3.08	3.11	3.15	3.17
1.86	0.00	0.00	0.00	0.92	1.72	2.27	2.66	2.75	2.81	2.85	2.91	2.95	3.01	3.04	3.08	3.10
1.87	0.00	0.00	0.00	0.86	1.65	2.20	2.59	2.68	2.74	2.78	2.84	2.88	2.94	2.97	3.01	3.03
1.88	0.00	0.00	0.00	0.79	1.58	2.13	2.52	2.61	2.67	2.71	2.77	2.81	2.87	2.90	2.94	2.96
1.89	0.00	0.00	0.00	0.73	1.51	2.06	2.45	2.54	2.60	2.64	2.71	2.75	2.81	2.84	2.87	2.89

Q_U or Q_L	Sample Size															
	3	4	5	7	10	15	25	30	35	40	50	60	85	115	175	230
1.90	0.00	0.00	0.00	0.67	1.45	1.99	2.38	2.47	2.53	2.57	2.64	2.68	2.74	2.77	2.81	2.83
1.91	0.00	0.00	0.00	0.62	1.38	1.93	2.32	2.41	2.47	2.51	2.58	2.61	2.67	2.70	2.74	2.76
1.92	0.00	0.00	0.00	0.56	1.32	1.86	2.25	2.34	2.41	2.45	2.51	2.55	2.61	2.64	2.68	2.70
1.93	0.00	0.00	0.00	0.51	1.26	1.80	2.19	2.28	2.34	2.38	2.45	2.49	2.55	2.58	2.61	2.63
1.94	0.00	0.00	0.00	0.46	1.20	1.74	2.13	2.22	2.28	2.32	2.39	2.43	2.49	2.52	2.55	2.57
1.95	0.00	0.00	0.00	0.42	1.15	1.68	2.07	2.16	2.22	2.26	2.33	2.37	2.43	2.46	2.49	2.51
1.96	0.00	0.00	0.00	0.37	1.09	1.62	2.01	2.10	2.16	2.20	2.27	2.31	2.37	2.40	2.43	2.45
1.97	0.00	0.00	0.00	0.33	1.04	1.57	1.95	2.04	2.10	2.14	2.21	2.25	2.31	2.34	2.38	2.40
1.98	0.00	0.00	0.00	0.30	0.99	1.51	1.90	1.99	2.05	2.09	2.15	2.19	2.25	2.28	2.32	2.34
1.99	0.00	0.00	0.00	0.26	0.94	1.46	1.84	1.93	1.99	2.03	2.10	2.14	2.20	2.23	2.26	2.28
2.00	0.00	0.00	0.00	0.23	0.89	1.41	1.79	1.88	1.94	1.98	2.05	2.08	2.14	2.17	2.21	2.23
2.01	0.00	0.00	0.00	0.20	0.84	1.36	1.74	1.83	1.89	1.93	1.99	2.03	2.09	2.12	2.16	2.18
2.02	0.00	0.00	0.00	0.17	0.80	1.31	1.69	1.78	1.83	1.87	1.94	1.98	2.04	2.07	2.10	2.12
2.03	0.00	0.00	0.00	0.14	0.75	1.26	1.64	1.73	1.78	1.82	1.89	1.93	1.99	2.02	2.05	2.07
2.04	0.00	0.00	0.00	0.12	0.71	1.21	1.59	1.68	1.73	1.77	1.84	1.88	1.94	1.97	2.00	2.02
2.05	0.00	0.00	0.00	0.10	0.67	1.17	1.54	1.63	1.69	1.73	1.79	1.83	1.89	1.92	1.95	1.97
2.06	0.00	0.00	0.00	0.08	0.63	1.12	1.49	1.58	1.64	1.68	1.74	1.78	1.84	1.87	1.91	1.93
2.07	0.00	0.00	0.00	0.06	0.60	1.08	1.45	1.54	1.59	1.63	1.70	1.74	1.79	1.82	1.86	1.88
2.08	0.00	0.00	0.00	0.05	0.56	1.04	1.40	1.49	1.55	1.59	1.65	1.69	1.75	1.78	1.81	1.83
2.09	0.00	0.00	0.00	0.03	0.53	1.00	1.36	1.45	1.50	1.54	1.61	1.64	1.70	1.73	1.77	1.79
2.10	0.00	0.00	0.00	0.02	0.49	0.96	1.32	1.41	1.46	1.50	1.56	1.60	1.66	1.69	1.72	1.74
2.11	0.00	0.00	0.00	0.01	0.46	0.92	1.28	1.36	1.42	1.46	1.52	1.56	1.61	1.64	1.68	1.70
2.12	0.00	0.00	0.00	0.00	0.43	0.88	1.24	1.32	1.38	1.42	1.48	1.52	1.57	1.60	1.64	1.66
2.13	0.00	0.00	0.00	0.00	0.40	0.85	1.20	1.28	1.34	1.38	1.44	1.48	1.53	1.56	1.60	1.62
2.14	0.00	0.00	0.00	0.00	0.38	0.81	1.16	1.25	1.30	1.34	1.40	1.44	1.49	1.52	1.56	1.58
2.15	0.00	0.00	0.00	0.00	0.35	0.78	1.13	1.21	1.26	1.30	1.36	1.40	1.45	1.48	1.52	1.54
2.16	0.00	0.00	0.00	0.00	0.32	0.75	1.09	1.17	1.22	1.26	1.32	1.36	1.41	1.44	1.48	1.50
2.17	0.000	0.00	0.00	0.00	0.30	0.71	1.06	1.13	1.18	1.22	1.29	1.32	1.38	1.41	1.44	1.46
2.18	0.00	0.00	0.00	0.00	0.28	0.68	1.02	1.10	1.15	1.19	1.25	1.28	1.34	1.37	1.40	1.41
2.19	0.00	0.00	0.00	0.00	0.26	0.65	0.99	1.06	1.11	1.15	1.22	1.25	1.30	1.33	1.37	1.39
2.20	0.000	0.000	0.000	0.000	0.236	0.625	0.954	1.030	1.083	1.122	1.178	1.214	1.267	1.299	1.330	1.346
2.21	0.000	0.000	0.000	0.000	0.217	0.597	0.922	0.997	1.058	1.089	1.144	1.180	1.233	1.265	1.295	1.311
2.22	0.000	0.000	0.000	0.000	0.199	0.570	0.891	0.966	1.018	1.056	1.111	1.147	1.199	1.231	1.261	1.277
2.23	0.000	0.000	0.000	0.000	0.182	0.544	0.861	0.935	0.986	1.025	1.079	1.115	1.167	1.197	1.228	1.244
2.24	0.000	0.000	0.000	0.000	0.166	0.519	0.831	0.905	0.956	0.994	1.048	1.083	1.135	1.165	1.195	1.211
2.25	0.000	0.000	0.000	0.000	0.150	0.495	0.802	0.875	0.926	0.964	1.018	1.052	1.104	1.134	1.163	1.179
2.26	0.000	0.000	0.000	0.000	0.136	0.471	0.775	0.847	0.897	0.935	0.987	1.022	1.073	1.103	1.132	1.148
2.27	0.000	0.000	0.000	0.000	0.123	0.449	0.748	0.819	0.869	0.906	0.958	0.993	1.043	1.073	1.103	1.118
2.28	0.000	0.000	0.000	0.000	0.111	0.427	0.722	0.792	0.841	0.878	0.930	0.964	1.014	1.044	1.073	1.088
2.29	0.000	0.000	0.000	0.000	0.099	0.406	0.697	0.766	0.814	0.851	0.902	0.936	0.986	1.015	1.044	1.059

Table 7-24 *(Continued)* **Table for Estimating the Lot Percent Defective Using Range Method.**

Q_U or Q_L	Sample Size															
	3	4	5	7	10	15	25	30	35	40	50	60	85	115	175	230
2.30	0.000	0.000	0.000	0.000	0.089	0.386	0.672	0.741	0.789	0.825	0.875	0.909	0.959	0.988	1.016	1.031
2.31	0.000	0.000	0.000	0.000	0.079	0.367	0.648	0.716	0.763	0.799	0.849	0.882	0.931	0.960	0.988	1.003
2.32	0.000	0.000	0.000	0.000	0.070	0.348	0.624	0.691	0.739	0.774	0.823	0.856	0.905	0.934	0.962	0.976
2.33	0.000	0.000	0.000	0.000	0.061	0.330	0.601	0.668	0.715	0.750	0.798	0.831	0.879	0.908	0.935	0.950
2.34	0.000	0.000	0.000	0.000	0.054	0.313	0.579	0.645	0.691	0.720	0.774	0.807	0.854	0.882	0.909	0.924
2.35	0.000	0.000	0.000	0.000	0.047	0.296	0.558	0.623	0.669	0.703	0.750	0.782	0.829	0.857	0.884	0.899
2.36	0.000	0.000	0.000	0.000	0.040	0.280	0.538	0.602	0.646	0.680	0.728	0.759	0.806	0.833	0.860	0.874
2.37	0.000	0.000	0.000	0.000	0.035	0.265	0.518	0.580	0.624	0.658	0.705	0.736	0.782	0.809	0.836	0.850
2.38	0.000	0.000	0.000	0.000	0.029	0.250	0.496	0.560	0.604	0.637	0.683	0.714	0.759	0.787	0.813	0.827
2.39	0.000	0.000	0.000	0.000	0.025	0.236	0.479	0.541	0.584	0.616	0.662	0.693	0.737	0.764	0.791	0.804
2.40	0.000	0.000	0.000	0.000	0.021	0.223	0.461	0.521	0.564	0.596	0.641	0.671	0.715	0.742	0.769	0.782
2.41	0.000	0.000	0.000	0.000	0.017	0.210	0.443	0.503	0.545	0.577	0.621	0.651	0.695	0.721	0.747	0.760
2.42	0.000	0.000	0.000	0.000	0.014	0.198	0.426	0.485	0.526	0.557	0.601	0.631	0.674	0.701	0.726	0.739
2.43	0.000	0.000	0.000	0.000	0.011	0.186	0.410	0.467	0.508	0.539	0.582	0.611	0.654	0.679	0.705	0.718
2.44	0.000	0.000	0.000	0.000	0.009	0.175	0.393	0.450	0.491	0.521	0.564	0.593	0.635	0.660	0.685	0.698
2.45	0.000	0.000	0.000	0.000	0.007	0.165	0.378	0.434	0.473	0.503	0.545	0.573	0.616	0.641	0.665	0.678
2.46	0.000	0.000	0.000	0.000	0.005	0.154	0.362	0.417	0.456	0.486	0.528	0.556	0.597	0.622	0.646	0.659
2.47	0.000	0.000	0.000	0.000	0.004	0.145	0.348	0.403	0.441	0.470	0.511	0.538	0.579	0.604	0.627	0.640
2.48	0.000	0.000	0.000	0.000	0.003	0.136	0.333	0.387	0.425	0.454	0.494	0.522	0.562	0.586	0.609	0.622
2.49	0.000	0.000	0.000	0.000	0.002	0.127	0.321	0.372	0.409	0.438	0.478	0.504	0.545	0.569	0.593	0.605
2.50	0.000	0.000	0.000	0.000	0.001	0.118	0.307	0.358	0.395	0.423	0.463	0.489	0.528	0.552	0.575	0.587
2.51	0.000	0.000	0.000	0.000	0.001	0.111	0.294	0.345	0.381	0.409	0.447	0.473	0.512	0.536	0.558	0.570
2.52	0.000	0.000	0.000	0.000	0.000	0.103	0.282	0.331	0.367	0.394	0.432	0.458	0.497	0.519	0.542	0.553
2.53	0.000	0.000	0.000	0.000	0.000	0.096	0.270	0.319	0.354	0.381	0.418	0.444	0.481	0.503	0.526	0.537
2.54	0.000	0.000	0.000	0.000	0.000	0.089	0.258	0.306	0.340	0.367	0.404	0.428	0.466	0.488	0.510	0.522
2.55	0.000	0.000	0.000	0.000	0.000	0.083	0.247	0.294	0.328	0.354	0.390	0.415	0.451	0.473	0.495	0.506
2.56	0.000	0.000	0.000	0.000	0.000	0.077	0.237	0.283	0.316	0.341	0.377	0.401	0.437	0.459	0.480	0.491
2.57	0.000	0.000	0.000	0.000	0.000	0.071	0.227	0.272	0.304	0.328	0.364	0.388	0.424	0.445	0.466	0.477
2.58	0.000	0.000	0.000	0.000	0.000	0.066	0.217	0.261	0.292	0.317	0.352	0.376	0.411	0.432	0.452	0.463
2.59	0.000	0.000	0.000	0.000	0.000	0.061	0.207	0.251	0.282	0.305	0.340	0.363	0.397	0.418	0.439	0.449
2.60	0.000	0.000	0.000	0.000	0.000	0.056	0.198	0.240	0.271	0.294	0.328	0.351	0.385	0.406	0.426	0.436
2.61	0.000	0.000	0.000	0.000	0.000	0.052	0.189	0.231	0.260	0.283	0.317	0.339	0.372	0.393	0.413	0.423
2.62	0.000	0.000	0.000	0.000	0.000	0.048	0.181	0.221	0.250	0.273	0.306	0.327	0.360	0.381	0.400	0.410
2.63	0.000	0.000	0.000	0.000	0.000	0.044	0.173	0.212	0.241	0.263	0.295	0.316	0.349	0.368	0.388	0.398
2.64	0.000	0.000	0.000	0.000	0.000	0.040	0.164	0.203	0.232	0.253	0.285	0.306	0.338	0.357	0.376	0.386
2.65	0.000	0.000	0.000	0.000	0.000	0.037	0.157	0.195	0.223	0.244	0.274	0.295	0.327	0.346	0.365	0.375
2.66	0.000	0.000	0.000	0.000	0.000	0.034	0.149	0.186	0.213	0.234	0.265	0.285	0.316	0.335	0.353	0.363
2.67	0.000	0.000	0.000	0.000	0.000	0.031	0.143	0.179	0.205	0.225	0.255	0.275	0.305	0.324	0.342	0.352
2.68	0.000	0.000	0.000	0.000	0.000	0.028	0.136	0.171	0.197	0.217	0.246	0.266	0.296	0.314	0.332	0.342
2.69	0.000	0.000	0.000	0.000	0.000	0.025	0.129	0.164	0.190	0.209	0.238	0.257	0.286	0.304	0.321	0.331

Table 7-24 (*Continued*) *Table for Estimating the Lot Percent Defective Using Range Method.*

Q_U or Q_L	Sample Size															
	3	4	5	7	10	15	25	30	35	40	50	60	85	115	175	230
2.70	0.000	0.000	0.000	0.000	0.000	0.023	0.123	0.156	0.182	0.201	0.228	0.248	0.277	0.295	0.311	0.321
2.71	0.000	0.000	0.000	0.000	0.000	0.021	0.117	0.150	0.174	0.193	0.220	0.239	0.267	0.285	0.302	0.311
2.72	0.000	0.000	0.000	0.000	0.000	0.019	0.111	0.143	0.167	0.185	0.212	0.231	0.259	0.275	0.292	0.301
2.73	0.000	0.000	0.000	0.000	0.000	0.017	0.106	0.137	0.160	0.178	0.205	0.222	0.250	0.266	0.283	0.292
2.74	0.000	0.000	0.000	0.000	0.000	0.015	0.101	0.131	0.153	0.171	0.197	0.215	0.241	0.258	0.274	0.282
2.75	0.000	0.000	0.000	0.000	0.000	0,014	0.096	0.125	0.147	0.164	0.189	0.207	0.233	0.248	0.266	0.274
2.76	0.000	0.000	0.000	0.000	0.000	0.012	0.091	0.120	0.141	0.158	0.182	0.200	0.225	0.241	0.257	0.265
2.77	0.000	0.000	0.000	0.000	0.000	0.011	0.086	0.114	0.135	0.152	0.175	0.192	0.217	0.232	0.249	0.257
2.78	0.000	0.000	0.000	0.000	0.000	0.010	0.081	0.109	0.130	0.146	0.169	0.185	0.210	0.226	0.241	0.249
2.79	0.000	0.000	0.000	0.000	0.000	0.008	0.077	0.103	0.124	0.140	0.163	0.179	0.202	0.218	0.233	0.241
2.80	0.000	0.000	0.000	0.000	0.000	0.007	0.074	0.099	0.118	0.134	0.156	0.172	0.196	0.210	0.225	0.233
2.81	0.000	0.000	0.000	0.000	0.000	0.007	0.070	0.094	0.113	0.129	0.150	0.165	0.189	0.204	0.218	0.226
2.82	0.000	0.000	0.000	0.000	0.000	0.006	0.066	0.090	0.109	0.123	0.144	0.159	0.183	0.194	0.211	0.219
2.83	0.000	0.000	0.000	0.000	0.000	0.005	0.062	0.085	0.103	0.118	0.139	0.154	0.176	0.190	0.204	0.212
2.84	0.000	0.000	0.000	0.000	0.000	0.004	0.059	0.082	0.099	0.113	0.134	0.148	0.170	0.184	0.197	0.205
2.85	0.000	0.000	0.000	0.000	0.000	0.004	0.055	0.078	0.095	0.109	0.128	0.143	0.164	0.178	0.191	0.198
2.86	0.000	0.000	0.000	0.000	0.000	0.003	0.053	0.074	0.091	0.104	0.124	0.137	0.159	0.172	0.185	0.192
2.87	0.000	0.000	0.000	0.000	0.000	0.003	0.050	0.070	0.087	0.100	0.119	0.132	0.152	0.166	0.179	0.185
2.88	0.000	0.000	0.000	0.000	0.000	0.002	0.047	0.067	0.082	0.095	0.114	0.127	0.147	0.160	0.173	0.179
2.89	0.000	0.000	0.000	0.000	0.000	0.002	0.044	0.064	0.079	0.091	0.109	0.122	0.142	0.155	0.167	0.173
2.90	0.000	0.000	0.000	0.000	0.000	0.002	0.042	0.061	0.075	0.088	0.105	0.117	0.138	0.149	0.161	0.168
2.91	0.000	0.000	0.000	0.000	0.000	0.001	0.039	0.057	0.072	0.084	0.101	0.112	0.132	0.145	0.156	0.162
2.92	0.000	0.000	0.000	0.000	0.000	0.001	0.037	0.055	0.069	0.080	0.097	0.107	0.127	0.140	0.151	0.157
2.93	0.000	0.000	0.000	0.000	0.000	0.001	0.035	0.052	0.066	0.077	0.093	0.104	0.123	0.134	0.146	0.151
2.94	0.000	0.000	0.000	0.000	0.000	0.001	0.033	0.049	0.062	0.073	0.089	0.100	0.118	0.129	0.141	0.146
2.95	0.000	0.000	0.000	0.000	0.000	0.001	0.031	0.047	0.059	0.070	0.086	0.096	0.114	0.125	0.136	0.142
2.96	0.000	0.000	0.000	0.000	0.000	0.001	0.029	0.044	0.056	0.067	0.082	0.092	0.110	0.121	0.132	0.137
2.97	0.000	0.000	0.000	0.000	0.000	0.000	0.027	0.042	0.054	0.064	0.079	0.088	0.105	0.116	0.127	0.132
2.98	0.000	0.000	0.000	0.000	0.000	0.000	0.025	0.039	0.051	0.061	0.075	0.085	0.101	0.112	0.123	0.128
2.99	0.000	0.000	0.000	0.000	0.000	0.000	0.024	0.038	0.049	0.058	0.072	0.082	0.098	0.108	0.119	0.124
3.00	0.000	0.000	0.000	0.000	0.000	0.000	0.022	0.036	0.047	0.056	0.069	0.078	0.094	0.105	0.115	0.120
3.01	0.000	0.000	0.000	0.000	0.000	0.000	0.022	0.034	0.044	0.053	0.066	0.075	0.091	0.101	0.111	0.116
3.02	0.000	0.000	0.000	0.000	0.000	0.000	0.020	0.032	0.042	0.050	0.063	0.072	0.087	0.097	0.107	0.112
3.03	0.000	0.000	0.000	0.000	0.000	0.000	0.019	0.030	0.040	0.048	0.061	0.069	0.084	0.094	0.103	0.108
3.04	0.000	0.000	0.000	0.000	0.000	0.000	0.017	0.028	0.038	0.045	0.058	0.066	0.081	0.090	0.099	0.104
3.05	0.000	0.000	0.000	0.000	0.000	0.000	0.016	0.027	0.036	0.043	0.056	0.064	0.078	0.086	0.096	0.101
3.06	0.000	0.000	0.000	0.000	0.000	0.000	0.015	0.025	0.034	0.041	0.053	0.061	0.075	0.083	0.092	0.097
3.07	0.000	0.000	0.000	0.000	0.000	0.000	0.014	0.024	0.032	0.039	0.051	0.059	0.072	0.080	0.089	0.094
3.08	0.000	0.000	0.000	0.000	0.000	0.000	0.013	0.022	0.030	0.037	0.049	0.056	0.069	0.077	0.086	0.091
3.09	0.000	0.000	0.000	0.000	0.000	0.000	0.012	0.021	0.029	0.036	0.046	0.054	0.067	0.075	0.083	0.088

Q_U or Q_L	Sample Size															
	3	4	5	7	10	15	25	30	35	40	50	60	85	115	175	230
3.10	0.000	0.000	0.000	0.000	0.000	0.000	0.011	0.020	0.027	0.034	0.044	0.051	0.064	0.072	0.080	0.085
3.11	0.000	0.000	0.000	0.000	0.000	0.000	0.011	0.019	0.026	0.032	0.042	0.050	0.061	0.069	0.077	0.082
3.12	0.000	0.000	0.000	0.000	0.000	0.000	0.010	0.018	0.025	0.031	0.041	0.048	0.060	0.067	0.074	0.079
3.13	0.000	0.000	0.000	0.000	0.000	0.000	0.010	0.017	0.024	0.029	0.039	0.046	0.057	0.064	0.072	0.075
3.14	0.000	0.000	0.000	0.000	0.000	0.000	0.009	0.015	0.022	0.028	0.037	0.044	0.055	0.062	0.069	0.073
3.15	0.000	0.000	0.000	0.000	0.000	0.000	0.008	0.014	0.021	0.026	0.036	0.042	0.053	0.060	0.067	0.070
3.16	0.000	0.000	0.000	0.000	0.000	0.000	0.008	0.014	0.020	0.025	0.034	0.040	0.051	0.057	0.064	0.067
3.17	0.000	0.000	0.000	0.000	0.000	0.000	0.007	0.013	0.019	0.024	0.033	0.038	0.049	0.056	0.062	0.065
3.18	0.000	0.000	0.000	0.000	0.000	0.000	0.006	0.012	0.017	0.022	0.031	0.036	0.046	0.053	0.060	0.063
3.19	0.000	0.000	0.000	0.000	0.000	0.000	0.006	0.012	0.017	0.021	0.030	0.034	0.044	0.052	0.057	0.060
3.20	0.000	0.000	0.000	0.000	0.000	0.000	0.005	0.011	0.016	0.020	0.028	0.033	0.043	0.049	0.055	0.058
3.21	0.000	0.000	0.000	0.000	0.000	0.000	0.005	0.010	0.015	0.019	0.027	0.032	0.041	0.047	0.053	0.056
3.22	0.000	0.000	0.000	0.000	0.000	0.000	0.004	0.009	0.014	0.018	0.025	0.031	0.040	0.045	0.051	0.054
3.23	0.000	0.000	0.000	0.000	0.000	0.000	0.004	0.009	0.013	0.017	0.024	0.029	0.037	0.043	0.049	0.052
3.24	0.000	0.000	0.000	0.000	0.000	0.000	0.004	0.009	0.013	0.016	0.023	0.028	0.037	0.042	0.047	0.050
3.25	0.000	0.000	0.000	0.000	0.000	0.000	0.003	0.008	0.012	0.015	0.022	0.027	0.035	0.040	0.046	0.049
3.26	0.000	0.000	0.000	0.000	0.000	0.000	0.003	0.007	0.011	0.015	0.021	0.025	0.033	0.039	0.044	0.047
3.27	0.000	0.000	0.000	0.000	0.000	0.000	0.003	0.007	0.011	0.014	0.021	0.024	0.032	0.037	0.042	0.045
3.28	0.000	0.000	0.000	0.000	0.000	0.000	0.003	0.006	0.010	0.013	0.019	0.023	0.031	0.036	0.040	0.043
3.29	0.000	0.000	0.000	0.000	0.000	0.000	0.003	0.006	0.009	0.012	0.018	0.023	0.029	0.034	0.039	0.042
3.30	0.000	0.000	0.000	0.000	0.000	0.000	0.003	0.005	0.009	0.012	0.017	0.021	0.028	0.033	0.037	0.040
3.31	0.000	0.000	0.000	0.000	0.000	0.000	0.003	0.005	0.008	0.011	0.017	0.021	0.027	0.032	0.036	0.039
3.32	0.000	0.000	0.000	0.000	0.000	0.000	0.002	0.004	0.007	0.010	0.016	0.020	0.026	0.030	0.034	0.037
3.33	0.000	0.000	0.000	0.000	0.000	0.000	0.002	0.004	0.007	0.010	0.015	0.019	0.025	0.029	0.033	0.036
3.34	0.000	0.000	0.000	0.000	0.000	0.000	0.002	0.004	0.007	0.009	0.014	0.018	0.024	0.028	0.032	0.035
3.35	0.000	0.000	0.000	0.000	0.000	0.000	0.002	0.004	0.006	0.009	0.014	0.017	0.023	0.027	0.031	0.033
3.36	0.000	0.000	0.000	0.000	0.000	0.000	0.002	0.004	0.006	0.008	0.013	0.016	0.022	0.026	0.030	0.032
3.37	0.000	0.000	0.000	0.000	0.000	0.000	0.002	0.004	0.006	0.008	0.012	0.015	0.021	0.024	0.028	0.031
3.38	0.000	0.000	0.000	0.000	0.000	0.000	0.001	0.003	0.005	0.007	0.012	0.014	0.019	0.024	0.027	0.030
3.39	0.000	0.000	0.000	0.000	0.000	0.000	0.001	0.003	0.005	0.007	0.011	0.014	0.019	0.022	0.027	0.029
3.40	0.000	0.000	0.000	0.000	0.000	0.000	0.001	0.003	0.005	0.007	0.010	0.013	0.018	0.021	0.026	0.028
3.41	0.000	0.000	0.000	0.000	0.000	0.000	0.001	0.002	0.004	0.006	0.010	0.012	0.018	0.021	0.025	0.027
3.42	0.000	0.000	0.000	0.000	0.000	0.000	0.001	0.002	0.004	0.006	0.009	0.012	0.017	0.020	0.024	0.026
3.43	0.000	0.000	0.000	0.000	0.000	0.000	0.001	0.002	0.004	0.005	0.009	0.011	0.016	0.019	0.023	0.025
3.44	0.000	0.000	0.000	0.000	0.000	0.000	0.001	0.002	0.004	0.005	0.008	0.011	0.015	0.018	0.022	0.024
3.45	0.000	0.000	0.000	0.000	0.000	0.000	0.001	0.002	0.004	0.005	0.008	0.011	0.014	0.017	0.021	0.023
3.46	0.000	0.000	0.000	0.000	0.000	0.000	0.001	0.002	0.003	0.005	0.008	0.010	0.014	0.017	0.020	0.022
3.47	0.000	0.000	0.000	0.000	0.000	0.000	0.001	0.002	0.003	0.004	0.007	0.010	0.014	0.016	0.019	0.021
3.48	0.000	0.000	0.000	0.000	0.000	0.000	0.001	0.002	0.003	0.004	0.007	0.009	0.013	0.015	0.018	0.020
3.49	0.000	0.000	0.000	0.000	0.000	0.000	0.000	0.001	0.003	0.004	0.006	0.009	0.012	0.015	0.018	0.020

Q_U or Q_L	Sample Size															
	3	4	5	7	10	15	25	30	35	40	50	60	85	115	175	230
3.50	0.000	0.000	0.000	0.000	0.000	0.000	0.000	0.001	0.002	0.003	0.006	0.008	0.012	0.014	0.017	0.019
3.51	0.000	0.000	0.000	0.000	0.000	0.000	0.000	0.001	0.002	0.003	0.006	0.008	0.011	0.014	0.016	0.018
3.52	0.000	0.000	0.000	0.000	0.000	0.000	0.000	0.001	0.002	0.003	0.006	0.007	0.010	0.013	0.016	0.017
3.53	0.000	0.000	0.000	0.000	0.000	0.000	0.000	0.001	0.002	0.003	0.005	0.007	0.010	0.013	0.015	0.016
3.54	0.000	0.000	0.000	0.000	0.000	0.000	0.000	0.001	0.002	0.003	0.004	0.007	0.010	0.012	0.014	0.015
3.55	0.000	0.000	0.000	0.000	0.000	0.000	0.000	0.001	0.002	0.003	0.004	0.006	0.009	0.012	0.014	0.015
3.56	0.000	0.000	0.000	0.000	0.000	0.000	0.000	0.001	0.001	0.002	0.004	0.006	0.009	0.011	0.013	0.014
3.57	0.000	0.000	0.000	0.000	0.000	0.000	0.000	0.001	0.001	0.002	0.004	0.005	0.008	0.011	0.012	0.013
3.58	0.000	0.000	0.000	0.000	0.000	0.000	0.000	0.001	0.001	0.002	0.003	0.005	0.008	0.010	0.012	0.013
3.59	0.000	0.000	0.000	0.000	0.000	0.000	0.000	0.001	0.001	0.002	0.003	0.005	0.008	0.010	0.011	0.012
3.60	0.000	0.000	0.000	0.000	0.000	0.000	0.000	0.001	0.001	0.002	0.003	0.005	0.007	0.009	0.011	0.012
3.61	0.000	0.000	0.000	0.000	0.000	0.000	0.000	0.001	0.001	0.002	0.003	0.004	0.007	0.009	0.011	0.011
3.62	0.000	0.000	0.000	0.000	0.000	0.000	0.000	0.001	0.001	0.002	0.003	0.004	0.007	0.009	0.010	0.011
3.63	0.000	0.000	0.000	0.000	0.000	0.000	0.000	0.001	0.001	0.001	0.003	0.004	0.006	0.008	0.010	0.010
3.64	0.000	0.000	0.000	0.000	0.000	0.000	0.000	0.001	0.001	0.001	0.003	0.003	0.006	0.008	0.009	0.010
3.65	0.000	0.000	0.000	0.000	0.000	0.000	0.000	0.001	0.001	0.001	0.003	0.003	0.006	0.008	0.009	0.010
3.66	0.000	0.000	0.000	0.000	0.000	0.000	0.000	0.000	0.000	0.001	0.003	0.003	0.005	0.007	0.009	0.009
3.67	0.000	0.000	0.000	0.000	0.000	0.000	0.000	0.000	0.000	0.001	0.003	0.003	0.005	0.007	0.008	0.009
3.68	0.000	0.000	0.000	0.000	0.000	0.000	0.000	0.000	0.000	0.001	0.002	0.003	0.005	0.006	0.008	0.008
3.69	0.000	0.000	0.000	0.000	0.000	0.000	0.000	0.000	0.000	0.001	0.002	0.002	0.004	0.006	0.008	0.008
3.70	0.000	0.000	0.000	0.000	0.000	0.000	0.000	0.000	0.000	0.001	0.002	0.002	0.004	0.006	0.007	0.008
3.71	0.000	0.000	0.000	0.000	0.000	0.000	0.000	0.000	0.000	0.001	0.001	0.002	0.004	0.006	0.007	0.007
3.72	0.000	0.000	0.000	0.000	0.000	0.000	0.000	0.000	0.000	0.001	0.001	0.002	0.004	0.006	0.007	0.007
3.73	0.000	0.000	0.000	0.000	0.000	0.000	0.000	0.000	0.000	0.001	0.001	0.002	0.004	0.006	0.007	0.007
3.74	0.000	0.000	0.000	0.000	0.000	0.000	0.000	0.000	0.000	0.001	0.001	0.002	0.004	0.005	0.006	0.007
3.75	0.000	0.000	0.000	0.000	0.000	0.000	0.000	0.000	0.000	0.001	0.001	0.002	0.003	0.005	0.006	0.006
3.76	0.000	0.000	0.000	0.000	0.000	0.000	0.000	0.000	0.000	0.001	0.001	0.002	0.003	0.005	0.006	0.006
3.77	0.000	0.000	0.000	0.000	0.000	0.000	0.000	0.000	0.000	0.001	0.001	0.002	0.003	0.005	0.006	0.006
3.78	0.000	0.000	0.000	0.000	0.000	0.000	0.000	0.000	0.000	0.000	0.001	0.002	0.003	0.004	0.005	0.005
3.79	0.000	0.000	0.000	0.000	0.000	0.000	0.000	0.000	0.000	0.000	0.001	0.002	0.003	0.003	0.005	0.005
3.80	0.000	0.000	0.000	0.000	0.000	0.000	0.000	0.000	0.000	0.000	0.001	0.002	0.003	0.003	0.005	0.005
3.81	0.000	0.000	0.000	0.000	0.000	0.000	0.000	0.000	0.000	0.000	0.001	0.002	0.003	0.003	0.005	0.005
3.82	0.000	0.000	0.000	0.000	0.000	0.000	0.000	0.000	0.000	0.000	0.001	0.002	0.003	0.003	0.005	0.005
3.83	0.000	0.000	0.000	0.000	0.000	0.000	0.000	0.000	0.000	0.000	0.001	0.002	0.003	0.003	0.004	0.004
3.84	0.000	0.000	0.000	0.000	0.000	0.000	0.000	0.000	0.000	0.000	0.001	0.001	0.002	0.003	0.004	0.004
3.85	0.000	0.000	0.000	0.000	0.000	0.000	0.000	0.000	0.000	0.000	0.001	0.001	0.001	0.002	0.004	0.004
3.86	0.000	0.000	0.000	0.000	0.000	0.000	0.000	0.000	0.000	0.000	0.000	0.001	0.001	0.002	0.004	0.004
3.87	0.000	0.000	0.000	0.000	0.000	0.000	0.000	0.000	0.000	0.000	0.000	0.001	0.001	0.002	0.004	0.004
3.88	0.000	0.000	0.000	0.000	0.000	0.000	0.000	0.000	0.000	0.000	0.000	0.001	0.001	0.002	0.004	0.004
3.89	0.000	0.000	0.000	0.000	0.000	0.000	0.000	0.000	0.000	0.000	0.000	0.001	0.001	0.002	0.003	0.003
3.90	0.000	0.000	0.000	0.000	0.000	0.000	0.000	0.000	0.000	0.000	0.000	0.001	0.001	0.002	0.003	0.003

7-5 EXAMPLES—MIL-STD-414

EXAMPLE 7-12   (Standard Deviation Method, Single-Specification Limit)

Given   A manufacturer produces a lot of 35 thermostats. The maximum temperature at which the thermostat must actuate is 100°C.

Question   Determine the acceptability of the lot using an AQL = 1.5%.

Solution

(a) Look up the specified AQL value in Table 7-16 (p. 179) and determine the AQL range and the applicable AQL value in the right-hand column. In this case, AQL = 1.5% appears in the right-hand column and is therefore used.

(b) From Table 7-17 (p. 179), determine the sample-size code letter. Since inspection level IV applies unless otherwise specified, the sample-size code letter is D.

(c) Refer to Table 7-19 (p. 181) and locate the sample-size code letter D in column 1. The corresponding sample size is read from column 2. Following this, determine the factor k for the sample-size code letter and the specified acceptable quality level (AQL). In this case, the sample size is 5 and the factor k is 1.40.

(d) Pick a random sample of five thermostats from the lot and measure and record the actuation temperature. Assume that the five values were 95°, 99°, 100°, 97°, and 98°.

(e) Compute the mean $\bar{X}$ and the estimated standard deviation s:

$$\bar{X} = \frac{\sum X}{n} = \frac{95 + 99 + 100 + 97 + 98}{5} = 97.8, \text{ say } 98.$$

$$s = \sqrt{\frac{\sum (X - \bar{X})^2}{n - 1}}$$

X	$X - \bar{X}$	$(X - \bar{X})^2$
95	−3	9
99	1	1
100	2	4
97	−1	1
98	0	0

$$\sum (X - \bar{X})^2 = 15 \qquad s = \sqrt{\frac{15}{4}} = 1.936$$

(f) Determine the number of standard deviations Q_U that the mean $\bar{X}$ is away from the upper specification limit U:

$$Q_U = \frac{U - \bar{X}}{s} = \frac{100 - 98}{1.936} = 1.033$$

(g) Compare the Q_U value to the factor k which was determined in step 3. If the value of Q_U is equal or greater than k, the lot is acceptable.

In our case, $Q_U = 1.033$ and $k = 1.40$; hence, we reject the lot. Note, that the lot is rejected even though none of the 5 samples exceed the specified 100°C limit. The reason is the inherent variability of the production lot.

EXAMPLE 7-13 (Standard Deviation Method, Single-Specification Limit).

Given Use the same specification as for Example 7-12, except that instead of the maximum actuation temperature of 100°C, a minimum actuation temperature of 95°C applies. The AQL and the readings of the five sample units are likewise unchanged.

Question Determine the acceptability of the lot, using an AQL = 1.5%.

Solution As before, $n = 5$, $k = 1.4$, $s = 3.75$, and $\bar{X} = 97.8$, say 98

$$Q_L = \frac{\bar{X} - L}{s} = \frac{98 - 95}{1.936} = 1.549$$

Compare Q_L to k. If Q_L is equal to or greater than k, accept the lot.
 Since 1.549 is greater than 1.4, the lot is acceptable.

EXAMPLE 7-14 (Standard Deviation Method, Double Specification Limit).

Given A lot of 40 transformers has a critical specification for excitation current. The latter must fall between 12 and 16 milliamperes. An AQL = 2% and inspection level IV applies.

Question Should the lot be accepted?

Solution

   (a) Look up the AQL range in Table 7-16 (p. 179) and determine the associated AQL value. AQL = 2.5, from the right-hand column.
   (b) From Table 7-17 (p. 179), determine the sample-size code letter. For inspection level IV, the code letter is D.
   (c) Refer to Table 7-20 (p. 182) and read off sample size $n = 5$ for sample-size code letter D. Then determine factor M for AQL = 2.50 and sample-size code letter D. The value is $M = 9.80$.
   (d) Pick a random sample of 5 transformers and measure and record their excitation current. Assume that the five readings were 15, 16, 13, 14, and 15 milliamperes.
   (e) Compute the mean $\bar{X}$ and the estimated standard deviation

$$\bar{X} = \frac{\sum X}{n} = \frac{15 + 16 + 13 + 14 + 15}{5} = 14.6$$

$$s = \sqrt{\frac{\sum (X - \bar{X})^2}{n - 1}}$$

X	$X - \bar{X}$	$(X - \bar{X})^2$
15	0.4	0.16
16	1.4	1.96
13	−1.6	2.56
14	−0.6	0.36
15	0.4	0.16

$$\sum (X - \bar{X})^2 = 5.20$$

$$s = \sqrt{\frac{5.20}{4}} = 1.14$$

(f) Determine Q_U and Q_L.

$$Q_U = \frac{U - \bar{X}}{s} = \frac{16 - 14.6}{1.14} = 1.23$$

$$Q_L = \frac{\bar{X} - L}{s} = \frac{14.6 - 12}{1.14} = 2.28$$

(g) Refer to Table 7-21 (p. 183) and determine P_U, the estimated lot percent defective above U, and P_L, the estimated lot percent defective below L for the sample size $n = 5$.

$$\text{For } Q_U = 1.23, P_U = 9.97$$
$$\text{For } Q_L = 2.28, P_L = 0.00$$

(h) The total estimated percent defective in the lot is

$$p = P_U + P_L$$
$$= 9.97 + 0.00 = 9.97$$

(i) Compare p to M. The lot meets the acceptability criterion when p is equal to or less than M.

In our case, $p = 9.97$ and $M = 9.80$; hence, the lot is to be rejected.

EXAMPLE 7-15 (Range Method, Single-Specification Limit). Repeat Example 7-12, except use the range method.

Solution

(a) In Example 7-12 the highest and lowest actuation temperature in the sample drawn was 100° and 95°, respectively. Therefore, range $R = 5$. Since there is only one subgroup of 5 units, the average range of the sample is $\bar{R} = 5$.

(b) Determine Q_U.

$$Q_U = \frac{U - \bar{X}}{\bar{R}} = \frac{100 - 98}{5} = 0.4$$

(c) From Table 7-22 (p. 192), determine the value of factor k for the particular sample-size code letter and AQL.

$$\text{sample-size code letter} = D$$
$$\text{AQL} = 1.5\%$$
$$k = 0.565$$

(d) If Q_U is greater than k, the lot is acceptable. Since $Q_U = 0.4$ and $k = 0.565$ the lot is rejected.

EXAMPLE 7-16 (Range Method, Double-Specification Limit). Repeat Example 7-14, except use the range method.

Solution

(a) In Example 7-14 the highest and lowest excitation current readings were 16 and 12 mA, respectively. Therefore,

range $R = 4$. Since there is only one subgroup of 5 units, the average range of the sample is $R = 4$.

(b) Find constant c from Table 7-23 (p. 193) for sample-size code letter D:

$$c = 2.474$$

Also, determine constant M from the same table for AQL $= 2.50$:

$$M = 9.90$$

(c) Determine values for Q_U and Q_L. From Example 7-14, $U = 16$, $L = 12$, $\bar{X} = 14.6$

$$Q_U = \frac{(U - \bar{X})c}{\bar{R}} = \frac{(16 - 14.6)2.474}{4} = 0.87$$

$$Q_L = \frac{(\bar{X} - L)c}{\bar{R}} = \frac{(14.6 - 12)2.474}{4} = 1.61$$

(d) Determine the total estimated percent defective:

$$p = P_U + P_L$$

where P_U and P_L, corresponding to the computed values of Q_U and Q_L, are determined from Table 7-24 (p. 194) for the sample size $n = 5$.

$$P_U = 20.31$$
$$P_L = 1.87$$

Therefore, $p = 20.31 + 1.87 = 22.18$

(e) If p is equal to or smaller than M, the lot is accepted.

Since p is 22.18 and M is 9.90, the lot is rejected.

EXAMPLE 7-17 (Standard Deviation Method, Single Specification Limit)

Given   A lot of 65 special jigs are to be subjected to inspection by variables to an AQL $= 2\%$. The angle between two positioning fingers of this jig is important and is limited to a maximum of 35°.

Question Draw the necessary size sample and determine whether the readings obtained from this sample warrant acceptance of the lot of 65 units.

Solution

(a) Per Table 7-16 (p. 179), the specified AQL = 2% falls within the range of AQL values from 1.65 to 2.79 (see the left column of the table). The corresponding AQL value to be used, from the right-hand column of the table, is AQL = 2.5.

(b) Refer to Table 7-17 (p. 179). Since no specific inspection level was specified, inspection level IV will be used. For a lot size of 65 pieces and inspection level IV, the table shows sample size code letter E.

(c) Turn to Table 7-19 (p. 181) and determine the sample size n for sample-size code letter E and constant k for the applicable AQL value. The size of the sample $n = 7$. The constant k for sample-size code letter E and AQL = 2.5 is $k = 1.33$.

(d) Since the size of the sample is $n = 7$, choose 7 units randomly from the production lot of 65 pieces and measure and record the quality characteristic of interest. Let us assume that the critical angles between the positioning fingers of the jig as determined on the 7 units are as follows:

$$20°, 25°, 29°, 28°, 23°, 24°, \text{ and } 26°$$

(e) Compute the mean $\bar{X}$ and the estimated deviation.

$$\bar{X} = \frac{\sum X}{n} = \frac{20 + 25 + 29 + 28 + 23 + 24 + 26}{7}$$

$$\bar{X} = 25$$

$$s = \sqrt{\frac{\sum (X - \bar{X})^2}{n - 1}}$$

X	$X - \bar{X}$	$(X - \bar{X})^2$
20	−5	25
25	0	0
29	4	16
28	3	9
23	−2	4
24	−1	1
26	1	1
		$\sum (X - \bar{X})^2 = 56$

$$s = \sqrt{\frac{56}{6}} = 3.05$$

(f) Determine the number of standard deviations Q_U that the mean $\bar{X}$ is away from the upper specification limit U.

$$Q_U = \frac{U - \bar{X}}{s} = \frac{35 - 25}{3.05} = 3.28$$

(g) Compare the computed Q_U value to the factor k determined in step 4. If the value of Q_U is equal to or greater than k, the lot is acceptable.

In this case, $Q_U = 3.28$ and $k = 1.33$. Hence, Q_U exceeds k, and the lot of 65 jigs may be accepted.

EXAMPLE 7-18   (Standard Deviation Method, Double Specification Limit).

Given   A production lot of one hundred 150-W amplifiers is to be subjected to inspection by variables. The acceptable quality level is AQL $= 0.65\%$. The quality characteristic to be inspected is power output in watts for a constant voltage input of 0.5-V rms 400 hertz (Hz). The required output wattage range specified is 145 to 155 W.

Question   Draw the necessary size sample and determine the amplification of each unit so that the acceptability of the entire production lot may be determined. Inspection level IV is to be used.

Solution

(a) Per Table 7-16 (p. 179), AQL $= 0.65$ appears in the right-hand column of the table and is therefore used.
(b) Per Table 7-17 (p. 179), for a lot size of 100 and inspection level IV, the sample size code letter is F.
(c) Refer to Table 7-20 (p. 182) and read off sample size $n = 10$ for sample-size code letter F. Then determine constant M for AQL $= 0.65$ and sample-size code letter F.

$$M = 2.17$$

(d) Output wattage readings on 10 amplifiers, drawn randomly, were found to be 151, 153, 150, 154, 152, 147, 149, 150, 155, and 149 W.
(e) The mean $\bar{X}$ and the standard deviation are

$$\bar{X} = \frac{151 + 153 + 150 + 154 + 152 + 147 + 149 + 150 + 155 + 149}{10}$$

$$= 151$$

X	$X - \bar{X}$	$(X - \bar{X})^2$
151	0	0
153	2	4
150	-1	1
154	3	9
152	1	1
147	-4	16
149	-2	4
150	-1	1
155	4	16
149	-2	4
		$\sum (X - \bar{X})^2 = \overline{56}$

$$s = \sqrt{\frac{\Sigma (X - \bar{X})^2}{n - 1}}$$

$$= \sqrt{\frac{56}{10 - 1}} = 2.5$$

(f) Determine Q_U and Q_L.

$$Q_U = \frac{U - \bar{X}}{s} = \frac{155 - 151}{2.5} = 1.6$$

$$Q_L = \frac{\bar{X} - L}{s} = \frac{151 - 145}{2.5} = 2.4$$

(g) Refer to Table 7-21 (p. 183) and determine P_U and P_L for $n = 10$ and the computed Q_U and Q_L values.

$$P_U = 4.54$$

$$P_L = 0.109$$

(h) The total estimated percent defective in the lot is

$$p = P_U + P_L$$

$$= 4.54 + 0.109 = 4.649$$

(i) Compare p to M; p exceeds the value of M. Since the lot meets the acceptability criterion when p is equal to or less than M, this lot has to be rejected.

EXAMPLE 7-19 (Range Method, Double Specification Limit).

Given The required overall length of a piston rod is 40 centimeters (cm). The latter is the critical quality characteristic and is specified with a tolerance of $\pm 2\%$. A total of 100 units are subject to inspection by variables with an AQL $= 3\%$. Inspection level IV applies.

Question After drawing the necessary size sample and measuring the lengths of the piston rods, determine whether to accept or reject the whole lot.

Solution

(a) The minimum length of the piston rod is 39.2 cm; the maximum, 40.8 cm.
(b) Per Table 7-16 (p. 179), the specified AQL $= 3\%$ falls in the range 2.80 to 4.39, with an equivalent AQL $= 4\%$ (see the right-hand column in the table).
(c) Also, from Table 7-17, the sample-size code letter for a lot size of 100 and an inspection level IV is F.
(d) From Table 7-23 (p. 193) the sample size is 10 and the value of M, determined for sample-size code letter F and AQL $= 4\%$, is $M = 10.79$.
(e) Draw 10 sample units randomly from the lot of 100 units. Assume that the 10 sample units rendered the following piston rod lengths (2 subgroups of 5 measurements are listed in the order taken):

Subgroup 1	Subgroup 2
39.5 cm	40.1 cm
39.4 cm	38.3 cm
39.6 cm	39.7 cm
39.7 cm	39.8 cm
40.0 cm	39.6 cm

(f) Determine the average $\bar{X}$.

$$\bar{X} = \frac{39.5 + 39.4 + 39.6 + 39.7 + 40.0 + 40.1 + 38.3 + 39.7 + 39.8 + 39.6}{10}$$

$$= 39.57$$

(g) Determine the average range $\bar{R}$: There are two subgroups of five readings each. The range of subgroup 1 covers 39.5, 39.4, 39.6, 39.7, and 40. Hence, $R = 40 - 39.4 = 0.6$. The range of subgroup 2 covers 40.1, 38.3, 39.7, 39.8, and 39.6. Hence $R = 40.1 - 38.3 = 1.8$, and the average range of the sample is:

$$\bar{R} = \frac{0.6 + 1.8}{2} = 1.2$$

(h) Determine constant c from Table 7-23 (p. 193) for sample-size code letter F:

$$c = 2.405$$

(i) Determine values for Q_U and Q_L:

$$Q_U = \frac{(U - \bar{X})c}{\bar{R}} = \frac{(40.8 - 39.57)2.405}{1.2} = 2.46$$

$$Q_L = \frac{(\bar{X} - L)c}{\bar{R}} = \frac{(39.57 - 39.2)2.405}{1.2} = 0.74$$

(j) Determine the total estimated percent defective.

$$p = P_U + P_L$$

From Table 7-24 (p. 194), the following values of P_U and P_L (based on the corresponding values of Q_U and Q_L) are determined:

$$P_U = 0.005$$
$$P_L = 23.54$$
$$p = 23.545$$

(k) The criterion for lot acceptance is a comparison of p and M. If p is equal to or smaller than M, the lot is to be accepted.

Since $p = 23.545$ and $M = 10.79$, the lot is rejected.

PROBLEMS

7-1 List several of the areas of applications of MIL-STD-105D sampling plans.

7-2 Define defects and defectives.

7-3 (a) State the definition given by MIL-STD-105D for AQL.
    (b) Define lot or batch.

7-4 What are the rules for changing
    (a) From normal to tightened inspection?
    (b) From tightened to normal inspection?
    (c) From normal to reduced inspection?
    (d) From reduced to normal inspection?

7-5 (a) Discuss the inspection levels of MIL-STD-105D.
    (b) What are the three types of sampling plans offered by MIL-STD-105D?

7-6 Discuss the basic differences between the three types of plans referred to in 7-5(b).

7-7 What information is given by average-sample-size curves?

7-8 (a) What is limiting quality protection?
    (b) Give values of commonly used risks in connection with LQ plans.

7-9 A manufacturer submits a production lot of 1000 pieces for inspection by attributes to an AQL $= 1.5\%$. Single sampling and normal inspection, level II, apply.
    (a) What is the sample size?
    (b) What are the accept/reject criteria?
    (c) What are the accept/reject criteria for tightened inspection?

7-10 A lot of 2000 resistors is subjected to sampling by attributes. Single sampling, normal inspection, and an AQL $= 1.5\%$ applies. Determine the sampling plan per MIL-STD-105D and identify the accept/reject criteria.

7-11 Parts for a contract are bought in lots of 2000 pieces. What AQL must be specified so that lots are rejected when 4 or more defective parts are found in the sample drawn? Attribute sampling, general inspection level II, and single sampling, and normal inspection apply.

7-12 Refer to Problem 7-11. Suppose that seven consecutive lots of 2000 pieces each were subjected to inspection. Lots 1 through 3 were accepted, lots 4 and 5 were rejected, and lots 6 and 7 were accepted. What were the accept and reject numbers for each of the seven lots?

7-13 A production lot of 1250 units is subjected to attribute sampling. General inspection level II, single sampling and normal inspection apply. The AQL $= 1.0\%$. If the production lot contains 3% defectives, what is the probability that the production lot will be accepted by the sampling plan?

7-14 A production lot of 1000 units is subjected to attribute sampling. A single-sampling plan, normal inspection, and general inspection level III apply. The AQL $= 0.40\%$. If the probability that the production lot will be accepted is 70%, what is the quality of the production lot, expressed in percent defectives?

7-15 A double-sampling plan, using normal inspection, inspection level II, with an AQL $= 1.5\%$ is applied to a production lot of 2850 units. The first sample drawn contained 3 defective units. Assuming that the second sample contains 5 defective units, what decision as to the acceptance or rejection of the production lot is to be made?

7-16 (a) Discuss the acceptability criterion as delineated by MIL-STD-414.
    (b) What is the range of acceptable quality levels provided by MIL-STD-414?
    (c) How is the sample size determined by MIL-STD-414?
    (d) If the specified AQL value is 7.5%, what AQL value is used for a MIL-STD-414 sampling plan?

7-17 A production lot of 100 spring devices is subjected to inspection by variables. Variability is not known, AQL $= 1.5\%$, and the single-specification limit applies. The length of the device, fully compressed, is the critical quality characteristic and should not exceed 0.75 in. A sample of 10 units was drawn randomly. The quality characteristic was measured and the following readings recorded:

Unit No.	Length Measured (in)
1	0.65
2	0.66
3	0.68
4	0.70
5	0.69
6	0.72
7	0.64
8	0.67
9	0.69
10	0.70

(a) Was the size of the sample correct?

(b) Should the lot of 100 devices be accepted or rejected?

7-18 Inspection by variables is applied to a production lot of 39 precision heating elements. With the precise voltage applied, the temperature developed by the heating elements must not be less than 146°C and not more than 152°C. Variability is unknown, $AQL = 0.65\%$, and a double-specification limit applies. A sample of 5 units, drawn randomly, furnishes the following temperature readings:

Unit No.	Temperature (°C)
1	149
2	148
3	150
4	149
5	152

Should the production lot of 39 heating elements be accepted? Compute the result using the Standard Deviation Method.

7-19 Determine the average outgoing quality limit (AOQL) for production lots of 5000 units each. The AQL is 4% and a single-sampling plan is used.

7-20 What is the average sample number (ASN) for single sampling, double sampling, and multiple sampling for production quantities of 850 units each with an $AQL = 1.0\%$.

7-21 Using an inspection plan by variables per MIL-STD-414, determine the acceptability of a production lot of 65 miniature devices. The weight of these units is critical and must be limited to 85 grams. The $AQL = 4\%$ and inspection level IV applies. Seven sample units are drawn randomly, and the following weights recorded: 65 g, 68 g, 70 g, 75 g, 80 g, 67 g, and 79 g.

7-22 Refer to Problem 7-21. Determine the acceptability of the production lot if the criterion of a critical weight of 85 g maximum is changed to a minimum weight of 65 g.

7-23 The sensitivity of an electronic device must be such as to assure at least 10-mW output for an input signal voltage range from 75 to 85 mV. A MIL-STD-414 inspection plan by variables is applied to a production lot of 100 units. The AQL = 1.5% and inspection level IV is used. A sample lot of 10 units is drawn randomly and tested. The input signal level applied to each unit is raised slowly until the minimum wattage of 10 mW is produced on the output. The following readings are recorded:

Unit No.	Output (mW)	Input (mV)
1	10	81
2	10	79
3	10	80
4	10	80
5	10	75
6	10	79
7	10	83
8	10	75
9	10	81
10	10	79

Should the lot of 100 units be accepted?

7-24 A production lot of 40 units is subjected to a MIL-STD-414 inspection plan by variables. The AQL = 4.0% and inspection level IV applies. The quality characteristic of interest is the speed of the device in feet per second. The maximum speed is 15 ft/s. Readings on 5 units were as follows:

Unit No.	Speed (ft/s)
1	12
2	11
3	14
4	10
5	15

Should the lot of 40 units be accepted?

8

PRODUCT RELIABILITY

8-1 BACKGROUND AND IMPORTANCE OF PRODUCT RELIABILITY

In recent decades the complexity of equipments used for military, space, and commercial applications has increased manifold. By the time of the Korean war we were faced with a real problem of electronic equipment failure on the front, because of the equipment's inability to withstand the rigors of field operation. The United States and Russia have also had costly failures in space which were due to malfunctions of electronic component parts. In commercial applications, such as computers, data processing, automated production processes, communication networks, medical electronics, and so on, we have also pushed equipment complexity far beyond expectation. The more complex a piece of equipment, the less reliable it is and the more likely it is that one or more of its parts may fail.

Military electronic gear which fails, when needed, may adversely affect the outcome of an engagement and cost lives. Downtime on commercial computers or process control devices could cause extensive financial losses, while failures in medical electronics or safety equipments could endanger life.

In 1952 the Department of Defense created an advisory group dealing with the reliability of electronic equipment. Its studies and reports formed the foundation for today's product reliability technology.

8-2 DEFINITION OF RELIABILITY

The definition for reliability encompasses four basic elements:

1. Probability.

2. Performance.
3. Time of operation.
4. Operating conditions.

A definition of reliability is as follows: *Product reliability is the probability of a product performing its intended function over its intended life and under the operating conditions encountered.* The term "probability" is quantitative, since it is expressed in percent or fraction. To say, for example, that the probability of survival of an equipment is 85% for 200 h of operation means that, on the average, 85 of 100 such pieces of equipments will last for 200 h.

8-3 TIME DEPENDENCY

Our definition for reliability involves a time constraint. This is not unusual. No product lasts forever; therefore, its reliability, under fully specified conditions of use, should be defined in terms of time. A manufacturer could, for example, state that his product is guaranteed to be reliable for one year's continuous use when operated as prescribed, or intended. Obviously, before a manufacturer would be willing to give such a guarantee he would want to be reasonably certain that the chances that his product will last one year are indeed very good—that is, that the probability that the product will survive for 1 year is indeed high. Such confidence in the product may stem from past experience or tests conducted. In any event, the manufacturer is willing to assume the risk that a small number of units might still fail in use and that he would consequently have to make good for them.

8-4 RELATIONSHIP TO QUALITY CONTROL

The definition of product reliability clearly links the probability of product survival under stated operating conditions to a defined period of time. Quality control is concerned with conformance of the product to specifications. The distinction between reliability and quality control was discussed in greater detail in Section 4.5 (see p. 68).

8-5 ACHIEVING RELIABILITY

(a) Difficulty of Achieving Reliability

The difficulty of achieving high reliability in modern electronic equipment is due to the complexity and sophistication of the equipment and the requirement that large numbers of parts must work without malfunctioning for long periods of time. In attempting to improve a product's reliability we could overdesign by using parts having higher ratings and greater safety margins, but this would enormously increase size, weight, and cost of the equipment. Another difficulty often encountered is that data concerning the reliability of parts under the desired conditions of use are not always available. Trade-offs among size, weight, cost, and reliability are frequently necessary to strike at the proper compromise. In many instances, reliability verification tests are conducted after the equipment is built to check on the closeness of the initial estimate.

(b) Estimation of Reliability

A preliminary reliability prediction, using realistic failure rates for the parts in the equipment, is an important initial step in the design of the product. It furnishes the designer an estimate of the theoretically achievable reliability and points up areas of high-failure-rate concentration in the equipment.

The latter may be eliminated by judicious use of redundancy techniques or by reducing the circuit complexity. The reliability prediction is then repeated to ascertain whether the reliability goal has now been met. Several parts, circuit, or configuration changes may be necessary in this process.

(c) Redundancy

One of the techniques that can reduce the hazard of high-failure-rate concentrations is redundancy. Using this method we can, for example, assemble two parts in parallel, such that the failure of one part will not cause failure of the equipment, as there would still be one part left to perform the necessary function. Redundancy is, however, not always the panacea to equipment reliability, since two (or more) parts are needed to perform the function of one part, with an accompanying increase in size, cost, and weight. Also, we need to consider the following: If two parts are connected in parallel and one fails by burning open, there is no interference with the action of the remaining good part; if, on the other hand, the failure of the one part is a short, it will also interfere with the functioning of the alternate part. In cases where this kind of failure mode is anticipated, some switching device has to be incorporated to remove the failed component. If the reliability requirement for the switching device is now considered in addition to the redundant part, the advantage of the redundancy scheme may be lost.

As is probably apparent by now, a reliable design does not just happen but requires careful planning and trade-off studies. Redundancy techniques will be discussed in more detail later.

(d) Parts Engineering

1. Choosing, Specifying, and Procuring Parts

The reliability of a system depends largely on the reliability of its parts. A part failure may precipitate a systems failure, or it may not be critical to system success; it depends on the specific circuit application and circuit function. There are military specifications for the great majority of electronic and electromechanical parts. These parts are built to rigid specifications and are tested rigorously prior to shipment.

The choice of parts must be responsive to the requirements of the particular application. Merely to select reliable parts will not guarantee a reliable system; the application of the parts in the circuit is of prime importance. Certain constraining factors need to be considered, such as electrical stress, temperature range and other environmental conditions, and part tolerances and drift characteristics. Table 8-1 shows failure rates of fixed-film resistors per MIL-R-10509 and MIL-R-55182 as influenced by electrical stress (ratio of operating to rated wattage) for various ambient temperatures.

Table 8-1 *Failure Rates λ_b (failures/10^6 h) for MIL-R-10509 and MIL-R-55182 Fixed-Film Resistors. (Source: MIL-HDBK-217B, Table 2.5.2-5.)*

T (°C)	RATIO OF OPERATING TO RATED WATTAGE									
	0.1	0.2	0.3	0.4	0.5	0.6	0.7	0.8	0.9	1.0
0	0.0012	0.0013	0.0014	0.0016	0.0018	0.0020	0.0022	0.0024	0.0027	0.0029
10	0.0013	0.0014	0.0016	0.0018	0.0020	0.0022	0.0024	0.0027	0.0030	0.0033
20	0.0014	0.0016	0.0018	0.0020	0.0022	0.0025	0.0027	0.0031	0.0034	0.0038
30	0.0016	0.0017	0.0020	0.0022	0.0025	0.0027	0.0031	0.0034	0.0038	0.0043
40	0.0017	0.0019	0.0022	0.0024	0.0027	0.0031	0.0034	0.0039	0.0044	0.0049
50	0.0019	0.0021	0.0024	0.0027	0.0030	0.0034	0.0039	0.0044	0.0049	0.0055
55	0.0020	0.0022	0.0025	0.0028	0.0032	0.0036	0.0041	0.0046	0.0052	0.0059
60	0.0021	0.0023	0.0026	0.0030	0.0034	0.0038	0.0043	0.0049	0.0056	0.0063
65	0.0022	0.0025	0.0028	0.0032	0.0036	0.0041	0.0046	0.0052	0.0059	0.0067
70	0.0023	0.0026	0.0029	0.0033	0.0038	0.0043	0.0049	0.0055	0.0063	0.0071
75	0.0024	0.0027	0.0031	0.0035	0.0040	0.0045	0.0052	0.0059	0.0067	0.0076
80	0.0025	0.0028	0.0032	0.0037	0.0042	0.0048	0.0055	0.0062	0.0071	0.0081
85	0.0026	0.0030	0.0034	0.0039	0.0044	0.0051	0.0058	0.0066	0.0075	0.0086
90	0.0027	0.0031	0.0036	0.0041	0.0047	0.0054	0.0061	0.0070	0.0080	0.0092
95	0.0029	0.0033	0.0038	0.0043	0.0049	0.0057	0.0065	0.0074	0.0085	0.0097
100	0.0030	0.0034	0.0040	0.0045	0.0052	0.0060	0.0069	0.0079	0.0090	0.010
105	0.0031	0.0036	0.0042	0.0048	0.0055	0.0063	0.0073	0.0084	0.0096	0.011
110	0.0033	0.0038	0.0044	0.0050	0.0058	0.0067	0.0077	0.0089	0.010	0.011
115	0.0034	0.0040	0.0046	0.0053	0.0061	0.0071	0.0082	0.0094	0.010	0.012
120	0.0036	0.0042	0.0048	0.0056	0.0065	0.0075	0.0086	0.010	0.011	0.013
125	0.0038	0.0044	0.0051	0.0059	0.0068	0.0079	0.0091	0.010	0.012	0.013
130	0.0040	0.0046	0.0053	0.0062	0.0072	0.0083	0.0097	0.011	0.013	
135	0.0041	0.0048	0.0056	0.0065	0.0076	0.0088	0.010	0.011	0.013	
140	0.0043	0.0051	0.0059	0.0069	0.0080	0.0093	0.010	0.012		
145	0.0046	0.0053	0.0062	0.0072	0.0084	0.0098	0.011	0.013		
150	0.0048	0.0056	0.0065	0.0076	0.0089	0.010	0.012			
155	0.0050	0.0058	0.0069	0.0080	0.0094	0.011	0.012			
160	0.0052	0.0061	0.0072	0.0084	0.0099	0.011	0.013			
165	0.0055	0.0064	0.0076	0.0089	0.010	0.012				
170	0.0057	0.0068	0.0080	0.0094	0.011	0.013				
175	0.0060	0.0071	0.0084	0.0099	0.011	0.013				

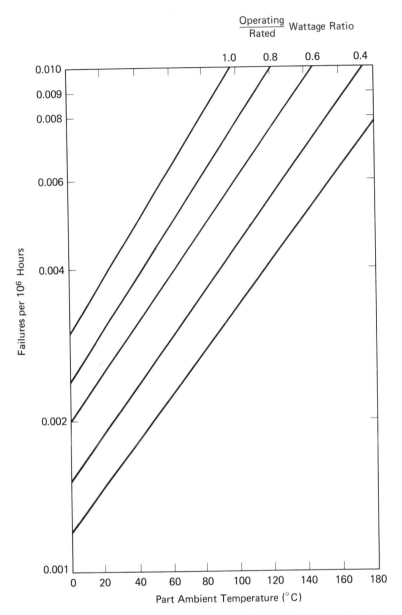

Figure 8-1 *Plot of base failure rates (in failures per 10⁶ h) for typical stress values for fixed film resistors per MIL-R-10509 and MIL-R-55182.*

Figure 8-1 is a plot of these data for typical stress values. Table 8-2 is a similar tabulation for MIL-C-11015 and MIL-C-39014 ceramic capacitors. Table 8-3 represents typical values of initial and end-of-life tolerances for several resistors and capacitors. It is important that parts be selected which have a known history of good performance in similar applications and that adequate controls for the manufacture, quality con-

Table 8-2 *Base Failure Rates λ_b (for T = 85°C max. rated)* (failures/10^6 h) for MIL-C-11015 and MIL-C-39014 Fixed-Ceramic Capacitors. (Source: MIL-HDBK-217B Table 2.6.4-3.)*

T (°C)	\multicolumn{10}{c}{S, RATIO OF OPERATING TO RATED VOLTAGE}

T (°C)	0.1	0.2	0.3	0.4	0.5	0.6	0.7	0.8	0.9	1.0
0	0.0019	0.0024	0.0038	0.0064	0.010	0.017	0.026	0.038	0.053	0.072
5	0.0020	0.0025	0.0038	0.0065	0.010	0.017	0.026	0.038	0.054	0.073
10	0.0020	0.0025	0.0039	0.0066	0.011	0.017	0.026	0.039	0.054	0.074
15	0.0020	0.0025	0.0039	0.0067	0.011	0.017	0.027	0.039	0.055	0.075
20	0.0020	0.0026	0.0040	0.0068	0.011	0.018	0.027	0.040	0.056	0.076
25	0.0021	0.0026	0.0040	0.0068	0.011	0.018	0.028	0.040	0.057	0.077
30	0.0021	0.0026	0.0041	0.0069	0.011	0.018	0.028	0.041	0.058	0.078
35	0.0021	0.0027	0.0042	0.0070	0.011	0.018	0.028	0.042	0.058	0.080
40	0.0022	0.0027	0.0042	0.0071	0.012	0.019	0.029	0.042	0.059	0.081
45	0.0022	0.0028	0.0043	0.0072	0.012	0.019	0.029	0.043	0.060	0.082
50	0.0022	0.0028	0.0043	0.0073	0.012	0.019	0.030	0.043	0.061	0.083
55	0.0023	0.0028	0.0044	0.0074	0.012	0.020	0.030	0.044	0.062	0.084
60	0.0023	0.0029	0.0045	0.0076	0.012	0.020	0.030	0.045	0.063	0.085
65	0.0023	0.0029	0.0045	0.0077	0.012	0.020	0.031	0.045	0.064	0.087
70	0.0024	0.0030	0.0046	0.0078	0.013	0.020	0.031	0.046	0.064	0.088
75	0.0024	0.0030	0.0047	0.0079	0.013	0.021	0.032	0.046	0.065	0.089
80	0.0024	0.0030	0.0047	0.0080	0.013	0.021	0.032	0.047	0.066	0.090
85	0.0025	0.0031	0.0048	0.0081	0.013	0.021	0.033	0.048	0.067	0.092

* Applicable to styles CKR 13, 48, 64, 72 of MIL-C-39014. Applicable to A-rated temperature of MIL-C-11015, as shown in type designation (e.g., CK61AW222M).

Table 8-3 *Typical Initial and End-of-Life Design Tolerances for Resistors and Capacitors.*

	Initial Tolerance	End-of-Life Tolerance
Resistors		
Composition, carbon (RCR)	±5%	±20%
Film:		
Insulated (RLR)	±2%	±7%
High stability (RNC)	±1%	±2%
Power (RD)	±2%	±7%
Wire-wound:		
Accurate (RBR)	±0.1%	±0.3%
Power (RW)	±10%	±15%
Capacitors		
Ceramic (characteristic X)	±10%	+25%, −35%
Aluminum (35 MF 100 V dc)	−10%, +50%	−30%, +75%
Tantalum, solid	±10%	+10%, −20%
Tantalum, foil	±20%	+20%, −50%
Mica (purchase tolerance J)	±5%	±7%
Glass	±2%	±4%
Paper	±10%	±25%

trol, and testing exist. The procurement of the parts should be based on specifications that detail the performance requirements, environmental conditions, tolerances, and test requirements.

For military procurement there are standard MIL specifications for a vast variety of electrical, electronic, electromechanical, and mechanical components which adequately define the requirements for parts discussed above. In recent years the "Established Reliability Specifications" for military parts have been introduced which require the manufacturer to prove the reliability level of the parts by extended life test and statistical evaluation of the results. Several levels of reliability may be proved by the manufacturer in succession provided he continues his program of life testing. Failure rate levels of 5.0 to 0.001 %/1000 h may be certified.

Parts produced to MIL specifications require qualification inspection. Qualification inspection consists of subjecting groups of devices to various electrical, mechanical, and environmental tests. Table 8-4 shows a typical qualification inspection for transformers and inductors.

Upon successful completion of these tests, the vendor's name is placed on a qualified products list (QPL). Procurement of parts from vendors on the QPL is usually mandatory on military contracts.

When parts are required for which no MIL specifications are in existence, it is usually necessary to prepare control drawings with adequate definition of the performance, tests, and qualification requirements. The purchase of parts merely by vendor part number is not desirable since no controls exist and the vendor may institute changes in the manufacture, quality, or performance of the parts without making special announcement thereof. For his own protection, the equipment manufacturer should subject all procured lots of parts to an incoming sampling inspection test.

Great effort has been expended by both industry associations and the military to limit and standardize sizes and tolerances of parts. For example, nearly all manufacturers of capacitors and resistors abide by standard values. Many government procuring agencies stipulate in their contracts that parts be chosen from standard parts lists and require that requests for deviation from this list be submitted for review. Approval for the use of the desired nonstandard part is only granted when it is established that no standard part is available that could be substituted. In the commercial equipment field, established reliability parts, or part procurement to MIL specifications, although just as desirable, may be too costly for the competitive market.

2. Parts Derating

When parts are used at their full rating (e.g., a 1-W resistor used in a 1-W application) they are not as reliable as derated parts. To illustrate the above further, a 400-V capacitor will operate quite satisfactorily in a 400-V application, but it will probably last several times as long in a 200-V application. Similarly, resistors, transformers, transistors, and other electronic parts last longer when operated at lower power or voltage levels than those they are rated for. As before, there are again trade-offs among size, weight, and cost and increased parts reliability. Table 9-13 gives typical derating factors for a group of part types.

3. Part Burn-in

All equipments go through an initial phase which exhibits an excessive number of early failures due to weak parts and manufacturing deficiencies. This is the equipment debugging phase and is characterized by a decreasing failure rate. That is, as more and more of the weak parts drop out, the failure rate of the remaining parts in the equipment becomes progressively lower.

The function of part burn-in is to expose parts to an extended test for the purpose of detecting and removing those units which have some inherent weakness. These parts would probably have quickly failed in the equipment following turn-on. The burn-in test for parts eliminates most of the early parts failures by operating the parts for a period of time (168 h in many instances) under conditions simulating, or sometimes exceeding, the intended use. Electrical measurements are performed on each part before and after the burn-in test and if, as a result of the exposure, certain specified characteristics or parameters degrade beyond the limits of the specification, this is interpreted as an indication of a potentially weak or defective part. This part is eliminated from further use, as are parts that fail catastrophically during the burn-in test.

(e) Environmental Factors

1. Environmental Stresses

Both the mechanical and electrical design must take into account the expected environment in which the equipment will operate. This includes the environmental conditions resulting from the anticipated geographical and physical location of the equipment, such as the prevailing temperature, humidity, altitude, solar radiation, and so on. To these are added the stresses due to the particular use and application of the equipment, such as shock and vibration, internally generated heat, and similar factors. As an example, electronic and mechanical components in a missile must withstand severe shock, vibration, acceleration, and deceleration stresses.

Equipment that is not hermetically sealed and protected by climatic control may undergo repeated severe cyclic changes in temperature and humidity when operated in humid climates. When the equipment is inactive, moisture covers the surfaces of its parts and assemblies. Some of these moisture particles also penetrate into the interior of assemblies by a process called "breathing." Wherever moisture has penetrated, there is a lower resistance to voltage breakdown. In the case of low-voltage equipment this may not be very important, but when equipments operate at high voltages, it is imperative that the design allow for sufficient "creepage distances" and insulation.

Consider, for example, an electrolytic capacitor exposed to high-vacuum space environment. The capacitor is usually packaged in a tubular aluminum case. The terminals protrude through some kind of insulated end-cap assembly. Inside the case is the electrolyte, which is a conductive compound and which is instrumental in bringing about the capacitor action. When the capacitor is exposed to deep-space vacuum conditions, the stress resulting from the differential pressure between the environment prevailing inside and outside the capacitor has often been found to be responsible for rupture of the case seal and splattering of the conductive electrolyte onto adjacent parts and wiring, thereby causing electrical shorts and equipment failure. It is obvious that the simultaneous presence of two or more adverse environmental factors increases the risk of failure. For example, if severe vibration or shock were present while a capacitor seal is under severe internal pressure stress, rupture of the seal would be hastened.

Under vacuum conditions, distances between voltage-carrying conductors and ground and between conductors themselves must be many times larger than for ground use. For example, whereas a spacing of 0.1-in breaks down when exposed to a 3000-V stress at ground-level conditions, as stress of only 350 V will cause breakdown over the same distance at an altitude of 100,000 ft. This is a ratio of nearly 9 : 1, and sufficient protective margin must be built into the design to permit operation at either extreme. Adverse effects of space environment were also noted due to radiation. Transistors, for example, unless shielded properly, will suffer an increase in their leakage current and various insulation materials may become brittle and lose some of their insulating ability. Lubricants are equally vulnerable in space, and there are instances on record of locked bearings due to lubricant deterioration.

The above-cited examples relating to the impact of environmental conditions on the reliability of exposed equipments are indicative of the scope and variety of environmental influences to be considered. For ground-based equipment, the effects of humidity, shock, vibration and temperature are most significant. The type of vibration that equipment could be subjected to may vary in accordance with regular periods (sinusoidal vibrations) or may be composed of a whole spectrum of random vibrational components. The stress experienced by the equipment undergoing vibration are definable in terms of multiple g-level acceleration. The g levels depend on the frequency of the vibration and the amplitude of the excursions. The relationship is expressed by the equation

$$g_{pk} = 0.051 \times \text{DA} \times f^2 \tag{58}$$

where g_{pk} = peak gravity-level acceleration
$\quad\text{DA}$ = double amplitude, in
$\quad\quad f$ = frequency

The following example will illustrate the use of the equation.

EXAMPLE 8-1 What is the effective peak g-level acceleration when an assembly is exposed to a sinusoidal vibration having a frequency of 70 Hz and physical excursion of the assembly of 0.01 in to either side of its original position at rest?

Solution

$$g_{pk} = 0.051 \times \text{DA} \times f^2 \quad\quad \text{DA} = 2 \times 0.01 = 0.02 \text{ in total excursion}$$
$$= 0.051 \times 0.02 \times 70^2$$
$$= 5$$

Equation (58) also reveals that the effects of frequency of a vibration on the effective g level are in a quadrature relationship. This means that for every doubling of frequency, the g level increases four times.

Shock is a sudden application of force transmitted directly to equipment by mechanical impact or through the air by shock waves. For example, the former could occur if equipment is intentionally or unintentionally hit or dropped. The latter could occur due to the proximity of an explosion or a shot.

Shock is measured in g levels of acceleration and is usually limited to a number of milliseconds of duration. During this short period of application shocks may reach high g levels of acceleration, with magnitudes of 100 to 500 g's for periods of 5 to 50 ms not unusual. Since the g levels during shock exposure only last a very short period of time, the fact that they are 200 to 300 times higher than for the periodic (sinusoidal) vibration is not significant. What is important is the combination of g level and the exposure time.

Chassis, cabinets, and sensitive components are the principal areas of shock and vibration failures and require special attention in design. Shock mounts may be used in certain applications, but care must be exercised since they may tend to cause an increase of the vibration amplitude [DA in equation (58)]. Stress on equipment in motor vehicles ranges from 30 to 90 Hz at an acceleration level of up to 7 g's. Shocks encountered by equipment due to ordinary freight handling may range up to 10 g's.

A detailed analysis of the effects of shock and vibration on the equipment is difficult and frequently not economical. The designer will often be forced to make simplifying assumptions in his analysis.

2. Environmental Testing

Environmental testing plays a major role in the development of reliable equipments. It is an effective method for the early detection of weaknesses in the design, thereby permitting effective corrective action. Environmental testing has assumed increased importance in recent years for several reasons: first, the customer rightly expects his equipment to perform adequately for its lifetime. Second, changes or modifications to the equipment, once marketed, are costly and time-consuming. Also, maintenance costs could be excessive if problems only show up in the field when the equipment is exposed to actual environmental stresses. And finally, there is the legal problem of product safety and liability.

As a result, environmental simulation and testing of the equipment in the laboratory on a prototype model has become not only good business but an economic necessity. Traditionally, the objective of product testing has always been to demonstrate the prescribed performance. These tests failed to give insight into the possible modes of equipment failure and were soon recognized to be inadequate for equipment designed and procured for military purposes. A system of qualification inspection was instituted which requires the product to withstand a very comprehensive test program.

Table 8-4 illustrates the qualification inspection imposed on transformers and inductors. An examination of the table shows a listing of the required tests for grades

Table 8-4 *Qualification Inspection for Transformers and Inductors. (Source: MIL-T-39013A.)*

Examination or Test	Grades 4	Grades 5	Require-ment Paragraph	Method Paragraph	Number of Sample Units To Be Inspected	Number of Defectives Permitted
Group I (screening test)						
Thermal shock	X	X	3.22	4.8.17	All	Not
Winding continuity	X	X	3.21	4.8.16	units	applicable
Group II						
Visual and mechanical examination (external)	X	X	3.1, 3.5 to 3.5.3 incl, 3.6 to 3.6.2.3 incl, 3.6.3, 3.6.7, 3.28, 3.29	4.8.1.1		
Electrical characteristics	X	X	3.7	4.8.2		
Resistance to soldering heat*	X	X	3.8	4.8.3		
Terminal strength	X	X	3.9	4.8.4	All	0
Seal	X	X	3.10	4.8.5	units	
Dielectric withstanding voltage:						
At atmospheric pressure	X	X	3.11	4.8.6.1 or 4.8.6.1.1		
At barometric pressure (when applicable)	X	X	3.11	4.8.6.2		
Induced voltage	X	X	3.12	4.8.7		
Insulation resistance	X	X	3.13(a)	4.8.8		
Fungus†	X	X	3.14	4.8.9		
Group III						
Solderability (two sample units)‡	X	X	3.27	4.8.22		
Life	X	X	3.16	4.8.11		
Dielectric withstanding voltage:						
At reduced voltage	X	X	3.11	4.8.6.3		
Insulation resistance	X	X	3.13(c)	4.8.8	22	0
Induced voltage	X	X	3.12	4.8.7		
Visual and mechanical examination (external) (post-test)	X	X	3.30	4.8.1.1.1		
Electrical characteristics	X	X	3.7	4.8.2		

* Through-hole-type terminals and printed-circuit-type units only.

† Test should not be performed if the manufacturer provides certification that all external materials are fungus-resistant.

‡ Printed-circuit-type transformers and inductors only.

Table 8-4 *Qualification Inspection.* (Continued)

Examination or Test	Grades 4	Grades 5	Require-ment Paragraph	Method Paragraph	Number of Sample Units To Be Inspected	Number of Defectives Permitted
Group IV						
Corona discharge (when specified)	X	X	3.17	4.8.12		
Temperature rise (two sample units)	X	X	3.18	4.8.13		
Vibration	X	X	3.19	4.8.14		
Shock	X	X	3.20	4.8.15		
Dielectric withstanding voltage:						
At reduced voltage	X	X	3.11	4.8.6.3		
Induced voltage	X	X	3.12	4.8.7		
Winding continuity	X	X	3.21	4.8.16		
Immersion	X	X	3.23	4.8.18		
Moisture resistance	X	X	3.24	4.8.19		
Dielectric withstanding voltage:						
At reduced voltage	X	–	3.11	4.8.6.3		
Induced voltage	X	–	3.12	4.8.7		
Insulation resistance	X	–	3.13(b)	4.8.8		
Winding continuity	X	–	3.21	4.8.16		
Overload	–	X	3.25	4.8.20		
Dielectric withstanding voltage:						
At reduced voltage	–	X	3.11	4.8.6.3	6	0
Induced voltage	–	X	3.12	4.8.7		
Insulation resistance	–	X	3.13(b)	4.8.8		
Winding continuity	–	X	3.21	4.8.16		
Electrical characteristics	X	X	3.7	4.8.2		
Visual and mechanical examination (external) (post-test)	X	X	3.30	4.8.1.1.1		
Resistance to solvents (three sample units)§	X	X	3.15	4.8.10		
Flammability (two sample units)	–	X	3.26	4.8.21		
Visual and mechanical examination (internal) (three sample units; same units used for resistance to solvents test)	X	X	3.1, 3.5 to 3.5.3 incl, 3.6.4, 3.6.5, and 3.6.7	4.8.1.2		

§ For back-to-back testing, a minimum of 22 samples is required.

4 and 5 transformers or inductors (the grades represent different levels of mechanical design). Associated with each test is a requirement and a method paragraph and an indication of the number of units to be exposed to the test and the number of defects allowed. The examinations or tests are conducted in 4 groups.

The group I tests involve thermal shock and winding continuity examinations.

The group II tests involve visual and mechanical examinations and electrical performance tests. The group III tests involve life tests on 22 units for 2016 h as well as solderability tests. The group IV tests involve all the environmental testing (shock, vibration, moisture resistance, water immersion), electrical, as well as flammability and solvent-resistance tests. Zero failures are permitted on any of the tests.

MIL-STD-202E[1] is a military specification for environmental testing and is used extensively for the specification of environmental testing. Tables A1 through A8, which are contained in Appendix A, are taken from MIL-STD-202E as typical examples of environmental tests. Table A1 is the numerical index of test methods and details the types of environmental tests, physical-characteristic tests, and electrical-characteristic tests described. The remaining tables cover the following tests:

Table A2	Temperature Cycling (Method 102A)
Table A3	Humidity (Steady State) (Method 103B)
Table A4	Moisture Resistance (Method 106D)
Table A5	Thermal Shock (Method 107D)
Table A6	Life (Method 108A)
Table A7	Vibration (Method 204C)
Table A8	Shock, High Impact (Method 207A)

The shock environment[2] is usually difficult to simulate. Many military specifications give mechanical details of the shock apparatus to be used. The most common shock test machines are the free-fall types, which subject the assembly under test to a predetermined height of fall and impact on a fixed anvil. The shape or amplitude of the shock wave is also frequently modified by providing damping of the anvil through the use of springs, lead pellets, air, or fluid means.

Equation (59) may be used to determine the number of g accelerations as a result of a free fall of the equipment from a certain height:

$$g = \sqrt{\frac{2Ch}{W_P + W_E}} \tag{59}$$

where g = number of g accelerations
$\quad C$ = spring constant of shock-absorbing spring, 1b/in.
$\quad h$ = height of equipment drop, in.
$\quad W_P$ = weight of platform on which equipment is mounted for the drop test, lb
$\quad W_E$ = weight of equipment to be tested, lb

Equation (60) permits the calculation of the duration of the shock pulse:

$$t = 159\sqrt{\frac{W_P + W_E}{C}} \tag{60}$$

where t = pulse duration (ms) and W_P, W_E, and C are as in equation (59).

[1] MIL-STD-202E, Military Standard, "Test Methods for Electronic and Electrical Component Parts," April 16, 1973.
[2] The shock-test description, shock and time-duration formulas, and the example are based on material in Chapter 8, R. H. Myers, K. L. Wong, and H. M. Gordy (eds.), *Reliability Engineering for Electronic Systems*, John Wiley & Sons, Inc., New York, 1964.

EXAMPLE 8-2 A shock machine has a mounting platform that weighs 20 lb and a spring constant of 4000 lb/in. What drop height is required if a piece of equipment weighing 30 lb is to be exposed to 40 g's of acceleration? What will the shock duration be, in milliseconds?

Solution

Using equation (59),

$$40 = \sqrt{\frac{2(4000)h}{20 + 30}}$$

Solving for h,

$$1600 = \frac{8000h}{50}$$

$$h = 10 \text{ in}$$

To calculate the shock duration, we use equation (60);

$$t = 159\sqrt{\frac{20 + 30}{4000}}$$

$$= 159\sqrt{0.0125} = 17.8 \text{ ms}$$

3. Step-Stress Testing

A technique that is particularly useful in the initial design phase is the step-stress test method. When applying this method, the stress level is raised in steps until failure occurs. The type of stress (temperature, voltage, vibration, shock, etc.) is selected on the basis of its ability to bring about failure due to a suspected failure mechanism. The stress level is raised until all units in the test sample have failed. To be sure that the failures obtained are really due to the suspect failure mode, a failure analysis of the failed parts has to be carried out.

In general, step-stress tests and steady life tests using rated stress give the same results. The step-stress tests are considerably faster than regular life testing. This is especially true of high-reliability components when regular life tests could proceed without failure for many thousands of hours. Not infrequently, both steady life tests and step-stress tests are run concurrently to verify the results of the step-stress test. The knowledge gained from step-stress testing may be applied to selecting burn-in levels for parts, so that failure-prone units are removed without approaching the failure range of the unit as determined during step-stress testing.

(f) Worst-Case Analysis

Worst-case analysis is concerned with examining how the performance of a circuit will change in time as a result of drift in parts characteristics. If parts-characteristic changes are severe as a result of environmental stresses or aging, the circuit in which these parts are used may no longer adequately perform its required function. For all practical purposes, this circuit would differ little from one that has failed catastrophically because of a burned-out part.

The important fact is that random catastrophic failures cannot be eliminated in a circuit even though it has been properly designed, that is, where the parts, stresses, and environmental conditions have been adequately considered; Drift failures, on the other hand, can be avoided by allowing for part parameter variations in the design.

Taking these variations into account is not always simple to accomplish, especially in complex circuitry. At times circuits are designed with nominal part characteristic values and tested experimentally in the laboratory to determine the effects of intentionally introduced variations in parameters.

> **EXAMPLE 8-3** The resistance tolerance of a 100-Ω resistor, as purchased, is $\pm 10\%$. This type of resistor may drift up to $\pm 15\%$ in resistance value after 2 years of operation. What are the worst-case considerations for the application of this resistor?
>
> **Solution**
>
> 1. The lowest resistance value, as purchased, could be
> $$100 - 0.1(100) = 90\ \Omega$$
> 2. The lowest resistance value after 2 years could be
> $$90 - 0.15(90) = 76.5\ \Omega$$
> 3. The highest resistance value as purchased could be
> $$100 + 0.1(100) = 110\ \Omega$$
> 4. The highest resistance value after 2 years could be
> $$110 + 0.15(110) = 126.5\ \Omega$$
>
> Therefore, unless the circuit is designed to be sufficiently tolerant to a variation of this 100-Ω resistor, ranging from a possible 76.5 to 126.5 Ω, there could be a circuit malfunction. It is obvious that if this resistor is part of a circuit which contains parts having characteristics which drift the opposite way, there could be some compensating situations. The complexity of determining the total interaction of a group of parts on circuit performance is evident and it is not infrequent that computer programs have to be written to obtain the desired answers. In less complex situations, graphic plots may simplify the analysis.

Worst-case analysis is one of the more sophisticated tools used for achieving product reliability and is mostly employed in the design of complex electronic equipments.

(g) Failure-Modes-and-Effects Analysis

The reporting of failures during the fabrication, testing, and field use of equipments, followed by analysis of the failures and corrective action, is an extremely important aid in producing reliable equipment. This is discussed in more detail later in the chapter. Knowledge of the failure modes of parts, circuits, and subsystems is of great help to the designer, as is knowledge of the effect(s) of a given failure on the next-higher level of assembly in the equipment.

There are two approaches to obtaining failure-modes-and-effects information:

1. Failure-mode discovery by testing and laboratory investigation.
2. Failure-mode discovery by analysis.

Both methods are effective in pointing up hidden modes of failure in equipment. The analysis method is preferred since it forces the designer to analyze the system on a systematic, step-by-step basis.

The failure-modes-and-effect analysis is an analytical procedure for critically examining the possible ways in which a system, component, circuit, or part may fail and the resultant effect upon performance.

One of the objectives of the analysis is to detect areas of high-failure-rate concentration (low probability of survival) to permit the reduction of circuit complexity in that area, the introduction of redundancy techniques, or other remedial actions. The analysis can be conducted on a system, subsystem, subassembly, component, circuit, or part level. Typical items to be included are as follows:

1. Item designation: the block diagram is shown, with each part, component, and circuit subassembly designated by an item number.
2. Assumed failure: list all conceivable failures, both degradative and catastrophic —everything that could possibly happen.
3. Possible causes: list all causes that are believed capable of giving rise to the failure.
4. Symptoms and effects: the immediate result of each assumed failure at the next-higher level of assembly.
5. Compensating provisions: compensating provisions embodied in the design (such as a redundant component) should be listed. It should be stated whether the compensation is total or partial, and whether resorting to the compensating provisions would limit performance or reduce efficiency.
6. Effect on system: the effect of the postulated failure on the overall system (or equipment) should be identified.

Further refinements of the failure-modes-and-effects-analysis technique are possible by incorporating failure rates with each of the assumed failures, thereby determining the probability of occurrence of each of the assessed modes of failure. As in worst-case analysis, the use of failure-modes-and-effects analysis is reserved for complex systems, such as space satellite, missile, radar, and computer electronics equipment.

(h) Design Reviews

Design reviews are used to assess whether a proposed design is capable of meeting all the performance requirements specified. Alternative design options are examined, when necessary, and the effects of changes in design philosophy, complexity, and configuration analyzed with respect to manufacturability, reliability, safety, cost, styling, and so on. Liberman[3] describes the following four distinguishing characteristics of the modern design review:

Competence: More than one reviewer is used, and collectively the reviewers must be at least as competent as the designer, and in more areas.
Objectivity: There must be sufficient breadth to the review team so that it is not a review by design supervision of a design they are responsible for.
Systematicness: The review must be oriented to specified objectives, must consider a predetermined body of data, and must follow a reasoned plan and schedule.
Formality: There must be sufficient formality to maintain the review as a disciplined activity. This requires existence of a review policy, leadership by a formally constituted body, and documentation of the proceedings and decisions.

Design reviews may cover the entire piece of equipment or any portions thereof, depending on specific needs. Participants in design reviews include the designers, manufacturing and management personnel, and competent specialists in the specific areas reviewed. The recommendations produced during the review are documented and specific action items are assigned to individuals. These action items, as well as the minutes of the design review, are published for the record. After all action items have been implemented, the design review chairman officially closes out the review.

On relatively simple equipment, one design review may suffice. On more complex equipment for military application, it is customary to have a preliminary design review for the design proposal developed during the study phase. The second design review is the "major review." This review is concerned with the application of the parts used, specifications, and trade-offs among performance, reliability, weight, size, and cost.

The third design review is the final, preproduction review. All design aspects are presented, parts choices, specifications, worst-case and failure-modes-and-effects analyses are reviewed, the reliability prediction is approved, and the design, for all practical purposes, is approved and "frozen" for production release.

(i) Failure-Analysis Program

1. Failure Occurrence

During the initial phases of testing, the failure rate of equipment is high, owing to the infant-mortality failures of the weaker parts and the usual need for debugging of the equipment. Quite a few explainable, nonrelevant failures occur during this period.

2. Failure Reporting

A comprehensive failure reporting, analysis, and corrective-action program is extremely important to the achievement of reliability in equipments. First, it is necessary to have a detailed program which assigns responsibilities to appropriate personnel.

---

[3] D. S. Liberman, "Design Review—A Tool for All Seasons," *Proceedings of 1972 Annual Reliability and Maintainability Symposium* p. 438.

Usually, training sessions are desirable to instruct test personnel in the necessary details required by the failure report form. This includes the recording of the exact location of failed items in the equipment, drawing numbers, serial numbers, observations about the anomaly, action taken (such as replacement of failed parts), retest information, and related facts.

There have to be adequate controls to ensure that all failures are reported and analyzed. This can be accomplished by instituting a system by which replacement for a failed part can only be obtained from stock if the failed part is surrendered with a failed part ticket attached. The latter should reference the failure report number. In many companies it is the responsibility of the cognizant quality-control inspector to assure that all failures are reported. Surveillance by a QC inspector during the testing phases of equipment is very desirable since, not infrequently, failures resulting from human error or negligence may be reported as random equipment failures. Failures due to human error or workmanship are not considered relevant failures, since they are caused by assignable and preventable factors. Human error can be prevented by greater care or double checking of an operation by a second person. Workmanship errors are preventable by proper care, training, and inspection. Relevant failures occur randomly after all these precautions have been taken.

After the debugging period, the equipment settles down to the useful life period with relatively constant failure rate. The failures occurring during this period are of a nonpredictable, random nature.

3. Analysis and Corrective Action

All failure reports are usually sent to the reliability department for analysis and corrective action recommendations. The failed parts are normally sent to the laboratory for detail analysis. Such analysis may include optical, metallurgical, chemical, electrical, or X-ray analysis and may involve dissection of the failed item. Photographic evidence of the failure is often included in the analysis report. The reliability department will also investigate previous occurrences of this type of failure and evaluate the possibility that one bad lot may have been responsible for several failures. If such a situation has been confirmed, all units involved will be purged from the equipment and from stock.

All failures which are judged relevant count against the reliability of the equipment, and it is the responsibility of the reliability department to keep account of the reported failures and close them out on the basis of the analysis conducted, the failure classification assigned, and the corrective action (if any) implemented. Figure A1, Appendix A, is an example of a typical failure-analysis report form.

4. Integrated-Circuit-Analysis Techniques

Failure-analysis techniques have in recent years become more sophisticated. This is especially true of integrated circuit analysis. Failures in integrated circuits may range from simple mechanical lead bonding problems to sophisticated chemical or metallurgical anomalies. Figures 8-2 and 8-3 illustrate simple lead bonding problems

which should be detected by an alert QC inspector. Figure 8-2 shows an emitter doughnut bond with aluminum pad area showing, indicative of an excessive combination
of the pressure-time-temperature bonding variables. Additionally, the collector lead

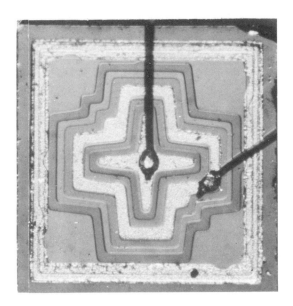

Figure 8-2 *Bond problems (magnification 100X).*

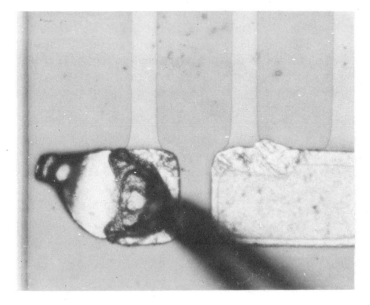

Figure 8-3 *Bond problem with damage to resistive element (magnification 500X).*

is not positioned on its assigned pad. This is an example of a workmanship deficiency and inadequate QC surveillance. Figure 8-3 illustrates a "doughnut bond" where the pressure-time-temperature combination was so excessive as to force the bonder through the gold and aluminum pad into the nichrome resistive element. The bond strength is seriously impaired at the junction. Additionally, the proper functioning of the resistive element is impaired.

The success or failure of an analysis depends primarily on the method of approach and the sequencing of nondestruct methods before those of a destructive nature. The degree of investigation is determined by the complexity of the failure. The analysis is terminated when the failure mechanism is located or sufficient information is derived to generate corrective action or preventive steps to preclude this type of failure in the future. A systematic approach, and a plan for an effective analysis flow, is essential for a complex integrated-circuit analysis.

As an illustration, the following failure-analysis technique for transistor–transistor logic (TTL) integrated circuits is presented. This material is based on a 1972 NASA report.[4] First some details regarding basic construction, processes, and materials used in TTL integrated-circuit construction are presented. This is followed by a comprehensive listing of problem codes and definitions (see Table 8-5). These problem codes

Table 8-5 *Problem Codes and Definitions.*

PROBLEM CODE		
Area	*Cause*	*Definition*
I		Open
	A	Die metallization
	1	Microcracks at window cutouts
	2	Scratches across metallization stripes or pads
	3	Voids
	4	Incorrect/insufficient alloying
	5	Formation of insulating layer under metallization at cutouts due to faulty oxide removal
	6	Peeling, lifting, or flaking metal
	7	Mask misalignment
	8	Masking defect
	9	Corrosion
	B	Wiring/bonding
	1	Lifted bond
	2	Cracked bond wire at heel of bond
	3	Broken bond wire, or nicked or necked-down wire
	4	Corroded wire
	5	Misbond—insufficient pressure or temperature/excessive pressure or temperature
	6	Missing bonds or jumpers
	7	Plague (intermetallics)

[4] National Aeronautics and Space Administration, "Parts Materials, and Processes Experience Summary," Report CR114391, February 1972.

Table 8-5 *Problem Codes and Definitions* (Continued).

PROBLEM CODE		
Area	*Cause*	*Definition*
	8	Incorrect bond sizes
	9	Misplaced bonds
	C	Chip or substrate
	1	Cracked chip or substrate
	2	Lifted chip or substrate
	3	Lifted molytab(s)
	4	Diffusion defect: bad mask, contamination, incorrect, etc.
	5	Thin-film defect: contact, scratch, omissions, lifting, peeling, voids, and corrosion
	6	Thick-film defects: design error, silk screening, scratches, voids, omissions, contacts, and contamination
	7	Oxide defects
II		Short
	A	Chip
	1	Smeared metallization
	2	Unetched metallization
	3	Lack of/or too thin oxide insulation
	4	Pinholes
	5	Mask(s) misalignment: diffusion masks, metallization masks overlapping
	6	Diffusion shorts due to faulty masks
	7	Particulate contamination embedded during diffusion
	8	Chemical contamination as result of faulty cleaning, rinsing, etc.
	9	Metallic or conductive particle short due to extraneous material
	10	Metallization pads placed too close to chip edge
	11	Chip-edge chipouts, exposing bare silicon to metallization scribing
	12	Excess flow of eutectic during die attach
	13	Short due to moisture or faulty hermetic seal—contamination
	14	Junction punch-through overstress
	B	Wiring
	1	Leads too long
	2	Lead dressing—too close to chip edge or adjacent chip or wire, or crossed over
	3	Sagging leads
	4	Leads in contact with package lid
	5	Bond tails too long
	6	Improper bond sizes, causing adjacent shorts
	7	Improperly terminated leads
	8	Extraneous and/or loose wires or bonds
	C	Substrates
	1	Smeared or scratched thin- or thick-film depositions
	2	Improperly deposited metallization patterns (e.g., misalignments)
	3	Conductive contamination from inks or chemicals

Table 8-5 *Problem Codes and Definitions* (Continued).

PROBLEM CODE		
Area	*Cause*	*Definition*
	4	Shorts due to bonding agents, whether eutectic or conductive epoxies
	5	Extraneous intraconnect wires or ribbons
	6	Misplaced bonds
	7	Extraneous particulate matter—bond melt, lid solder melt, or silicon chipouts, wire slivers, or contamination
	D	Package
	1	Eutectic lid solder melt in contact with package leads
	2	Package lead misaligned in package, causing package to lead short
	3	Loss of hermetic seal
	4	Short due to external contamination
III		Operational degradation
	A	Surface inversion N to P, or vice versa
	B	Surface channeling
	C	High leakage
	D	Loss of beta
	E	Oversensitivity to voltage variation
	F	Oversensitivity to temperature variation
	G	Die misorientation
	H	Design change
	I	Design deficiency
	J	Sequence of power application
	K	Misapplication of device
	L	Electrostatic discharge in MOS devices
IV		Mechanical anomaly
	A	Marking or identification (misidentification or illegibility)
	B	Mislocation of index mark/tab
	C	Damage to package
	D	Damage to package leads or lead to package seals
	E	Loss of hermeticity
	F	X-ray anomalies
	G	Incorrect package
	H	Missing leads
	I	Dimensional errors

list the various causes for failure resulting from the following conditions:

1. Open.
2. Short.
3. Operational degradation.
4. Mechanical anomaly.

Figure 8-4 shows the die topography for a typical TTL integrated circuit identifying

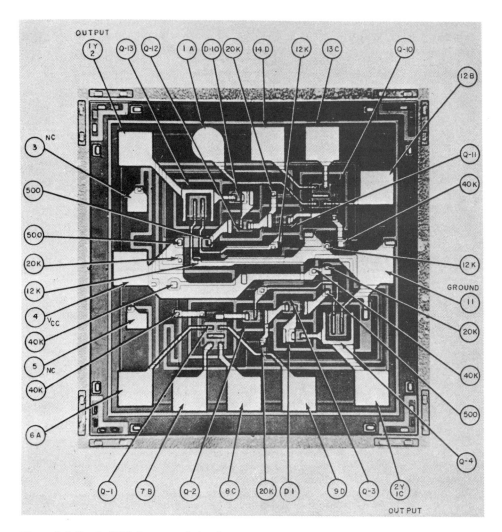

Figure 8-4 *Typical TTL integrated circuit.*

the die elements, terminations, and pin locations. The symbols located around the die periphery depict the die active elements and the various inputs and outputs with associated pin locations. These may be referenced to Figure 8-5, which is the corresponding schematic diagram locating the symbols used in the die topography illustration in the actual circuit locations. Finally, Figure 8-6 depicts the sequential failure-analysis flow for the defined problem codes. The use of this flow chart is self-evident by examination.

The basic construction processes and materials are as follows: The basic TTL gate on a monolithic silicon chip consists of a diffused multiemitter input transistor

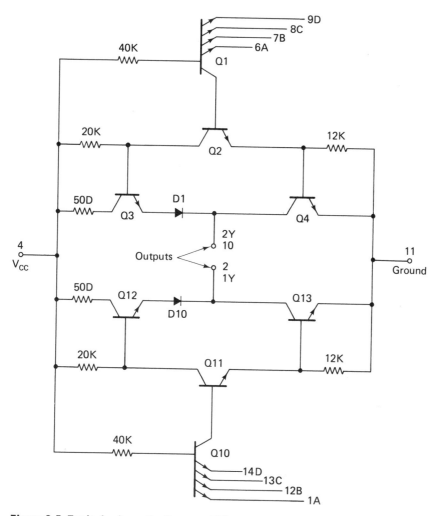

Figure 8-5 *Typical schematic diagram—TTL.*

and its associated current-gain drive circuitry. The multiemitter transistor and the other drive transistors are fabricated by diffusing an isolated collector region. The base region is then diffused into this collector region and the several emitter regions are diffused as separate areas within the base region. Between diffusions, there are repetitive steps consisting of oxide growth, masking, etching, isolation, and deposition, then diffusion.

The conducting paths in this TTL integrated circuit are accomplished by use of deposited metals (e.g., molybdenum–gold). Gold wires (0.001 in diameter) are then utilized for external connections through use of thermocompression gold ball bonds and wedge bonds.

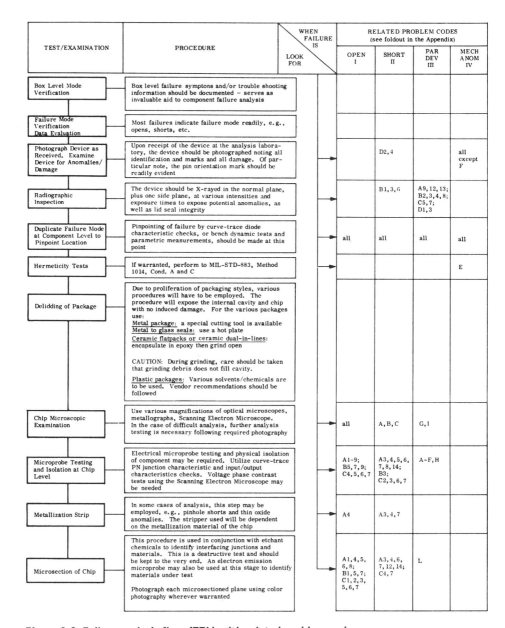

TEST/EXAMINATION	PROCEDURE	WHEN FAILURE IS LOOK FOR	RELATED PROBLEM CODES (see foldout in the Appendix)			
			OPEN I	SHORT II	PAR DEV III	MECH ANOM IV
Box Level Mode Verification	Box level failure symptons and/or trouble shooting information should be documented — serves as invaluable aid to component failure analysis					
Failure Mode Verification Data Evaluation	Most failures indicate failure mode readily, e.g., opens, shorts, etc.					
Photograph Device as Received. Examine Device for Anomalies/ Damage	Upon receipt of the device at the analysis laboratory, the device should be photographed noting all identification and marks and all damage. Of particular note, the pin orientation mark should be readily evident			D2,4		all except F
Radiographic Inspection	The device should be X-rayed in the normal plane, plus one side plane, at various intensities and exposure times to expose potential anomalies, as well as lid seal integrity			B1,3,6	A9,12,13; B2,3,4,8; C5,7; D1,3	
Duplicate Failure Mode at Component Level to Pinpoint Location	Pinpointing of failure by curve-trace diode characteristic checks, or bench dynamic tests and parametric measurements, should be made at this point		all	all	all	all
Hermeticity Tests	If warranted, perform to MIL-STD-883, Method 1014, Cond. A and C					E
Delidding of Package	Due to proliferation of packaging styles, various procedures will have to be employed. The procedure will expose the internal cavity and chip with no induced damage. For the various packages use: Metal package: a special cutting tool is available Metal to glass seals: use a hot plate Ceramic flatpacks or ceramic dual-in-lines: encapsulate in epoxy then grind open CAUTION: During grinding, care should be taken that grinding debris does not fill cavity. Plastic packages: Various solvents/chemicals are to be used. Vendor recommendations should be followed					
Chip Microscopic Examination	Use various magnifications of optical microscopes, metallographs, Scanning Electron Microscope. In the case of difficult analysis, further analysis testing is necessary following required photography		all	A,B,C	G,I	
Microprobe Testing and Isolation at Chip Level	Electrical microprobe testing and physical isolation of component may be required. Utilize curve-trace PN junction characteristic and input/output characteristics checks. Voltage phase contrast tests using the Scanning Electron Microscope may be needed		A1-9; B5,7,9; C4,5,6,7	A3,4,5,6, 7,8,14; B3; C2,3,6,7	A-F,H	
Metallization Strip	In some cases of analysis, this step may be employed, e.g., pinhole shorts and thin oxide anomalies. The stripper used will be dependent on the metallization material of the chip		A4	A3,4,7		
Microsection of Chip	This procedure is used in conjunction with etchant chemicals to identify interfacing junctions and materials. This is a destructive test and should be kept to the very end. An electron emission microprobe may also be used at this stage to identify materials under test Photograph each microsectioned plane using color photography wherever warranted		A1,4,5, 6,8; B1,5,7; C1,2,3, 5,6,7	A3,4,6, 7,12,14; C4,7	L	

Figure 8-6 *Failure analysis flow (TTL) with related problem codes.*

PROBLEMS

8-1 Define reliability.

8-2 What is the distinction between reliability and quality?

8-3 What are some of the difficulties in achieving reliability?

8-4 Discuss redundancy techniques.

8-5 What is the purpose of qualification and quality conformance tests?

8-6 What is a derating policy?

8-7 Discuss the purpose of burning-in electronic parts.

8-8 What is the purpose of a worst-case analysis?

8-9 What are the worst-case inductance values of a 20-H reactor purchased to a $+15$ % -10% tolerance and an inductance drift guaranteed not to exceed -20%?

8-10 Discuss the methods employed in a failure-modes-and-effects analysis and state its purpose.

8-11 Describe the elements of a comprehensive failure reporting and analysis program.

8-12 (a) State the purpose for holding design reviews.
  (b) Four distinguishing characteristics of an effective design review were given in the chapter. What were they?
  (c) What are the three kinds of reviews? Give details.

8-13 What are typical environmental stresses to which equipment is exposed in military ground applications?

8-14 What are the effects of space environment on electronic equipments?

8-15 An equipment is exposed to a sinusoidal vibration of 50 Hz. The amplitude is 0.03 in peak to peak. What is the peak g-level acceleration?

8-16 What is the typical g level of shocks encountered by ordinary freight?

8-17 What are some of the environmental tests contained in MIL-STD-202E?

8-18 A shock test machine is used to expose a piece of equipment to a free-fall shock. The equipment weighs 50 lb and is to be exposed to 50 g's of acceleration.
  (a) What is the height from which the equipment has to be dropped if the weight of the mounting platform is 50 lb and the spring constant is 3000 lb/in?
  (b) What is the duration of the shock pulse?

8-19 Describe the method of step-stress testing.

8-20 (a) What is parts engineering concerned with?
  (b) What is the base failure rate for a $\frac{1}{4}$-W MIL-R-55182 fixed-film resistor operating in an ambient temperature of 65°C, and dissipating 50% of rated wattage? (Use Table 8-1.)

8-21 With the aid of Figure 8-1, determine the maximum permissible ambient temperature for an $\frac{1}{8}$-W MIL-R-10509 film resistor, operating at 0.1 W, if the failure rate is limited to 0.008 failure/10^6 h.

8-22 With the aid of Table 8-3, determine the end-of-life range for a 470-pF ceramic capacitor (characteristic X).

8-23 A circuit requires a very stable 1.0-Ω resistor whose end-of-life resistance values must not exceed the range 0.997 to 1.003 Ω. From Table 8-3, what is the appropriate choice?

8-24 (a) What are established reliability parts specifications?
  (b) What is a QPL listing?

8-25 When parts are procured by vendor part number, what risk is taken?

9

EQUIPMENT SURVIVAL

9-1 PROBABILITY-OF-SURVIVAL CONCEPT

When the number of failures that occur in equipment in equal time intervals remains constant, its probability of survival (or its reliability) may be predicted by the exponential distribution. The probability of survival can thus be best expressed in terms of the random failure rate of the equipment.

The purpose of this chapter is to become more familiar with the principles of equipment survival as governed by failure rates and derating procedures. In Chapter 10 we will apply our knowledge of failure rate (or MTBF) determination to the prediction of the reliability of circuits and equipments.

9-2 FAILURE RATE

The *failure rate*, λ, is defined as the number of failures in a given time interval and is expressed as

$$\lambda = \frac{\text{number of failures}}{\text{total unit operating hours}} \tag{61}$$

EXAMPLE 9-1 Ten transformers were tested for 500 h each within prescribed operating conditions, and one transformer failed exactly at the end of the 500-h exposure. What is the failure rate for this type of transformer?

Solution

$$\lambda = \frac{1}{10 \times 500} = \frac{1}{5000} = 0.0002 \text{ failure/hour}$$

EXAMPLE 9-2 Refer to Example 9-1 except four transformers failed after the following test time periods:

transformer 1 fails after 50 h
transformer 2 fails after 150 h
transformer 3 fails after 250 h
transformer 4 fails after 450 h

Solution

$$\lambda = \frac{4}{1 \times 50 + 1 \times 150 + 1 \times 250 + 1 \times 450 + 6 \times 500}$$

$$= \frac{4}{3900} = 0.00102 \text{ failure/hour}$$

Failure rate can be expressed in different ways:

1. Failures/hour.
2. Percent failures/1000 h.
3. Failures/10^6 h.

Published tables may use any of the above.

The conversion from failures/hour, for example, to percent failures/1000 h is as follows:

$$\text{failures/hour} \times 10^5 = \% \text{ failures/1000 h} \tag{62}$$

To convert from failures/hour to failures/10^6 h, multiply by 10^6:

$$\text{failures/hour} \times 10^6 = \text{failures/}10^6 \text{ h} \tag{63}$$

In order to use the formula for probability of survival (exponential failure distribution), the failure rate needs to be expressed in failures per hour [see equation (70)]. Table 9-1 summarizes several useful conversions.

Table 9-1 *Failure-Rate Conversions.*

If Failure Rate Is Given As:	Multiply By:	To Obtain Failure Rate in:
% Failures/1000 h	10^{-5}	Failures/hour
Failures/10^6 h	10^{-6}	Failures/hour
Failures/hour	10^5	% Failures/1000 h
Failures/hour	10^6	Failures/10^6 h

9-3 MIL-HDBK-217B

(a) General Introduction

A great deal of research and testing has been done to determine failure rates of parts and products, and many of these are published. The failure rates for electronic and electromechanical parts are by far the most important and are published in a military standardization handbook, MIL-HDBK-217B.[1] A failure rate data bank is also maintained by a government-industry data exchange program (GIDEP).[2] Failure rates for representative mechanical parts are available from a Bureau of Naval Weapons[3] publication. Failure rates for new mechanical designs are based on the strength of the materials used, safety factors, and experience. In the case of complex mechanical components, such as valves, a failure rate can be derived from the failure rates of the basic parts, such as bearings, bushings, and seals.

Failure rates for purely structural mechanical parts (brackets, channels, hardware, etc.) are considered negligible for the life of the equipment and are therefore not counted. We will not concern ourselves with the mechanical failure rates since they are only occasionally used in reliability calculations, but will concentrate on the far more important failure rates for electronic and electromechanical parts. The material related to MIL-HDBK-217B, presented herein, has either been directly drawn from the handbook or is based on it.

To start, it is well worth relating several poignant points made in the introduction to MIL-HDBK-217B with respect to the reliability problem and limitations of reliability predictions.

1. The Reliability Problem

When it is proposed to design an electronic system to perform a complex and demanding job, it is assumed that the required investment will be justified in terms of equipment performance and ability to perform on demand repeatedly. In the design of these systems, considerable effort is made to obtain reliable performance, but because of the exploding electronics technology there is frequently no time for an orderly evaluation of systems, and the ratio of new to tried portions of the system is relatively high. As a result, a certain level of unreliability is inevitable and must be faced.

Reliability is a problem at all levels of electronics, from materials to operating systems, because materials go to make up parts, parts compose assemblies, and assemblies are combined in systems of ever-increasing complexity and sophistication.

2. Limitation of Reliability Prediction

The art of predicting the reliability of electronic equipment has fundamental limitations. To be of value, a prediction must be timely. However, the earlier it is needed, the more difficulties will be encountered to make it a meaningful prediction. For example, the reliability of a piece of electronic equipment is known with certainty only after it has been used in the field until it is worn out and its failure history has been faithfully recorded. Before this point, reliability cannot be known with certainty. Hence, the art of predicting the reliability of electronic equipment has fundamental

---

[1] MIL-HDBK-217B Military Standardization Handbook, "Reliability Prediction of Electronic Equipment," September 20, 1974.

[2] Government-Industry Data Exchange Program, GIDEP Operations Center, Corona, California, 91720.

[3] Bureau of Naval Weapons Failure Rate Data Handbook SP 63–470.

limitations such as those depending on data gathering and predictive technique complexity.

Considerable effort is required to generate sufficient failure-rate data on a part class to report a statistically valid reliability figure for that class. Casual data gathering on a part class occasionally accumulates data more slowly than the advance of technology in that class; consequently, a completely valid level of data is never attained. In the case of many part classes, the number of people participating in data gathering all over the industry is rather large with consequent varying methods and conditions which make exact coordination and correlation difficult.

The use of failure-rate data, obtained from field use of past systems, is applicable on future concepts, depending on the degree of similarity existing both in the hardware design and in the anticipated environments. Data obtained on a system used in one environment may not be applicable to use in a different environment, especially if the new environment substantially exceeds the design capabilities. Other variants that can affect the stated failure rate of a given system include different uses, different operators, and different measurement techniques or definitions of failure. Thus, a fundamental limitation of reliability prediction is the ability to accumulate data of known validity for the new application. Another fundamental limitation is the complexity of prediction techniques for very complicated systems. Oversimplified techniques may omit a great deal of distinguishing detail and the prediction suffers inaccuracy. Very sophisticated techniques, on the other hand, may become so time consuming and costly that the prediction may actually lag the principal hardware development effort.

MIL-HDBK-217B includes two methods of failure-rate prediction, "Part Stress Analysis" and "Parts Count." These methods vary in the degree of information needed to apply them. Part stress analysis requires the greatest amount of detail and is applicable during the later design phase, where actual hardware and circuits are being designed. The parts count method requires less information, generally that dealing with quantity of different part types, quality level of the parts, and the application environment. This method is applicable in the early design phase and during bid proposal formulation. The failure rates presented in the handbook represent the best available knowledge at the time of issue (September 1974).

(b) Part-Stress-Analysis Prediction

This method of prediction is most suited for instances when the electrical design has progressed to a point where detailed parts lists are available and the stresses which the parts will encounter have been determined. MIL-HDBK-217B provides failure rate models for a broad variety of electronic and electromechanical parts, connecting devices, printed wire boards, and miscellaneous parts.

The following factors influence the failure rates of parts:

1. Base Failure Rate

MIL-HDBK-217B assigns base failure rates, λ_b, to the various part categories. The factors that influence λ_b are temperature of operation and electrical stress.

TEMPERATURE

The higher the temperature at which an electronic or electromechanical part operates, the shorter its lifetime. Accordingly, the failure rate will be higher for higher temperatures of operation. The operating temperature of a part is composed of the sum of the surrounding (ambient) temperature and the temperature rise of the part itself. For example, the ambient temperature inside a certain television receiver is 50°C, and a resistor mounted on a chassis in the receiver has a temperature rise of 40°C. The operating temperature of the resistor is, consequently, 90°C.

The failure-rate tables in MIL-HDBK-217B only define the ambient temperature for the parts used. The temperature rise of the part itself is automatically considered in the failure rate given. The range of ambient part temperature in Table 8-1, for example, is from 0 to 175°C; it can be seen that the failure rate increases with temperature.

ELECTRICAL STRESS

Electronic parts are usually specified in terms of some rated voltage or power. For example, a capacitor may be rated at 400 V; or a resistor may be rated for 1 W of power. Experience has shown that the use of electronic or electromechanical parts below their rated voltage or power level increases their reliability. Accordingly, if we desire a low failure rate for a part, we "derate" it.

2. Part Quality

The quality of the part, π_0, enters in the failure-rate computation. If the particular part can be purchased to a multilevel quality specification,[4] the appropriate value for π_0 has to be selected for this category.

Table 9-2 lists quality designators for multilevel quality specifications for common capacitors and resistors. The detailed requirements for the quality levels are defined by the individual parts specification. Parts that are not covered by multilevel quality specifications (non-ER) are defined in their respective failure-rate models by one of three quality levels:

1. Upper.
2. MIL-Spec.
3. Lower.

The upper quality level is reserved for instances when a part can be procured to a specification that is more rigorous than the MIL-Spec such as an established reliability

Table 9-2 *Parts with Multilevel Quality Specifications.*

Part	Quality Designators
Capacitors, established reliability (ER)	L, M, P, R, S
Resistors, established reliability (ER)	M, P, R, S

[4] Multilevel quality specifications refer to established reliability (ER) levels (guaranteed failure-rate levels L, M, P, R, or S; see Tables 9-7 and 9-9).

(ER) specification. When the part is purchased to the MIL-Spec, the π_0-quality designator assigned to MIL applies.

When any part of a MIL-Spec is waived or a commercial part is used, the π_0 for lower quality is applicable.

3. Use Environment

All part-failure-rate models consider the effects of environmental stresses through the factor π_E. Table 9-3 defines the environments and gives a corresponding identifying symbol. The value for the latter is furnished by MIL-HDBK-217B for the specific part types.

Table 9-3 *Environmental Symbol Identification and Description.* (Source: *MIL-HDBK-217B*.)

Environment	π_E Symbol	Nominal Environmental Conditions
Ground, benign	G_B	Nearly zero environmental stress with optimum engineering operation and maintenance.
Space, flight	S_F	Earth orbital. Approaches ground, benign conditions without access for maintenance. Vehicle neither under powered flight nor in atmospheric reentry.
Ground, fixed	G_F	Conditions less than ideal to include installation in permanent racks with adequate cooling air, maintenance by military personnel, and possible installation in unheated buildings.
Ground, mobile (and portable)	G_M	Conditions more severe than those for G_F, mostly for vibration and shock. Cooling air supply may also be more limited, and maintenance less uniform.
Naval, sheltered	N_S	Surface-ship conditions similar to G_F but subject to occasional high shock and vibration.
Naval, unsheltered	N_U	Nominal surface shipborne conditions but with repetitive high levels of shock and vibration.
Airborne, inhabited	A_I	Typical cockpit conditions without environmental extremes of pressure, temperature, shock, and vibration.
Airborne, uninhabited	A_U	Bomb bay, tail, or wing installations where extreme pressure, temperature, and vibration cycling may be aggravated by contamination from oil, hydraulic fluid, and engine exhaust. Class I and Ia equipment of MIL-E-5400 should not be used in this environment.
Missile, launch	M_L	Severe conditions of noise, vibration, and other environments related to missile launch, and space vehicle boost into orbit, vehicle reentry, and landing by parachute. Conditions may also apply to installation near main rocket engines during launch operations.

4. Other π Factors

All part-failure-rate models contain the factors π_Q and π_E. Other π factors apply only to specific models and are defined in MIL-HDBK-217B for each part subsection. All the part-failure-rate models include both catastrophic and drift failures and are generally based upon a constant failure rate (some rotary devices show an increasing failure rate).

Table 9-4 presents an overall listing of π factors used in failure-rate models as applicable to specific part subsections, except microelectronics. Part-failure-rate

Table 9-4 π *Factors for Part-Failure-Rate Models (Except Microelectronics).* (**Source:** *MIL-HDBK-217B.*)

π Factor	Description
Common Factors—Used in All, or Many, Part Categories	
π_E	Environment—accounts for influence of environmental factors other than temperature. Related to application categories.
π_Q	Quality—accounts for effects of different quality levels.
Discrete Semiconductors	
π_A	Application—accounts for effect of application in terms of circuit function.
π_C	Complexity—accounts for effect of multiple devices in a single package.
π_{S2}	Voltage stress—adjusts model for a second electrical stress (application voltage) in addition to wattage included within λ_b.
Resistors	
π_R	Resistance—adjusts model for the effect of resistor ohmic values.
π_C	Construction class—accounts for influence of construction class of variable resistors as defined in individual part specifications.
π_V	Voltage—adjusts for effect of applied voltage in variable resistors in addition to wattage included within λ_b.
π_{TAPS}	Tap connections on potentiometers—accounts for effect of multiple taps on resistance element.
Capacitors	
π_{SR}	Series resistance—adjusts model for the effect of series resistance in circuit application of some electrolytic capacitors.
π_{CV}	Capacitance value—adjusts model for effect of capacitance related to case size.
Inductive Devices	
π_F	Family—adjusts model for influence of family type as defined by individual part specifications.
Rotating Devices	
π_F	Family—accounts for motor family and commutator vs. brushless construction.
π_N	Windings—Accounts for number of windings.
Relays	
π_C	Contacts—accounts for contact quantity and form.
π_{CYC}	Cycling—accounts for time rate of actuation.
π_L	Load—accounts for type of contact load.
π_F	Family—accounts for construction and application.
Switches	
π_C	Contacts—accounts for contact quantity and form.
π_{CYC}	Cycling—accounts for time rate of actuation.
Connectors	
π_P	Contacts—accounts for quantity of contacts.
λ_{CYC}	Cycling—accounts for time rate of mating and unmating.

models for microelectronic parts (monolithic bipolar and MOS digital or linear, monolithic bipolar and MOS memories, hybrids) differ significantly from the rest of the parts and are covered in a separate section of MIL-HDBK-217B

5. Typical Examples

Two examples of part-stress-analysis prediction will be demonstrated using the appropriate failure-rate models and π factors as required by MIL-HDBK-217B. For failure-rate models and π factors for other parts, refer to MIL-HDBK-217B.

EXAMPLE 9-3 Type RNR fixed-film resistor, established reliability (ER), 10,000 Ω per MIL-R-55182, failure-rate level P, rated at 0.25 W, is used in a mobile ground installation. The ambient temperature in which the resistor operates is 55°C. The resistor dissipates 0.125 W.

Solution The subsection of MIL-HDBK-217B on resistors contains the following failure-rate model for MIL-R-55182 resistors:

$$\lambda_P = \lambda_b(\pi_E \times \pi_R \times \pi_Q) \tag{64}$$

where λ_P = failure rate, failures/10^6 h
λ_b = base failure rate
π_E = environmental factor
π_R = resistance factor
π_Q = quality factor

Tables 9-5, 9-6, and 9-7, taken from the same subsection of MIL-HDBK-217B, give values for the π factors used above. For the base failure rate λ_b tabulation for MIL-R-55182 resistors, see Table 8-1.

Table 9-5 π_E, **Environmental Factor.**

Environment	π_E
G_B	1.0
S_F	1.0
G_F	2.5
A_I	5.0
N_S	7.5
G_M	10.0
N_U	11.0
A_U	12.0
M_L	18.0

Table 9-6 π_R, **Resistance Factor.**

Resistance Range	π_R
Up to 100,000 Ω	1.0
$>$ 0.1–1 MΩ	1.1
$>$ 1.0–10 MΩ	1.6
$>$ 10 MΩ	2.5

Table 9-7 π_Q, *Quality Factor.*

Failure-Rate Level	π_Q
M (1.0%/1000 h)	1.0
P (0.1%/1000 h)	0.3
R (0.01%/1000 h)	0.1
S (0.001%/1000 h)	0.03

Substituting the appropriate values from Tables 9-5, 9-6, and 9-7 into the failure-rate model will furnish the desired failure rate for this resistor. The ratio of operating to rated wattage is

$$\frac{0.125}{0.250} = 0.5$$

From Table 8-1 for an ambient temperature of 55°C, we obtain $\lambda_b = 0.0032$ failure/10^6 h. From Table 9-3, the π_E symbol for mobile ground operation is G_M. Entering Table 9-5 with symbol G_M renders $\pi_E = 10.0$. For a resistance value of 10,000 Ω, Table 9-6 furnishes $\pi_R = 1.0$. For a failure rate level P, Table 9-7 furnishes $\pi_Q = 0.3$. Hence

$$\lambda_P = 0.0032(10 \times 1.0 \times 0.3)$$

$$= \underline{0.0096 \text{ failure}/10^6 \text{ h}}$$

EXAMPLE 9-4 Type CKR ceramic capacitor, established reliability (ER), 680 pF per MIL-C-39014, failure rate level R rated at 200 V dc, 85°C, is used in an aircraft cockpit installation. The ambient temperature is 65°C. The capacitor is used at 140 V dc.

Solution Proceeding as in Example 9-3, we find the failure-rate model in the capacitor subsection of MIL-HDBK-217B:

$$\lambda_P = \lambda_b(\pi_E \times \pi_Q) \qquad (65)$$

The applicable tables from MIL-HDBK-217B are reproduced here as Tables 9-8 and 9-9.

Table 9-8 π_E, *Environmental Factor.*

Environment	π_E
G_B	1.0
S_F	1.0
G_F	2.0
A_I	4.0
N_S	4.0
G_M	4.0
N_U	8.0
A_U	10.0
M_U	15.0

Table 9-9 π_Q, *Quality Factor.*

Failure-Rate Level	π_Q
L (5%/1000 h)	1.5
M (1%/1000 h)	1.0
P (0.1%/1000 h)	0.3
R (0.01%/1000 h)	0.1
S (0.001%/1000 h)	0.03
For MIL-C-11015	10.0

The base failure rate λ_b for MIL-C-39014 capacitors is shown in Table 8-2. The ratio of operating voltage to rated voltage S is

$$S = \frac{140}{200} = 0.7$$

The value of λ_b, for $S = 0.7$ and 65°C ambient, is therefore 0.031 failure/10^6 h.

From Table 9-3 the π_E symbol for airborne, inhabited condition is A_I. Entering Table 9-8 with symbol A_I shows a value $\pi_E = 4.0$. For a failure-rate level R, Table 9-9 furnishes $\pi_Q = 0.1$. Hence,

$$\lambda_P = 0.031(4.0 \times 0.1)$$
$$= \underline{0.0124 \text{ failure}/10^6 \text{ h}}$$

(c) Parts Count Prediction

When the design of a piece of equipment has not progressed to the point where detailed parts lists and part stresses are known, a parts count failure-rate prediction is performed. The latter may also be useful in the bid (proposal) stage.

1. Information Required for a Parts Count Failure-Rate Prediction

1. Generic part types.
2. Quantities of particular parts used.
3. Part quality levels.
4. Equipment environment.

2. Calculation of the Total Part Failure Rate

Equation (66) relates these factors to λ_{PC}, the total part failure rate, for each part category.

$$\lambda_{PC} = n \times \lambda_G \times \pi_Q \tag{66}$$

where λ_{PC} = total part category failure rate (failures/10^6 h)
   n = quantity of parts in the category
   λ_G = generic failure rate
   π_Q = quality factor

3. Equipment Failure Rate

The total equipment failure rate, λ_{equip}, represents the arithmetic sum of the individual part category failure rates, provided the entire piece of equipment is subjected to the same environment. If portions of the equipment operate in different environments, individual calculations are necessary.

MIL-HDBK-217B contains a series of tables of generic failure rate λ_G for the various part categories. Tables 9-10 for resistors and 9-11 for capacitors are typical of

Table 9-10 *Generic Failure Rate, λ_G (failures/10^6 h) for Resistors.* (Source: *MIL-HDBK-217B.*)

Construction	Style	MIL-R-Spec.	G_B	S_F	G_F	A_I	N_S	G_M	A_U	N_U	M_L
Resistors, Fixed											
Composition	RCR	39008	0.00045	0.00045	0.002	0.0048	0.0077	0.0085	0.018	0.021	0.018
	RC	11	0.0023	0.0023	0.01	0.024	0.039	0.043	0.09	0.11	0.09
Film	RLR	39017	0.0024	0.0024	0.015	0.02	0.026	0.037	0.054	0.052	0.11
	RL	22684	0.013	0.013	0.075	0.1	0.13	0.19	0.27	0.26	0.55
	RNR	55182	0.0028	0.0028	0.017	0.023	0.03	0.042	0.063	0.062	0.12
	RN	10509	0.0028	0.0028	0.017	0.023	0.03	0.042	0.063	0.062	0.12
Film, power	RD	11804	0.18	0.18	0.96	1.3	1.5	2.3	3.1	2.8	6.8
Wire-wound, accurate	RBR	39005	0.0085	0.0085	0.056	0.15	0.18	0.19	0.32	0.26	0.68
	RB	93	0.043	0.043	0.28	0.75	0.9	0.95	1.6	1.3	3.4
Wire-wound, power	RWR	39007	0.009	0.009	0.033	0.066	0.085	0.11	0.26	0.16	0.33
	RW	26	0.046	0.046	0.17	0.33	0.43	0.55	1.3	0.8	1.7
Wire-wound, chassis mount	RER	39009	0.016	0.016	0.062	0.13	0.16	0.22	0.34	0.3	0.66
	RE	18546	0.080	0.080	0.31	0.65	0.8	1.1	1.7	1.5	3.3
Resistors, Variable											
Wire-wound, trimmer	RTR	39015	0.018	0.018	0.066	0.14	0.17	0.18	0.34	0.3	1.4
	RT	27208	0.09	0.09	0.33	0.7	0.85	0.9	1.7	1.5	9.0
Wire-wound, precision	RR	12934	0.43	0.43	2.7	5.8	6.1	5.8	11.0	9.0	70.
Wire-wound, semiprecision	RA	19	0.26	*	2.3	6.4	8.6	8.5	*	*	*
	RK	39002	0.26	*	2.3	6.4	8.6	8.5	*	*	*
Wire-wound power	RP	22	0.29	*	2.3	6.	7.5	8.1	*	*	*
Non-wire-wound, trimmer	RJ	22097	1.3	N/A	4.6	9.5	13.	16.	27.	23.	100.
Composition	RV	94	0.27	N/A	3.7	20.	22.	20.	34.	34.	40.

* Not normally used in these environments.

Table 9-11 *Generic Failure Rate, λ_G (failures/10^6 h) for Capacitors. (Source: MIL-HDBK-217B.)*

Dielectric	Style	MIL-C-Spec.	G_B	S_F	G_F	A_I	N_S	G_M	A_U	N_U	M_L
Paper/plastic	CHR	39022									
	CPV	14157	0.0002	0.0002	0.0006	0.0012	0.0016	0.0012	0.012	0.009	0.006
	CQR	19978									
	CQ	19978	0.002	0.002	0.006	0.012	0.016	0.012	0.12	0.09	0.06
Mica	CMR	39001	0.0003	0.0003	0.0032	0.006	0.0078	0.006	0.048	0.034	0.03
	CM	5	0.003	0.003	0.032	0.06	0.078	0.06	0.48	0.34	0.3
	CB	10950	0.12	0.12	0.58	0.93	0.99	0.93	4.8	3.7	4.7
Glass	CYR	23269	0.0013	0.0013	0.011	0.021	0.025	0.021	0.16	0.11	0.11
Ceramic	CKR	39014	0.01	0.01	0.022	0.044	0.044	0.044	0.11	0.096	0.17
	CK	11015	0.10	0.10	0.22	0.44	0.44	0.44	1.1	0.96	1.7
Tantalum, solid	CSR	39003	0.011	0.011	0.026	0.052	0.056	0.052	0.26	0.17	0.26
Tantalum, nonsolid	CLR	39006	0.015	0.015	0.034	0.11	0.11	0.11	0.44	0.34	0.54
	CL	3965	0.15	0.15	0.34	1.1	1.1	1.1	4.4	3.4	5.4
Aluminum oxide	CU	39018	0.073	0.073	0.23	1.6	1.9	1.6	6.9	5.3	5.4
Aluminum, dry Electrolyte	CE	62	0.11	0.11	0.41	3.0	3.8	3.0	16.	14.	10.
Ceramic, variable	CV	81	0.20	0.20	1.1	2.4	2.7	2.4	22.	12.	21.
Piston, variable	PC	14409	0.016	0.016	0.11	0.41	0.47	0.41	5.3	3.8	4.9

Table 9-12 *Quality Factors, π_Q, for Resistors and Capacitors.*

Failure-Rate Level	*π_Q
L	1.5
M	1.0
P	0.3
R	0.1
S	0.01

* For Non-ER parts (styles with only two letters in Tables 9-10 and 9-11), $\pi_Q = 1$, provided that parts are procured in accordance with the part specification. For ER parts (styles with three letters), use the π_Q value for the "letter" failure-rate level procured.

the generic failure-rate tables in MIL-HDBK-217B. Table 9-12, also taken from MIL-HDBK-217B, covers the quality factor π_Q for resistors and capacitors. The environmental symbols used in Tables 9-10 and 9-11 are as defined in Table 9-3.

4. Typical Example

EXAMPLE 9-5 A preliminary design for a large *RC* network consists of 50 film resistors, type RLR, per MIL-R-39017. The range of resistance values is from 1000 to 100,000 Ω. In addition, 35 mica capacitors, type CMR, per MIL-C-39001 are used ranging from 10 to 220 pF.

The failure-rate level for the resistors is $P(0.1\%/1000 \text{ h})$ and for the capacitors $R(0.01\%/1000 \text{ h})$. The equipment is to be designed for missile launch environment (M_L). Perform a parts-count-reliability prediction.

Solution For a parts-count-reliability prediction, the range of resistance values and capacitance values are of no consequence. The total equipment failure rate is

$$\lambda_{\text{equip}} = \lambda_{PC1} + \lambda_{PC2}$$

where λ_{equip} represents the total estimated failure rate for the *RC* network

λ_{PC1} = combined failure rate for part category 1 (50 RLR resistors)

λ_{PC2} = combined failure rate for part category 2 (35 CMR capacitors)

Per equation (66) for 50 resistors,

$$\lambda_{PC1} = 50 \times 0.11 \times 0.3 = 1.65 \text{ failures}/10^6 \text{ h}$$

where $\lambda_G = 0.11$ was determined from Table 9-10 and $\pi_Q = 0.3$ is per Table 9-9 for a failure-rate level *P*. Similarly, for 35 capacitors,

$$\lambda_{PC2} = 35 \times 0.03 \times 0.1 = 0.105 \text{ failure}/10^6 \text{ h}$$

where $\lambda_G = 0.03$ is per Table 9-11 and $\pi_Q = 0.1$ per Table 9-9 for a failure-rate level *R*. Hence,

$$\lambda_{\text{equip}} = 1.65 + 0.105 = \underline{1.755 \text{ failures}/10^6 \text{ h}}$$

5. Failure-Rate Summation Form

Most companies have developed their own standardized failure-rate computation forms which facilitate the determination of equipment failure rates. Figure A2, Appendix A, is an illustration of a commercially used failure-rate summation form. Operational system failure rates have to take into consideration any parallel channels or circuit redundancies (see Chapter 10).

9-4 DERATING PROCEDURE

Parts that are stressed beyond their maximum guaranteed rating will fail quickly. If the stress on the part is reduced, its probability of survival increases. Of course, size, weight, and cost enter into a derating consideration. If, in the interest of derating we were to use only 10% of a component's rated value, this would constitute an excessive overdesign, hence, a compromise has to be struck. Table 9-13 is typical of the derating policies for electronic and electromechanical parts. According to this table, a solid tantalum capacitor rated, for example, for 100 V dc, should not be exposed to more than 65 V dc in the circuit.

Table 9-13 *Derating Table. (Courtesy RCA Corporation.)*

Part	Maximum Acceptable Usage Level (% of Part Rating)	Rated Parameter
Capacitors		
Film		
Paper	60	Voltage
Plastic, paper plastic, metallized plastic, and metallized paper plastic	75	Voltage
Electrolytic		
Aluminum	75	Voltage
Nonsolid tantalum	75	Voltage
Solid tantalum	65	Voltage
Glass and mica		
Fixed mica	80	Voltage
Fixed glass	55	Voltage
Button mica	70	Voltage
Ceramic	70	Voltage
Variable		
Ceramic and air	65	Voltage
Tubular piston	70	Voltage
Connectors		
Connectors	90 (ambient + temp. rise)	Insert temperature
Contacts	60	Current
Magnetic devices		
Transformer, Inductors, and Coils	90 (ambient + temp. rise)	Class temperature
	80	Voltage
Microelectronics	125°C	Junction temperature
Relays		
Contacts	60	Current
No. of Poles	6	Quantity
Coil	(*See* Magnetic devices)	
Resistors		
Resistors	50	Wattage
Rotating devices		
Electrical	(*See* Magnetic devices)	
Mechanical	*Preference* should be shown to: 1. Decreasing the number of brushes (preferably none) 2. Reducing the rotational speed 3. Increasing the size of synchros and resolvers	
Semiconductor devices		
Semiconductor devices	125°C	Junction temperature
Diodes	70	Voltage
Transistors	60	Voltage
Switches		
Contacts	60	Current
No. of poles	4	Quantity

It should be noted that the mere derating of parts does not relieve the designer from assuring that the part will perform adequately in the operational functioning of the circuit for all specified conditions. Under no circumstances must the maximum ratings of parts be exceeded. This condition could occur when high operating temperatures (ambient temperature plus temperature rise) have not been taken into account.

9-5 MEAN TIME BETWEEN FAILURES (MTBF)

The reliability of a small computer has to be ascertained. Its parts complement consists of 1298 parts distributed in 17 parts categories. By the use of the parts-count-reliability prediction method, an aggregate failure rate of 948 failures/10^6 h has been computed. Since it was found that 948 failures are likely to occur during an operating period of 1 million hours, then 1 million hours divided by 948 will furnish the average number of hours between failures.

$$\text{MTBF} = \frac{10^6}{948} = 1054 \text{ h}$$

Generally speaking, the MTBF is the reciprocal of the failure rate with the appropriate conversion factor in the numerator. Table 9-14 summarizes three of the most frequently used conversions.

Table 9-14 *MTBF Calculation.*

Failure Rate	MTBF
failures/hour	$\dfrac{1}{\text{failures/hour}}$
% failures/1000 h	$\dfrac{10^5}{\text{% failures/1000 h}}$
failures/10^6 h	$\dfrac{10^6}{\text{failures/}10^6 \text{ h}}$

EXAMPLE 9-6 A piece of equipment has operated for 9000 h with three failures. What is the MTBF?

Solution

$$\text{MTBF} = \frac{1}{\text{failures/hour}}$$
$$= \frac{1}{3/9000} = \frac{9000}{3} = 3000 \text{ h}$$

EXAMPLE 9-7 The MTBF of an equipment is 500 h. What is the failure rate expressed in:

(a) % Failures/1000 h?
(b) Failures/10^6 h?
(c) Failures/hour?

Solution

(a) $\text{MTBF} = \dfrac{10^5}{\% \text{ failures}/1000 \text{ h}}$

   $\% \text{ failures}/1000 \text{ h} = \dfrac{10^5}{500} = \underline{200}$

(b) $\text{MTBF} = \dfrac{10^6}{\text{failures}/10^6 \text{ h}}$

   $\text{failures}/10^6 \text{ h} = \dfrac{10^6}{500} = \underline{2000}$

(c) $\text{MTBF} = \dfrac{1}{\text{failures}/\text{hour}}$

   $\text{failures}/\text{hour} = \dfrac{1}{500} = \underline{0.002}$

The MTBF figures is not a guaranteed failure-free period. It is a mean value of operating times between failures.

EXAMPLE 9-8   To help understand the MTBF concept, assume that the telephone company performs a life test on a large number of special voice amplifiers for a 1000-h period. The elapsed time from the start of the test to the occurrence of each failure is recorded. A total of 12 failures were experienced. The occurrence of the last failure was coincident with the termination of the test. The record of the failure times and the calculated failure-free periods is shown in Table 9-15.

Table 9-15 *Tabulation of Voice-Amplifier Failure Times.*

Elapsed Time (h)	Failure-Free Period (h)
50	50
200	150
210	10
280	70
300	20
490	190
530	40
540	10
630	90
710	80
880	170
1000	120
	Total 1000

The average failure-free period for the listed 12 periods is

$$\frac{1000}{12} = 83.3 \text{ h} \quad (\text{on the average})$$

Hence, the MTBF is 83.3 h, even though the failure-free periods range from 10 to 190 h. Translated into equipment service operations, service calls may be expected, on the average, once every 83.3 h during the 1000-h operating period.

9-6 CORRECTIVE MAINTENANCE EVENTS (SERVICE CALLS)

Since the mean time between failures is an indication of how many trouble-free hours of operation may, on the average, be expected from a piece of equipment, it may be used to approximate the average anticipated number of corrective maintenance events (service calls). For example, in the computer example presented in Section 9.5, operation for a period of 50,000 h may be desired. Assume that a large number of these computers are in operation. Since we anticipate approximately 948 failures in 1 million hours of operation, we may expect, on the average:

$$948 \times \frac{50,000}{10^6}$$

or approximately 48 failures in 50,000 h of operation per computer.

Since 48 failures represents approximately 5% of the 948 failures, the expected number of failures for each parts category will also be approximately 5% of the average number of failures that would have occurred in that part category in 10^6 h. It is apparent that parts categories that do not have an aggregate failure rate of at least 20 failures/10^6 h will, on the average, not even produce one failure in 50,000 h of operation (ignoring fractions of one failure). For example, assume the calculated failures/10^6 h for mica capacitors is 9.60. For 50,000 h of operation we use 5% of that value, or

$$9.60 \times 0.05 = 0.48 \text{ failure}/10^6 \text{ h}$$

This means that not even one failure of a mica capacitor is, on the average, anticipated.

Suppose that the equipment uses four parts categories, each with a total failure rate above 20 failures/10^6 h. The number of anticipated failure actions for these categories is arrived at by multiplying the corresponding failures/10^6 h by 0.05. All other parts categories with less than 20 failures/10^6 h comprise the group with less than one whole failure likely in 50,000 h of operation. Table 9-16 shows the schedule of anticipated corrective maintenance events likely to be needed for this computer.

Table 9-16 *Corrective Maintenance Events.*

Part Type	Failures/10^6 h	Average Anticipated Number of Maintenance Events per Computer
Diodes	150	8
Transistors	280	14
Resistors, fixed film	300	15
Integrated circuits	80	4
All other part categories	<20 each	<1 each

We can already predict from Table 9-16 that probably 8 of the anticipated 48 repairs will involve diodes, 14 repairs will involve transistors, 15 repairs will involve fixed-film resistors, 4 repairs will involve integrated circuits, and the remaining 7 repairs will involve the other part categories.

The value of this type of assessment is apparent. We can predict the average, approximate number of service calls (approximately 48 service calls per computer in 6 years of operation), and we can determine which part types are likely to fail during that period and use this information for planning purposes.

The ultimate goal for a spares-provisioning program is that there be a high probability (say, 95%) that spares will be available when and where needed. Mathematical procedures are used which define, with a desired probability of success, an appropriate number of spares based on the reliability of the particular part and the logistic sparing situation. Chapter 13, which deals with integrated logistic support, discusses some of these aspects of spares provisioning.

9-7 THE THREE PERIODS OF EQUIPMENT LIFE

The life of equipment may be divided into three major distinguishable periods:

1. Infant mortality period (also called burn-in or "debug" period).
2. Useful life period.
3. Wearout period.

Figure 9-1 illustrates these three periods with a plot of failure rate versus time. During the infant mortality period, the failure rate is high, owing to the presence of weak or substandard components. As these components drop out one by one, the failure rate keeps decreasing until a relatively low constant level is obtained at time t_1.

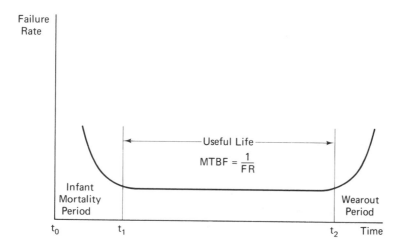

Figure 9-1 *Failure rate related to equipment operating time.*

This is the beginning of the useful life period, during which only random failures occur. During this period no weak components are assumed to have remained from the infant mortality period. Also, no deterioration of the components during the useful life period takes place, and there are no wearout conditions present. The failures that do occur during the period are truly random, unpredictable, and cannot be prevented by additional testing or burn-in of the components. The reason for the occurrence of these failures is not fully understood, but they are thought to be at least partially due to abrupt changes in stress distribution in the components.

As long as the components that fail during the useful life period are replaced, so that the population of components remains intact, the same number of components will, on the average, fail in any equal time period. It is during this period that the MTBF concept applies. The MTBF represents the reciprocal of the constant failure rate present during the useful-life period. During the infant mortality period and the wearout period, which follows the useful life, the MTBF concept is not applicable.

The wear-out period, beginning with time t_2, is characterized by a rapidly rising failure rate as more and more components break down.

9-8 MEAN LIFE

If we plot the number of failures that occur during the wearout period versus time, a normal curve is obtained. A plot of this curve is shown in Figure 9-2. Time t_2 represents the point in time when those components which are left at the end of the useful

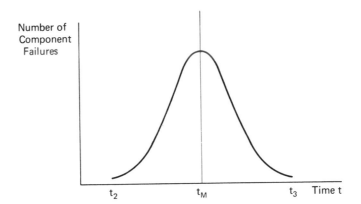

Figure 9-2 *Plot of wearout failures.*

life period enter the wear-out phase. During the wear-out phase, which lasts between times t_2 and t_3, all remaining components will have failed due to wear-out. During the time t_2 to t_M, half of the components, intact at time t_2, will have failed. In this book the time from the beginning of the infant mortality period to time t_M is called the "mean wear-out life" of the components, or simply the "mean life" of the components.

The difference between MTBF and mean life is noteworthy: the MTBF applies to the useful life period and designates the average failure-free period in hours. If the failure rate during this period is very low, the MTBF is correspondingly high, and this is independent of the length of the useful period of life. This is illustrated by the following example.

EXAMPLE 9-9   Certain components, such as ultra-high-speed motors, may only have a short useful life period of, say, about 500 h of continuous operation. After operating for this period of time, these motors have been shown to wear out somewhere during the next 100 h. The failure rate during the useful-life period of 500 h is extremely low; in other words, these motors perform extremely reliably during their useful-life period. Let us assume that the failure rate during this period is 10 failures/10^6 h. This means that there would be 10 failures in 1 million hours, or 1 failure in 100,000 h. Since the useful-life period is only 500 h, obviously we cannot operate the motors for 100,000 h; the wear-out stage would be reached before. The failure rate of 1 failure/100,000 h is merely the rate at which these motors are failing during the 500-h useful-life period, assuming scheduled motor replacement at, say, 400 hours.

There is far less than one motor failing on the average during this period. The fractional number of failures during the 500-h period is

$$\frac{1}{100,000} = \frac{X}{500}$$

$$X = \frac{500}{100,000} = \underline{0.005 \text{ failure/500 h}}$$

The MTBF for the 500-h period is correspondingly high:

$$\text{MTBF} = \frac{10^6}{\text{failures}/10^6 \text{ h}}$$

$$= \frac{10^6}{10} = 100,000 \text{ h}$$

As indicated above, since we are only operating for 500 h, an average failure-free period of 100,000 h is theoretical in significance.

Assume that in our application we are dealing with a quantity of 60 motors. There will in all likelihood be no failures during the 500-h life period. During the next 110 h each of the 60 motors will fail. Their failure times will be distributed in bell-shaped form (normal curve). Table 9-17 shows a typical distribution of these failures during the wear-out period.

Figure 9-3 is a plot of the failure data listed in Table 9-17. The mean of the bell-shaped curve is seen to occur 55 h after the onset of the wear-out period, or 555 h after the start of the useful-life period. The mean life for these motors is therefore 555 h.

It was stated earlier that as long as we keep replacing the failed components during the useful-life period of a piece of equipment, the failure rate will remain constant.

Table 9-17 *Distribution of Motor Failures During Wear-out.*

Wear-out Interval (h)	Number of Motor Failures
500–509.9	1
510–519.9	2
520–529.9	4
530–539.9	7
540–549.9	10
550–559.9	12
560–569.9	10
570–579.9	8
580–589.9	3
590–599.9	2
600–609.9	1
Total	60

We may ask the question: What if we choose not to replace the failed items during this period? Obviously, the population of components will shrink. If the failure rate happens to be high, it is not inconceivable that most components may fail before the end of the useful-life period is reached. In fact, it can be shown mathematically that on the average 63.2% of the components will have failed during that portion of the useful-life period which equals the MTBF period. This is illustrated by the following example.

EXAMPLE 9-10 The failure rate of a certain computer has been established as 20 failures during every 1000 h of operation. Five hundred computers are pressed into service and every computer that fails is removed but not replaced. How long will it be until only 36.8% of the original 500 computers will be potentially left in service?

Solution

$$20 \text{ failures/1000 h equals } \tfrac{1}{50} \text{ failure/h}$$

$$\text{MTBF} = \frac{1}{\frac{1}{50}} = 50 \text{ h}$$

It was stated above that without replacing failed components during the useful-life period, 63.2% of the total number of components will, on the average, have failed in a time period equaling the MTBF time. Hence, since the MTBF = 50 h, 63.2% of the 500 units may be expected to have failed during this period, leaving

$$
\begin{array}{r}
100\% \\
- \ 63.2\% \\
\hline
36.8\% \text{ of the units intact}
\end{array}
$$

We may conclude that in 50 h, 36.8% of the original number of computers will, on the average, be left. However, the actual percentage may vary considerably from this statistical average for any given 50-hour period.

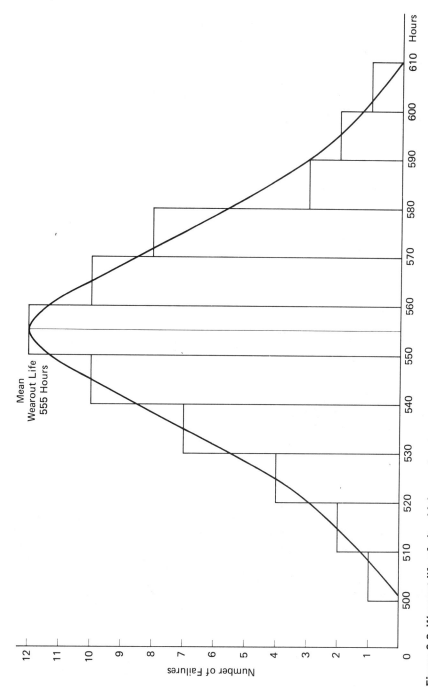

Figure 9-3 *Wearout life of ultra high speed motors.*

PROBLEMS

9-1 Define failure rate.

9-2 A 750-h life test is performed on six components. One component fails after 350 h of operation. All others survive the test. Compute the failure rate.

9-3 Determine the failure rate in %/1000 h for the following test:

test duration: 1000 h
number of units tested: 12
number of failed units: 5
failure history:

Unit No.	Hours to Failure
2	620
3	390
4	780
8	860
9	950

9-4 (a) Convert 0.000053 failure/hour to failures/10^6 h.
(b) Convert 12 failures/10^6 h to failures in %/1000 h.

9-5 Name and discuss the two methods of reliability prediction presented in MIL-HDBK-217-B.

9-6 What factors influence part failure rates? Discuss briefly.

9-7 What are established reliability specifications? Name the levels.

9-8 A type RNR 1/8-W 82,000-Ω fixed-film resistor per MIL-R-55182 is used in a Naval (sheltered) application. The ambient temperature is 65°C and the resistor dissipates 0.1 W. The resistor was purchased to failure-rate level R. What is the failure rate? The failure-rate model for RNR resistors is per equation (64) and the base failure rates are per Table 8-1. For π_E, π_R, and π_Q, use Tables 9-5, 9-6, and 9-7, respectively.

9-9 Determine the failure rate for a 4700-pF ceramic capacitor, type CK, per MIL-C-11015. The capacitor voltage rating is 100 V dc and it is used in a circuit that applies 40 V dc to the capacitor. The ambient temperature is 55°C and the equipment in which the resistor is used is designed for mobile ground use. The failure-rate level is P. For the failure-rate model, use equation (65), and for the base failure rate, use Table 8-2. For π_E and π_Q, use Tables 9-8 and 9-9, respectively.

9-10 Use the MIL-HDBK-217B parts-count-reliability-prediction method to compute the total failure rate of the following network used in a space flight environment:
(a) Resistors:

Type	Quantity	Resistance Range	Failure-Rate Level
RNR	21	10–10,000 Ω	R

(b) Capacitors:

Type	Quantity	Capacitance Range	Failure-Rate Level
CKR	15	47–47,000 pF	P

9-11 What is meant by parts derating and what is the purpose of this practice?

9-12 Using Table 9-13, determine the maximum dc voltage that should be applied in the particular circuit applications to the following types of capacitors:

Capacitor Type	Rated Voltage (dc)
Solid tantalum	50
Button mica	100
Ceramic	200
Tubular piston	1250

9-13 What is the maximum power which should be dissipated by a $\frac{1}{4}$-W fixed-film resistor, type RLR, per MIL-R-39017? Use Table 9-13.

9-14 (a) Define MTBF. Explain why MTBF is not a guaranteed failure-free period.
(b) What is the MTBF of a circuit whose total failure rate is 158 failures/10^6 h?
(c) Convert an MTBF of 2500 h to a failure rate in %/1000 h.

9-15 (a) The MTBF of a circuit is 1500 h. What is the corresponding failure rate expressed in failures/hour?
(b) A small computer has operated for 800 h with 4 failures. What is the MTBF of the computer?

9-16 (a) Discuss the three periods of equipment life.
(b) What is the bathtub curve?

9-17 Will the burning-in of components improve reliability during the useful portion of equipment life? Explain.

9-18 (a) What probability distribution is applicable during the wear-out period?
(b) Is the equipment failure rate presumed to be constant or variable during the useful portion of equipment life? Explain.

9-19 A complex electronic equipment has an aggregate failure rate of 650 failures/10^6 h. Semiconductors, resistors, capacitors, and inductive components are the four dominant part types. They represent the following percentages of the aggregate equipment failure rate:

Semiconductors	26%
Resistors	23%
Capacitors	21%
Inductive components	10%

How many corrective maintenance events (service calls) may, on the average, be expected for each of these four part types during a 40,000-h operational period?

9-20 (a) Explain mean life.

(b) Explain how components may have an extremely high MTBF and a comparatively short mean life.

9-21 The useful life period of a certain group of components is 850 h. The peak of the wear-out period occurs 100 h thereafter. What is the mean life of these components?

9-22 The MTBF of an electronic module was computed as 1250 h. If 50 of these modules are used in a piece of electronic equipment but failed modules are not replaced, how many modules can on the average be expected to be operational after 1250 h of equipment operation?

IO

RELIABILITY PREDICTION METHODS

In this chapter we use the exponential distribution to determine the reliability of various series and parallel circuits arrangements. We also examine the effects of redundancy on reliability. The investigation of redundant circuit arrangements will cover active parallel redundancy as well as standby redundancy arrangements with sensing and switching elements.

10-1 APPLYING THE POISSON DISTRIBUTION

In our discussion of the Poisson probability distribution we defined the quantity a or λt as the expected number of occurrences of an event and we tabulated the terms of the Poisson distribution in Table 2-7. The first Poisson term listed in the above mentioned table is referred to as the exponential distribution and is important in equipment reliability prediction. The probability of no equipment failure as expressed by this term is

$$P = e^{-a} \tag{67}$$

where P = probability of no failure (or the probability of survival)
e = base of the natural logarithm
a or λt = expected number of failures

Values of the quantity e^{-a}, expressed as $e^{-\lambda t}$ can be found from Table A10, Appendix A. The expected number of failures, as was already indicated, is the product of failure rate and the time of operation, assuming constant failure rate.

$$a = \text{failure rate} \times \text{time} = \lambda t \tag{68}$$

where a = expected number of failures
λ = failure rate, failures/hour
t = time of operation, hours

The expected number of failures may also be expressed in terms of MTBF. Since MTBF = $1/\lambda$, we can express a or λt as

$$\frac{1}{\text{MTBF}} \times t \tag{69}$$

Substituting equations (68) and (69) in equation (67), we obtain, for the probability of survival,

$$P = e^{-\lambda t} \tag{70}$$

and, in terms of the MTBF,

$$P = e^{-t/\text{MTBF}} \tag{71}$$

EXAMPLE 10-1   A piece of equipment has to operate for 10,000 h. Its calculated failure rate is 0.00001 failure/hour. What is the probability of survival of the equipment?

Solution

$$P = e^{-\lambda t}$$
$$= e^{-0.00001 \times 10,000}$$
$$= e^{-0.1}$$

From Table A10, Appendix A, for $\lambda t = 0.1$, we find $e^{-\lambda t} = 0.90484$. Hence, $P = e^{-0.1} = 0.90484$. The probability of survival is 90.48%.

Note that if the failure rate is not given in failures per hour, it must be converted to failures per hour to be used in equation (70) (see Table 9-1, p. 241).

EXAMPLE 10-2   An amplifier has a failure rate of 15%/1000 h (equivalent to 15 failures/10^5 hours) and is to operate for 1000 h. What is the probability of survival?

Solution

$$P = e^{-15 \times 10^{-5} \times 1000}$$
$$= e^{-0.15}$$

From Table A10, Appendix A,

$$P = \underline{0.8607}$$

EXAMPLE 10-3   Repeat Example 10-2 except for a failure rate of 15 failures/10^6 h.

Solution

$$P = e^{-15 \times 10^{-6} \times 1000}$$
$$= e^{-0.015}$$
$$= \underline{0.98511}$$

EXAMPLE 10-4   What is the highest failure rate, expressed in failures/10^6 h, that a piece of equipment can have if it is to operate with a probability of survival of 90% for 5000 h?

Solution

$$P = e^{-\lambda t}$$
$$= e^{-5000\lambda}$$

Since P is given as 90% (or 0.9), we look up in Table A10, Appendix A, to find the nearest value for $e^{-\lambda t} = 0.9$. The nearest value is

$$e^{-\lambda t} = 0.90032$$

corresponding to $\lambda t = 0.105$. Now we know that

$$\lambda t = 0.105$$

Since we know λt, we can calculate λ:

$$5000\,\lambda = 0.105$$
$$\lambda = \frac{0.105}{5000} = 0.000021 \text{ failure/hour}$$
$$= \underline{21 \text{ failures}/10^6 \text{ h}}$$

EXAMPLE 10-5   The MTBF of a piece of equipment is 1000 h. What is the probability of survival of this equipment for 200 h of operation?

Solution

$$P = e^{-t/\text{MTBF}}$$
$$= e^{-200/1000} = e^{-0.2}$$
$$= \underline{0.81873}$$

EXAMPLE 10-6   What is the minimum MTBF for a piece of equipment if the probability of survival is to be at least 85% for 4000 h of operation?

Solution

$$P = e^{-t/\text{MTBF}}$$
$$= e^{-4000/\text{MTBF}}$$

Use Table A10 Appendix A, to find nearest value for $e^{-\lambda t}$ which equals 0.85 and determine the corresponding value for λt. The nearest value for $e^{-\lambda t} = 0.85$ is $e^{-\lambda t} = 0.85044$, for which $\lambda t = 0.162$. Hence,

$$\lambda t = 0.162 = \frac{t}{\text{MTBF}}$$
$$\frac{4000}{\text{MTBF}} = 0.162$$
$$\text{MTBF} = \frac{4000}{0.162} = 24{,}691 \text{ h (min)}$$

10-2 PROBABILITY OF SURVIVAL OF SERIES SYSTEMS

In the discussion dealing with the probability of the occurrence of independent events, Example 2-8 considered the survival probability of a system consisting of four subsystems connected in series (all four subsystems critical to system survival). If the probabilities of success of the four subsystems are P_1, P_2, P_3, and P_4, respectively, the probability that all four components will survive was found to be

$$P_{\text{system}} = P_1 \times P_2 \times P_3 \times P_4$$

This equation applies for series systems in general.

EXAMPLE 10-7 A filter and an amplifier comprise a system that is to operate with a probability of success of 90% for 1000 h. The amplifier has a probability of survival of 95% for 1000 h. What probability of survival is required for the filter?

Solution

$$P_{\text{system}} = P_{\text{filter}} \times P_{\text{amplifier}}$$

$$0.90 = P_{\text{filter}} \times 0.95$$

$$P_{\text{filter}} = \frac{0.90}{0.95} = \underline{0.947}$$

EXAMPLE 10-8 An electronic system consists of a power supply, a receiver, and an amplifier. The equipment is to operate for 1000 h. The failure rates are

$$\lambda \text{ power supply} = 25 \text{ failures}/10^6 \text{ h}$$

$$\lambda \text{ receiver} = 30 \text{ failures}/10^6 \text{ h}$$

$$\lambda \text{ amplifier} = 15 \text{ failures}/10^6 \text{ h}$$

What is the probability of survival of this equipment?

Solution

$$P_{\text{power supply}} = e^{-25 \times 10^{-6} \times 1000}$$

$$= e^{-0.025} = 0.97531$$

$$P_{\text{receiver}} = e^{-30 \times 10^{-6} \times 1000}$$

$$= e^{-0.03} = 0.97045$$

$$P_{\text{amplifier}} = e^{-15 \times 10^{-6} \times 1000}$$

$$= e^{-0.015} = 0.98511$$

$$P_{\text{equipment}} = P_{\text{power supply}} \times P_{\text{receiver}} \times P_{\text{amplifier}}$$

$$= 0.97531 \times 0.97045 \times 0.98511$$

$$= \underline{0.93239}$$

The same answer could have been obtained by adding the expected number of failures λt of the three components of the system.

$$\lambda t_{\text{power supply}} = 25 \times 10^{-6} \times 1000$$

$$\lambda t_{\text{receiver}} = 30 \times 10^{-6} \times 1000$$

$$\lambda t_{\text{amplifier}} = \underline{15 \times 10^{-6} \times 1000}$$

$$\lambda t_{\text{total}} = 70 \times 10^{-6} \times 1000$$

$$P_{\text{equipment}} = e^{-\lambda t_{\text{total}}}$$

$$= e^{-0.07}$$

$$= \underline{0.93239}$$

In general terms, survival probability of a system containing a number of subsystems in a series configuration is

$$P = e^{-(\lambda_1 + \lambda_2 + \lambda_3 + \cdots)t} \tag{72}$$

EXAMPLE 10-9 Calculate the probability of survival of a piece of equipment that is to operate for 500 h and which consists of four subsystems having the following MTBF's:

Subsystem A: MTBF = 5000 h
Subsystem B: MTBF = 3000 h
Subsystem C: MTBF = 15,000 h
Subsystem D: MTBF = 15,000 h

Solution

$$\lambda_A = \frac{1}{5000} \text{ failure/hour}$$

$$\lambda_B = \frac{1}{3000} \text{ failure/hour}$$

$$\lambda_C = \frac{1}{15,000} \text{ failure/hour}$$

$$\lambda_D = \frac{1}{15,000} \text{ failure/hour}$$

$$\lambda_{\text{total}} = \frac{1}{5000} + \frac{1}{3000} + \frac{1}{15,000} + \frac{1}{15,000}$$

$$= \frac{3 + 5 + 1 + 1}{15,000}$$

$$= \frac{10}{15,000} = \frac{1}{1500}$$

$$= 0.00066 \text{ failure/hour}$$

$$P = e^{-(0.00066)500}$$

$$= e^{-0.33}$$

$$= \underline{0.71892}$$

Hence, the probability of survival of the system for 500 h is 71.9%.

10-3 PROBABILITY OF SURVIVAL OF PARALLEL REDUNDANT SYSTEMS- ALL SYSTEMS ARE ACTIVATED

Figure 10-1 depicts a parallel redundant system whose probability of survival is to be determined. There are three possible ways (also called "success states") in which this system is able to operate:

1. Subsystems A and B are working.
2. Subsystem A is working, B has failed.
3. Subsystem B is working, A has failed.

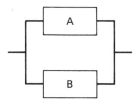

Figure 10-1 *Parallel redundant system.*

The survival of the system is therefore assured if either A or B, or both, are working. This condition exactly fits the case of not-mutually-exclusive events covered in Section 2.3. The reliability of the system is

$$P_{A \text{ and/or } B} = P_A + P_B - P_{A \text{ and } B}$$

or

$$P_{A \text{ and/or } B} = P_A + P_B - P_A P_B \tag{73}$$

EXAMPLE 10-10   Using the parallel system described above, let us assume that

$$P_A = 0.8$$
$$P_B = 0.6$$

What is the reliability of the system?

Solution

$$P_{A \text{ and/or } B} = 0.8 + 0.6 - 0.8 \times 0.6$$
$$= 1.4 - 0.48 = \underline{0.92}$$

Assuming the two subsystems A and B have identical probability of survival ($P_A = P_B$), equation (73) reduces from

$$P_{A \text{ and/or } B} = P_A + P_B - P_A P_B$$

to

$$P_{A \text{ and/or } B} = P_A + P_A - P_A^2$$
$$= 2P_A - P_A^2$$
$$= P_A(2 - P_A), \text{ or generally } P(2 - P) \tag{74}$$

EXAMPLE 10-11   A system consists of two identical subsystems in parallel redundancy. The probability of survival of each subsystem is $P = 0.96$. What is the reliability of the system?

Solution

$$P_{system} = P(2 - P)$$
$$= 0.96(2 - 0.96)$$
$$= \underline{0.9984}$$

The following is a general equation for identical subsystems in parallel redundancy:

$$P_{system} = 1 - Q^n \qquad (75)$$

where P_{system} = reliability of the system
Q = unreliability of each parallel subsystem ($Q = 1 - P_{subsystem}$)
n = number of parallel subsystems

EXAMPLE 10-12   A system consists of three identical subsystems in parallel redundancy. The probability of survival of a subsystem is $P = 0.8$. What is the probability of survival of the system?

Solution   Using equation (75),

$$P_{system} = 1 - Q^n$$
$$n = 3$$
$$Q_{subsystem} = 1 - P_{subsystem}$$
$$= 1 - 0.8$$
$$Q = 0.2$$
$$P_{system} = 1 - (0.2)^3$$
$$= 1 - 0.008$$
$$= \underline{0.992}$$

The increase in reliability of the system from $P = 0.8$ for a single subsystem compared to $P = 0.992$ for a total of three subsystems in parallel should be noted.

EXAMPLE 10-13   The failure rate of an electronic subsystem is 0.0005 failure/hour. If an operational period of 500 h with a probability of success of $P = 0.95$ is desired, what level of parallel redundancy is needed?

Solution   Using equation (75),

$$P_{system} = 1 - Q^n$$

we calculate the value of Q^n:

$$0.95 = 1 - Q^n$$
$$Q^n = 1 - 0.950$$
$$= 0.05$$

Next, we calculate the magnitude of Q:

$$P_{\text{subsystem}} = e^{-\lambda t}$$
$$= e^{-(0.0005)500}$$
$$= e^{-0.25}$$

From Table A10, Appendix A, we find that

$$P_{\text{subsystem}} = 0.77880$$

Therefore, since $P + Q = 1$,

$$Q = 1 - P$$
$$= 1 - 0.77880$$
$$= 0.22120$$

From the above we know that

$$Q^n = 0.05$$

Hence,

$$(0.22120)^n = 0.05$$

The exponent n is the unknown and represents the number of parallel subsystems required to obtain a system reliability of 0.95. We therefore have to ascertain how many times 0.22120 has to be multiplied by itself to obtain a value as close to 0.05 as possible, but no more than 0.05.

Let us try one subsystem in parallel redundancy with the original system (see Figure 10-2).

$$(0.22120)^2 = 0.048929$$

The answer $Q^n = 0.048929$ is satisfactory, since it is slightly less than the limit of 0.05.

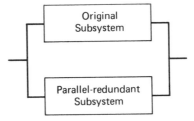

Figure 10-2 *Single parallel redundant subsystem.*

Trying two subsystems in parallel redundancy with the original subsystem would have rendered

$$(0.22120)^3 = 0.010823$$

This result is considerably lower than the required 0.05; hence, we surmise that one unit in parallel redundancy with the original unit is all that is needed to assure not less than $P = 0.95$ reliability.

Since $Q^n = (0.22120)^n$ turned out to be 0.048929 instead of the 0.05 the reliability of the system will be slightly above the desired $P = 0.95$. Let us calculate this reliability:

$$P_{system} = 1 - Q^n$$
$$= 1 - (0.22120)^2$$
$$= 1 - 0.048929 = 0.951071$$

Had we used an additional parallel-redundant subsystem, the system reliability would have been

$$P_{system} = 1 - (0.22120)^3$$
$$= 1 - 0.010823$$
$$= 0.989177$$

Clearly, this level of redundancy would have been in excess of the requirement.

Let us now calculate some examples involving configurations of parallel-redundant subsystems in series with single subsystems.

EXAMPLE 10-14   What is the probability of survival of the system shown in Figure 10-3?

Subsystem A: $P_A = 0.9$
Subsystem B: $P_B = 0.8$
Subsystem C: $P_C = 0.6$

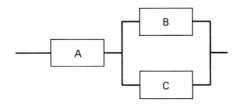

Figure 10-3 *System configuration involving a single and a parallel subsystem.*

Solution

$$P_{system} = P_A \times P_{parallel\ subsystem}$$

Using equation (73),

$$P_{parallel\ system} = P_B + P_C - P_B P_C$$

Hence,

$$P_{system} = P_A(P_B + P_C - P_B P_C)$$
$$= 0.9(0.8 + 0.6 - 0.8 \times 0.6)$$
$$= 0.9(1.4 - 0.48)$$
$$= 0.9(0.92) = \underline{0.828}$$

EXAMPLE 10-15 What is the reliability of the system shown in Figure 10-4?

$$P_A = P_B = P_C = 0.8 \text{ (parallel redundancy)}$$
$$P_D = 0.95$$
$$P_E = 0.85$$

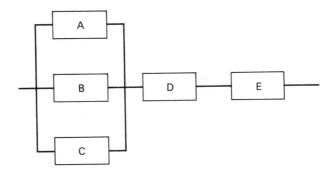

Figure 10-4 *Parallel-redundant subsystems followed by two single subsystems.*

Solution

$$P_{\text{system}} = P_{\text{parallel subsystems}} \times P_D \times P_E$$

$$
\begin{aligned}
P_{\text{parallel subsystems}} &= 1 - Q^3 \\
&= 1 - (1 - P)^3 \\
&= 1 - (1 - 0.8)^3 \\
&= 1 - 0.008 = 0.992 \\
P_{\text{system}} &= 0.992 \times 0.95 \times 0.85 \\
&= \underline{0.80104}
\end{aligned}
$$

How would the reliability improve further if we made subsystem E also parallel-redundant?

$$
\begin{aligned}
P_{\text{system}} &= 0.992 \times 0.95 \times P_{\text{parallel-redundant}} \\
P_{\text{parallel-redundant}} &= 1 - (1 - 0.85)^2 \\
&= 1 - 0.15^2 \\
&= 1 - 0.0225 \\
&= 0.9775 \\
P_{\text{system}} &= 0.992 \times 0.95 \times 0.9775 \\
&= \underline{0.92119}
\end{aligned}
$$

The system reliability is seen to improve from 0.80104 to 0.92119. The system configuration would look as shown in Figure 10-5.

Subsystems A, B and C are identical units, as are E and F. The subsystems connected in parallel are all under power throughout the operation of the system.

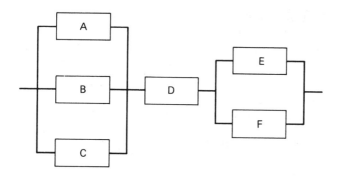

Figure 10-5 *Configuration of improved reliability system.*

In the more general case in which each subsystem of a parallel redundant system has a different reliability, we write for two subsystems

$$P_{\text{system}} = 1 - (1 - P_A)(1 - P_B) \tag{76}$$

and for three or more subsystems

$$P_{\text{system}} = 1 - (1 - P_A)(1 - P_B)(1 - P_C) \ldots . \tag{77}$$

10-4 APPLICATION OF BINOMIAL DISTRIBUTION TO PARALLEL-REDUNDANCY CASES

Equation (13) gave us the expansion for

$$(p + q)^3 = p^3 + 3p^2q + 3pq^2 + q^3$$

Consider Example 10-15, where identical subsystems A, B, and C were connected in parallel redundancy:

$$P_A = P_B = P_C = 0.8$$

and

$$P_{\text{parallel subsystem}} = 1 - (1 - 0.8)^3 = 0.992$$

Which of the terms of the binomial expansion apply to this example? Clearly, we have no system failure if one of the following conditions prevails, representing the situation where one or more subsystems are operating:

1. Subsystems A, B, and C are working.
2. Subsystems A and B are working, C failed [1]
3. Subsystems B and C are working, A failed.
4. Subsystems A and C are working, B failed.
5. Subsystem A is working, B and C failed.
6. Subsystem B is working, A and C failed.
7. Subsystem C is working, A and B failed.

---

[1] Parts are assumed to fail open.

p^3 represents A, B, C working

$3p^2q$ represents $\begin{cases} A, B \text{ working}; C \text{ failed} \\ B, C \text{ working}; A \text{ failed} \\ A, C \text{ working}; B \text{ failed} \end{cases}$

$3pq^2$ represents $\begin{cases} A \text{ working}; B, C \text{ failed} \\ B \text{ working}; A, C \text{ failed} \\ C \text{ working}; A, B \text{ failed} \end{cases}$

Therefore, $p^3 + 3p^2q + 3pq^2$ represent the desired condition of at least one subsystem operating. Since

$$p_{A,B,C} = 0.8$$

$$q_{A,B,C} = 1 - 0.8 = 0.2$$

$$p^3 + 3p^2q + 3pq^2 = 0.8^3 + 3(0.8)^2(0.2) + 3(0.8)(0.2)^2$$

$$= 0.512 + 0.384 + 0.096$$

$$= \underline{0.992}$$

This checks with the results of Example 10-15.

It now becomes apparent what the origin of the formula for probability of survival of identical parallel-redundant subsystems is. Since $p^3 + 3p^2q + 3pq^2 + q^3 = 1.0$, as it includes all possible combinations of success and failure of three parallel subsystems, we can state that

$$p^3 + 3p^2q + 3pq^2 = 1 - q^3$$

and $1 - (q)^3$ is exactly what we used to obtain

$$P_{\text{system}} = 0.992 \quad \text{in Example 10-15}$$

Therefore, the formula for parallel-redundancy $P = 1 - Q^n$ equals the value obtained when we subtract the binomial term Q^n from 1.0; we thereby obtain all the binomial terms representing success of one or more of the paralleled subsystems.

10-5 SERIES–PARALLEL REDUNDANCY

To improve reliability, units are sometimes assembled in series–parallel redundancy. Figure 10-6 represents one of the basic forms of a series–parallel arrangement. The reliability for this arrangement is

$$P_{\text{system}} = 1 - (1 - P_A P_B)(1 - P_C P_D) \tag{78}$$

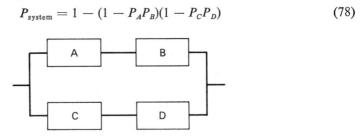

Figure 10-6 *Basic series-parallel redundancy.*

When the subsystems have equal reliability P, equation (78) reduces to

$$P_{\text{system}} = 1 - (1 - P^2)^2 \tag{79}$$

EXAMPLE 10-16   Assume that $P_A = P_B = P_C = P_D = 0.90$. What is the overall system reliability assuming a configuration as in Figure 10-6?

Solution

$$P_{\text{system}} = 1 - (1 - 0.9^2)^2 = 1 - (0.19)^2$$
$$= 0.9639$$

Generally, for n identical units connected in the series–parallel redundant configuration shown in Figure 10-7, the following equation applies:

$$P_{\text{system}} = 1 - (1 - P^n)^2 \tag{80}$$

where P is the reliability of one unit and n is the number of units per redundant subsystem (or "branch"). The exponent 2 in equation (80) is for two branches. The number of branches determines the value of this exponent (the assumption is made that only one branch has to survive).

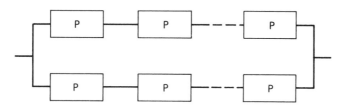

Figure 10-7 *Identical units in series-parallel redundancy—two branches.*

EXAMPLE 10-17   What is the reliability of the system shown in Figure 10-8?

Solution   Using equation (79) for $n = 3$, we have

$$P_{\text{system}} = 1 - (1 - 0.9^3)^2$$
$$= 1 - (1 - 0.729)^2$$
$$= 1 - 0.271^2$$
$$= 0.92656$$

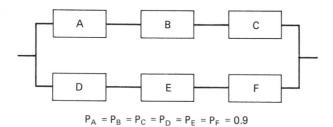

$$P_A = P_B = P_C = P_D = P_E = P_F = 0.9$$

Figure 10-8 *Series-parallel redundant system.*

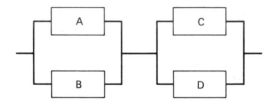

Figure 10-9 *Redundant units in series configuration.*

Figure 10-9 represents another common series–parallel configuration. The reliability for this arrangement is

$$P_{\text{system}} = [1 - (1 - P_A)(1 - P_B)][1 - (1 - P_C)(1 - P_D)] \tag{81}$$

when the subsystems have equal reliability P equation (81) reduces to

$$P = [1 - (1 - P)^2]^2 \tag{82}$$

EXAMPLE 10-18   Assuming that $P_A = P_B = P_C = P_D = 0.90$ in Figure 10-9, what is the overall reliability?

Solution

$$P_{\text{system}} = [1 - (1 - 0.9)^2]^2$$
$$= [1 - 0.1^2]^2$$
$$= (0.99)^2$$
$$= \underline{0.9801}$$

In the case depicted by Figure 10-10, which includes n identical pairs, the reliability of the system is expressed by

$$P_{\text{system}} = [1 - (1 - P)^2]^n \tag{83}$$

where P is the reliability of one unit.

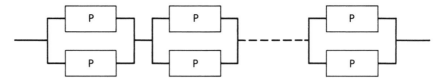

Figure 10-10 *Identical redundant pairs in series.*

EXAMPLE 10-19   What is the reliability of the system shown in Figure 10-11?

$$n = 3$$

Solution

$$P_{\text{system}} = [1 - (1 - 0.9)^2]^3$$
$$= [1 - (0.1)^2]^3$$
$$= (0.99)^3$$
$$= \underline{0.97029}$$

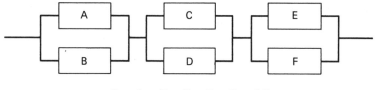

$$P_A = P_B = P_C = P_D = P_E = P_F = 0.9$$

Figure 10-11 *Three identical redundant pairs in series.*

EXAMPLE 10-20 Figure 10-12a illustrates a system consisting of two series elements. The reliability of both elements is the same.

$$P_A = P_B = 0.8$$

Determine the reliability improvement possible by the configurations shown in Figure 10-12b and 10-12c ($P_A = P_B = P_C = P_D = 0.80$).

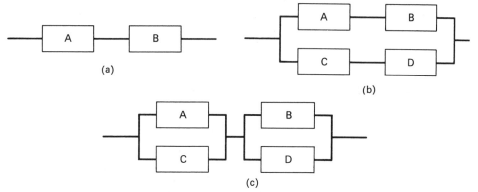

Figure 10-12 *(a) Simple series system; (b) series parallel redundancy; (c) redundant units in series configuration.*

Solution

(a) Reliability for the basic unimproved configuration per Figure 10-12a:

$$P = P_A \cdot P_B$$
$$= 0.8 \times 0.8 = \underline{0.64}$$

(b) Reliability for the configuration per Figure 10-12b:

$$P = 1 - (1 - 0.8^2)^2$$
$$= 1 - (1 - 0.64)^2$$
$$= 1 - 0.36^2$$
$$= \underline{0.8704}$$

(c) Reliability for the configuration per Figure 10-12c:

$$P = [1 - (1 - 0.8)^2]^2$$
$$= (1 - 0.04)^2$$
$$= (0.96)^2$$
$$= \underline{0.9216}$$

The improvement in case *c* is due to the more flexible redundancy, allowing A and D, or B and C to work together, if necessary. The spares provisioning models described in Chapter 13 for multiple operating units also apply to standby redundancy with multiple units backed up by one or more standby units (see examples 13-4 and 13-6 and text that precedes them).

10-6 LIMITATION OF REDUNDANT SUBSYSTEMS

When subsystems are used in parallel redundancy to increase reliability, it is often necessary to provide some means of disconnecting failed subsystems so that they do not interfere with the operation of the remaining good units. Figure 10-13a is the initial redundant system consisting of three parallel subsystems. Figure 10-13b illustrates symbolically a shorted subsystem *C*. Shorted subsystem *C* places a direct short across the remaining two subsystems, *A* and *B*, thereby disabling them also.

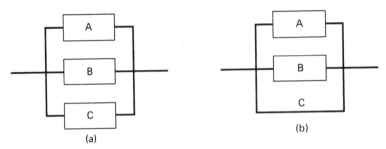

Figure 10-13 *(a) Redundant system—operational; (b) redundant system with shorted subsystem C.*

It is obvious that each subsystem which works in parallel redundancy with other subsystems must have some kind of failure sensing switch built in to disconnect the afflicted subsystem if the failure mode is a short. If the failure mode is an "open," the failed part ordinarily need not be disconnected from the rest of the system.

In the case of parallel power supplies sharing a common load, failure of one power supply must cause the load to be redistributed equally among the remaining supplies; necessary circuitry for accomplishing this must be provided.

Figure 10-14 schematically depicts an opened failure-sensing switch associated with failed subsystem C ("shorted" failure mode assumed). In the parallel-redundancy calculations we have assumed that the reliability of these failure-sensing switches is 100%. This is, of course, an idealized assumption; in practice, the reliability of these devices is less than 100%.

Furthermore, it is clear that there is a point of diminishing returns in using redundancy configurations; an increase in the level of parallel redundancy employed increases size, weight, cost, and volume of the equipment and often requires complicated failure-sensing devices whose reliabilities need to be considered.

It is usually necessary to perform trade-off calculations to determine the advis-

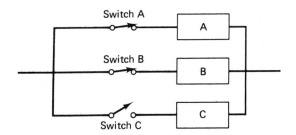

Figure 10-14 *Redundant system with failure-sensing switches.*

ability of parallel redundancy versus improvement of the reliability of the basic subsystem by other means. Methods for improvement could include the following considerations:

1. Reduce number of parts.
2. Simplify the system.
3. Improve reliability level of parts used.
4. Burn-in of parts.

In summary, the choice of parallel redundancy has to be made judiciously. Means must be provided to disconnect failed items from the redundant circuit, if their failure mode is such that they would interfere with the remaining unit(s). Furthermore, care must be taken not to change the basic circuit value represented by the redundant units beyond acceptable limits due to the elimination of one or more failed units. These disadvantages are to some extent alleviated by the standby redundancy arrangement.

10-7 PARALLEL REDUNDANT SYSTEMS IN STANDBY

In the standby configuration one or more subsystems are standing by to take over operation upon failure of the basic unit. The standby units are not operative until a failure-sensing device senses the failure of the basic unit and turns on and connects the standby unit. Owing to the fact that the unit in standby is not operating unless a failure of the basic unit occurs, system reliability for such a configuration is higher than for comparable systems with continuously operating units.

The terms of the Poisson distribution may be used to determine the reliability of standby systems since they display the constant λt characteristic of the Poisson process. Table 2-7 showed the probability of no failure to be $e^{-\lambda t}$. The probability of one failure is $\lambda t \cdot e^{-\lambda t}$. The probability of two failures is $(\lambda t)^2/2 \cdot e^{-\lambda t}$, and so on. These terms may be arranged to represent the reliability of various standby configurations.

For the configuration where one operating unit and one standby unit are grouped together, we have for the probability that no failure or one failure will occur (leaving one unit intact),

$$P_{\text{(with one standby unit)}} = e^{-\lambda t} + \lambda t e^{-\lambda t} \qquad (84)$$

where λt is expected number of failures.

For the configuration where one operating unit and two standby units are grouped together, the probability that no failure, or one failure, or two failures will occur (leaving at least one unit intact) is

$$P_{\text{(with two standby units)}} = e^{-\lambda t} + \lambda t e^{-\lambda t} + \frac{(\lambda t)^2}{2} e^{-\lambda t} \qquad (85)$$

For additional standby units, we add additional Poisson terms per Table 2-7. For example, for one operating unit and four standby units, the probability of success is

$$P_{\text{(with four standby units)}} = e^{-\lambda t} + \lambda t e^{-\lambda t} + \frac{(\lambda t)^2}{2} e^{-\lambda t} + \frac{(\lambda t)^3}{6} e^{-\lambda t} + \frac{(\lambda t)^4}{24} e^{-\lambda t}$$

$$= e^{-\lambda t}\left(1 + \lambda t + \frac{(\lambda t)^2}{2} + \frac{(\lambda t)^3}{6} + \frac{(\lambda t)^4}{24}\right)$$

EXAMPLE 10-21 Calculate the reliability of a standby system consisting of one operating unit and one identical standby unit operating for a period of 100 h. The failure rate for each unit is $\lambda = 0.001$ failure/hour. Assume that the failure sensing and connecting switch has 100% reliability.

Solution

$$P_{\text{(with one standby unit)}} = e^{-\lambda t} + \lambda t e^{-\lambda t}$$
$$= e^{-\lambda t}(1 + \lambda t)$$
$$\lambda t = 0.001 \times 100 = 0.1$$
$$P = e^{-0.1}(1 + 0.10)$$
$$= 1.10 e^{-0.1}$$

From Table A10, Appendix A,

$$e^{-0.10} = 0.90484$$
$$P = 1.10 \times 0.90484 = \underline{0.99532}$$

EXAMPLE 10-22 Refer to Example 10-21 except assume parallel redundancy, with the second unit operating throughout the mission instead of being on standby and compare the reliability of the two schemes.

Solution

$$P_{\text{parallel redundancy}} = 1 - Q^n$$
$$Q = 1 - P$$
$$= 1 - 0.90484 = 0.09516$$
$$P_{\text{parallel redundancy}} = 1 - (0.09516)^2$$
$$= 1 - 0.009055$$
$$= \underline{0.99094}$$

As should have been expected, the reliability of the standby system was found to be higher (0.99532) than that of the parallel-redundant system (0.99094).

(a) MTBF of Standby Systems[2]

The MTBF of a system consisting of the basic unit and one standby unit is

$$\text{MTBF} = \frac{2}{\lambda} \tag{86}$$

The MTBF of a system consisting of the basic unit and two standby units is

$$\text{MTBF} = \frac{3}{\lambda} \tag{87}$$

When $n - 1$ units are in standby, ready to take over if the basic unit should fail, the MTBF of the system is

$$\text{MTBF} = \frac{n}{\lambda} \tag{88}$$

EXAMPLE 10-23   The failure rate of an electronic device is $\lambda = 21$ failures/10^6 h. Two standby units are added to the system. What is the MTBF of the system?

Solution

$$\text{MTBF} = \frac{3}{\lambda}$$

The failure rate is converted to failures per hour.

$$21 \text{ failures}/10^6 \text{ h} = 21 \times 10^{-6} \text{ failure/hour}$$

$$\text{MTBF} = \frac{3}{21 \times 10^{-6}}$$

$$= \frac{3 \times 10^6}{21}$$

$$= \underline{143{,}000 \text{ h}}$$

(b) Fault-Sensing and Switching Reliability

When it cannot be assumed that the fault-sensing and switching element has 100% reliability, the following equation applies:

$$P = e^{-\lambda t} + P_{\text{switch}} \times (\lambda t)e^{-\lambda t}$$

or

$$P = e^{-\lambda t}(1 + \lambda t P_{\text{switch}}) \tag{89}$$

[2] The basic and the standby unit(s) are assumed to have the same failure rates while operating. No maintenance is performed until the last unit has failed. A much higher MTBF is produced by replacing failed units promptly, effectively reducing the chance of system failure to a negligible value. The policy of immediate or periodic maintenance may also be applied to parallel redundant systems with similar results. The mathematical models covering these cases which are beyond the scope of this book are treated in *Reliability Theory and Practice* by Igor Bazovsky (see bibliography).

The above relation is based on the following conditions:

1. Equal failure rates for the basic and the standby units.
2. Failure of the sensing unit does not affect the basic unit.
3. Switch reliability refers to the probability of a single successful switch actuation upon failure of the basic unit.

> **EXAMPLE 10-24** An electronic device has a failure rate of 500 failures/10^6 h. One identical standby unit is used to increase the reliability of the basic device. The operating time is 1000 h. The reliability of the failure-sensing and switching element is $P_{FSS} = 0.97$. See Figure 10-15 for the schematic of the system. What is the system reliability?

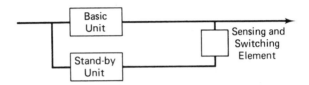

> **Figure 10-15** *Standby redundancy with sensing and switching element.*

Solution The expected number of failures is

$$\lambda t = 500 \times 10^{-6} \times 1000$$

where $500 \times 10^{-6} =$ failures/hour, so

$$\lambda t = 0.5$$

Using equation (89) we obtain

$$P = e^{-0.5}[1 + 0.5(0.97)]$$

From Table A10, Appendix A,

$$e^{-0.5} = 0.60653$$

$$P = 0.60653(1 + 0.485)$$

$$= 0.60653 \times 1.485$$

$$= \underline{0.90069}$$

Had the sensing and switching element been 100% reliable, the system reliability per equation (84) would have been

$$P = 0.60653 + 0.5(0.60653) = \underline{0.90979}$$

10-8 SYSTEM RELIABILITY

A system is a configuration of components designed to operate in unison to accomplish a desired mission. The reliability requirements of the system are usually given in the systems specification and frequently include limitations on downtime.

The system designer will lay out the integral parts of the system in the form of a block diagram with interconnecting lines showing the flow of signals and the supply of power. A preliminary estimate of the reliability of the individual subsystems or components making up the system is prepared, to ascertain whether the system reliability requirements can be met. If they cannot, trade-offs are performed involving redundancy of components or else other modifications to the system concept.

Once the overall configuration of the system is determined and compliance with the reliability requirements seems feasible (with the aid of a reliability mathematical model), a reliability goal is allocated to each subsystem or component of the system. As the design effort progresses, constant watch is necessary to assure that the total failure rate of each subsystem or component does not result in a reliability figure that exceeds the allocated value. The methods described in Chapters 9 and 10 or more complex models,[3] where necessary, may be used for the reliability analysis of the subsystems or components. If a reliability allocation is not met, design changes may be necessary unless a reallocation which does not compromise the overall reliability goal can be made. The design is also examined for compliance with the downtime or maintainability requirements, and necessary alterations in construction or configuration introduced to assure meeting these requirements.

PROBLEMS

10-1 (a) What is the expected number of failures for a piece of equipment having a failure rate of 12%/1000 h during a 850-h operating period?

   (b) What is the MTBF?

   (c) What is the probability of no failure occurring?

10-2 Perform the following conversions:

From	To
0.00003 failure/hour	failures/10^6 h
10.9% failures/1000 h	failures/hour
125 failures/10^6 h	failures/hour
6.3% failures/1000 h	failures/10^6 h

10-3 What is the maximum failure rate of a piece of equipment if a probability of survival of no less than 88% is desired for a 9000-h operating period? Express the failure rate in %/1000 h.

10-4 What is the MTBF for a piece of equipment if the probability of survival for a 1500-h operational period is 93.7%?

---

[3] Refer to *Reliability Theory and Practice* by Igor Bazovsky and *Mathematical Methods of Reliability Theory* by B. V. Guedenko, Yu. K. Belyayev and A. D. Solovyev—See Bibliography.

10-5 A stepdown transformer, rectifier, and filter comprise a series system. The following are the failure rates of these components:

Transformer	1.56% failures/1000 h
Rectifier	2.00% failures/1000 h
Filter	1.70% failures/1000 h

The equipment is to operate for 1500 h. What is the probability of survival?

10-6 An electronic system consists of five subsystems with the following MTBF's:

$$\text{Subsystem } A: \quad \text{MTBF} = 12{,}500 \text{ h}$$
$$\text{Subsystem } B: \quad \text{MTBF} = 2830 \text{ h}$$
$$\text{Subsystem } C: \quad \text{MTBF} = 11{,}000 \text{ h}$$
$$\text{Subsystem } D: \quad \text{MTBF} = 9850 \text{ h}$$
$$\text{Subsystem } E: \quad \text{MTBF} = 15{,}550 \text{ h}$$

The five subsystems are arranged in a series configuration. What is the probability of survival for an 800-h operating period?

10-7 Subsystems A and B are connected in a parallel system configuration. Both subsystems are activated during operation. If the probability of survival of subsystem A is 83% and of subsystem B is 91%, what is the probability of survival of the system?

10-8 (a) State the general equation for parallel redundant subsystems involving the unreliability of each parallel branch (both subsystems are operating).
(b) The unreliability of a subsystem is $Q = 0.5$. Five identical subsystems are connected in parallel redundancy. What is the probability of survival of the whole system?

10-9 How many subsystems, each having a failure rate of 30%/1000 h, need to be connected in parallel redundancy to assure a 90% probability of success of the whole system for an operational period of 1000 h?

10-10 (a) Refer to the circuit configuration of Figure 10-3. If $P_A = 0.7$, $P_B = 0.8$, and $P_C = 0.9$, determine the probability of survival of the system.
(b) Repeat the computation for part (a) but use two subsystems A in parallel.

10-11 Determine the probability of success for the following system, with all units operating.

$$P_A = 90\%$$
$$P_B = 85\%$$
$$P_C = 75\%$$
$$P_D = 80\%$$

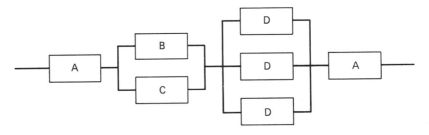

Problem 10-11.

10-12 Determine the probability of success for the following series-parallel system:

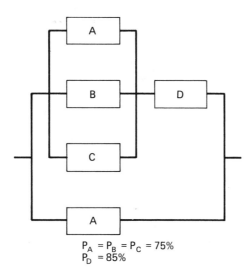

$$P_A = P_B = P_C = 75\%$$
$$P_D = 85\%$$

Problem 10-12.

10-13 For four subsystems connected in parallel redundancy, the binomial expansion is

$$(p + q)^4 = p^4 + 4p^3q + 6p^2q^2 + 4pq^3 + q^4$$

The meaning of p^4 is that all four subsystems are operating. What do the remaining four expressions in the expansion stand for?

10-14 Four subsystems are connected in series–parallel redundancy as shown in Figure 10-6.

$$P_A = P_B = P_C = P_D = 80\%$$

Determine the probability of success of this system.

10-15 (a) State the general formula for n identical units connected in series–parallel redundancy.

     (b) Refer to Figure 10-8. What is the reliability of each of the six identical subsystems to assure a probability of success of at least 68% for the overall series–parallel redundant system?

10-16 Four subsystems are connected as shown. The reliability of each of the four subsystems is 80%. What is the system probability of success?

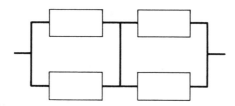

Problem 10-16.

10-17 A series circuit configuration consists of three series elements *A*, *B*, and *C*. The probability of success of this series system is 72.9%. Devise a redundancy scheme that will increase the probability of success to at least 92%.

10-18 (a) Discuss the limitations of ordinary parallel redundant circuits.
(b) What is the purpose of failure-sensing switches?

10-19 Four components are connected in series–parallel redundancy. In series with this arrangement are two components connected in parallel. Assuming each component has a reliability of 98%, what is the probability of success of the whole configuration?

10-20 (a) What is the formula for the reliability of a configuration consisting of an operating unit with one standby unit?
(b) Calculate the probability of survival of a configuration consisting of an operating module, with a reliability of 80%, and an identical standby unit which is switched into operation by a failure-sensing switch if the regular module should fail. The operating time is 800 h and the reliability of the failure-sensing switch is assumed to be 100%.

10-21 A circuit consists of a basic unit and two standby units. The operating failure rate of each unit is 1.2%/1000 h. What is the MTBF of the arrangement if the fault-sensing and switching device has an assumed 100% reliability and if no maintenance is performed until all three units have failed.

10-22 Refer to Problem 10-20(b). What is the probability of survival of the configuration if the reliability of the failure-sensing switch is 94% instead of the assumed 100%?

II

RELIABILITY TESTING

The ultimate purpose of reliability testing is to establish whether the equipment under test meets its reliability requirement. The criterion for acceptability is the demonstration of a minimum acceptable life. A variety of statistical procedures are available for reliability testing. Most of them are based on the exponential distribution.

In this chapter we will be concerned with the following three types of reliability tests:

1. Tests terminated upon occurrence of a number of preassigned failures.
2. Tests terminated at preassigned time.
3. Sequential tests.

Department of Defense (DoD) Handbook H108[1] is a useful source for fixed-time tests. Sections 11.1 and 11.2 are based on material from *Handbook H108*. The principles of sequential testing are introduced in Section 11.3.

Finally, MIL-STD-781B,[2] a military standard for reliability testing, is discussed and selected sections from the standard are reproduced. MIL-STD-781B contains both sequential and fixed-time reliability test plans and also introduces environmental stresses in the testing.

---

[1] Quality Control and Reliability Handbook H108, "Sampling Procedures and Tables for Life and Reliability Testing," April 29, 1960, Office of the Assistant Secretary of Defense, Washington, D.C.

[2] MIL-STD-781B, November 15, 1967, "Military Standard Reliability Tests: Exponential Distribution," Department of Defense, Washington, D.C.

11-1 RELIABILITY LIFE TESTS TERMINATED UPON OCCURRENCE OF PREASSIGNED NUMBER OF FAILURES

There are two procedures for use with reliability life tests terminated upon the occurrence of a preassigned number of failures: (1) testing with replacement of failed units, and (2) testing without replacement of failed units.

(a) Procedure for Testing with Replacement of Failed Items

The acceptability of a lot from which a random sample of size n was drawn and placed on life test is ascertained by comparing $\hat{\theta}$, the estimate of the mean life of the lot, with a constant k multiplied by θ_0, the mean life specified to be acceptable [see equation (90)]. The lot is acceptable when

$$\hat{\theta} \geq k \cdot \theta_0 \tag{90}$$

Values for k are given in Table 11-1. $\hat{\theta}$ is computed from the test results by using the equation

$$\hat{\theta} = \frac{nt_F}{r} \tag{91}$$

where $\hat{\theta}$ = estimated equipment mean life
n = sample size placed on test

Table 11-1 *Tabulation of Constant k Versus Termination Failures r for Two Values of Producer's Risk α. (Based on Table 2B-1 in DoD Handbook H108.)*

r	$k\ (\alpha = 0.05)$	$k\ (\alpha = 0.10)$
1	0.052	0.106
2	0.178	0.266
3	0.272	0.367
4	0.342	0.436
5	0.394	0.487
6	0.436	0.525
7	0.469	0.556
8	0.498	0.582
9	0.522	0.604
10	0.543	0.622
15	0.616	0.687
20	0.663	0.726
25	0.695	0.754
30	0.720	0.774
40	0.755	0.803
50	0.779	0.824

$t_F =$ time when the preassigned number of failures occurred
$r =$ number of failures which terminate the life test

The choice of the termination number r depends on the degree of protection desired against acceptance of material with unacceptable mean life. The larger the termination number r, the greater is the assurance against accepting material with unacceptable mean life.

Table 11-1 furnishes values for constant k for two values of producer's risk α. The definition for producer's risk as used here is as follows. The producer's risk, α, is the probability of rejecting lots with mean life θ_0, which is specified to be acceptable. The following example will illustrate the use of the method described.

EXAMPLE 11-1   A life-test plan is desired which will accept equipment having a specified acceptable mean life of 2000 h with a probability of 90%. Eight units are placed on test and the sixth failure occurs when 900 h test time is reached. The life test is terminated upon occurrence of the sixth failure. Determine whether the lot of equipments is acceptable.

Solution   Using the notation introduced earlier, we have

$$\theta_0 = 2000 \text{ h}$$

Since there is a probability of 0.90 of accepting θ_0, there is, conversely, a probability of

$$1 - 0.90 = 0.10$$

of rejecting θ_0. Since the probability of rejecting θ_0 was defined as the producer's risk, we have

$$\alpha = 0.10$$

The life test is terminated with the occurrence of $r = 6$ failures; the sample size is $n = 8$ and $t_F = 900$. Substituting these values in equation (91), we obtain

$$\hat{\theta} = \frac{8 \times 900}{6}$$

$$= 1200 \text{ h}$$

Next, determine the value of constant k from Table 11-1. Entering the table on the line for $r = 6$, we find $k = 0.525$ in the column for $\alpha = 0.10$. We can now test the acceptability of the results of the test with the aid of equation (90):

$$\hat{\theta} = 1200$$

$$k \cdot \theta_0 = 0.525 \times 2000 = 1050$$

The criterion for lot acceptance is that the value $\hat{\theta}$ be equal or greater than the value $k \cdot \theta_0$. Since

$$1200 > 1050$$

the above criterion is fulfilled and the lot is acceptable. The choice of the number of units n to be placed on test depends primarily on cost and test-time considerations.

(b) Procedure for Testing Without Replacement of Failed Items

When failed items are not replaced, the criterion for acceptability of a lot remains the same as before per equation (90), but $\hat{\theta}$ is calculated differently:

$$\hat{\theta} = \frac{1}{r}[t_1 + t_2 + t_3 + \cdots + (n - r)t_F] \tag{92}$$

where $\hat{\theta}$, n, r, t_F have the same meaning as before

t_1, t_2, t_3, etc., are the times when the first, second, third, etc, failures occur

> **EXAMPLE 11-2** Refer to Example 11-1, where the sixth failure, t_F, was 900 h. Assume that the first five failures occurred at 86, 199, 345, 550, and 770 h. The remainder of the specifications remain as before, except that failed units are not replaced.
>
> **Solution** Using equation (92),
>
> $$\hat{\theta} = \frac{1}{6}[86 + 199 + 345 + 550 + 770 + (8 - 6)900]$$
>
> $$= \frac{1}{6}(3750) = 625$$
>
> Since $k \cdot \theta_0$ remains the same value 1050, it follows that
>
> $$625 < 1050$$
>
> Since $\hat{\theta}$ is less than $k \cdot \theta_0$, we reject the lot.

11-2 RELIABILITY LIFE TESTS TERMINATED AT PREASSIGNED TIME

This kind of life test is stopped upon reaching a computed termination time T provided the specified number of failures did not occur before. The test is also stopped if the specified number of failures is reached before time T. This test procedure is applicable to testing with or without replacement of failed parts.

The acceptability of a lot is determined by the time required for a predetermined number of failures, r_F, to occur in a sample size of n, and a comparison of this time with the test termination time T multiplied by constant k. The value of constant k is obtained from Table 11-2 for $\alpha = 0.05$ or $\alpha = 0.10$, and for the following multiples of the predetermined number of failures: $2r_F, 3r_F, 4r_F, 5r_F, 6r_F$, and $7r_F$. A lot is accepted if the predetermined number of failures r_F has not yet occurred before termination time T is reached. T is determined from equation (93):

$$T = k \cdot \theta_0 \tag{93}$$

where T = test termination time

k = constant obtained from Table 11-2

θ_0 = mean life specified as acceptable

Table 11-2 *Values of Constant k Versus Termination Failures r_F for $\alpha = 0.05$ and $\alpha = 0.10$. Sample Size n Is Multiple of r_F.* [Based on DoD Handbook H108, Tables 2C-1(b), 2C-1(c), 2C-2(b), and 2C-2(c)].*

r_F	Yes $2r_F$	No $2r_F$	Yes $3r_F$	No $3r_F$	Yes $4r_F$	No $4r_F$	Yes $5r_F$	No $5r_F$	Yes $6r_F$	No $6r_F$	Yes $7r_F$	No $7r_F$	α
1	0.026	0.026	0.017	0.017	0.013	0.013	0.010	0.010	0.009	0.009	0.007	0.007	0.05
	0.053	0.053	0.035	0.035	0.026	0.026	0.021	0.021	0.018	0.018	0.015	0.015	0.10
2	0.089	0.104	0.059	0.065	0.044	0.048	0.036	0.038	0.030	0.031	0.025	0.026	0.05
	0.133	0.155	0.089	0.098	0.066	0.071	0.053	0.056	0.044	0.046	0.038	0.039	0.10
3	0.136	0.168	0.091	0.103	0.068	0.075	0.055	0.058	0.045	0.048	0.039	0.041	0.05
	0.184	0.226	0.122	0.139	0.092	0.101	0.073	0.079	0.061	0.065	0.052	0.055	0.10
4	0.171	0.217	0.114	0.132	0.085	0.095	0.068	0.074	0.057	0.061	0.049	0.052	0.05
	0.218	0.277	0.145	0.168	0.109	0.121	0.087	0.095	0.073	0.078	0.062	0.066	0.10
5	0.197	0.254	0.131	0.153	0.099	0.110	0.079	0.086	0.066	0.071	0.056	0.060	0.05
	0.243	0.314	0.162	0.189	0.122	0.136	0.097	0.106	0.081	0.087	0.070	0.074	0.10
6	0.218	0.284	0.145	0.170	0.109	0.122	0.087	0.095	0.073	0.078	0.062	0.066	0.05
	0.263	0.343	0.175	0.206	0.131	0.147	0.105	0.115	0.088	0.094	0.075	0.080	0.10
7	0.235	0.309	0.156	0.185	0.117	0.132	0.094	0.103	0.078	0.084	0.067	0.072	0.05
	0.278	0.366	0.185	0.219	0.139	0.157	0.111	0.122	0.093	0.100	0.079	0.085	0.10
8	0.249	0.330	0.166	0.197	0.124	0.141	0.100	0.110	0.083	0.090	0.071	0.076	0.05
	0.291	0.386	0.194	0.230	0.146	0.164	0.116	0.128	0.097	0.105	0.083	0.089	0.10
9	0.261	0.348	0.174	0.207	0.130	0.148	0.104	0.115	0.087	0.094	0.075	0.080	0.05
	0.302	0.402	0.201	0.239	0.151	0.171	0.121	0.133	0.101	0.109	0.086	0.092	0.10
10	0.271	0.363	0.181	0.216	0.136	0.154	0.109	0.120	0.090	0.098	0.078	0.083	0.05
	0.311	0.416	0.207	0.247	0.156	0.176	0.124	0.137	0.104	0.112	0.089	0.095	0.10
15	0.308	0.417	0.205	0.246	0.154	0.175	0.123	0.136	0.103	0.112	0.088	0.094	0.05
	0.343	0.465	0.229	0.275	0.172	0.196	0.137	0.152	0.114	0.124	0.098	0.105	0.10
20	0.331	0.451	0.221	0.266	0.166	0.189	0.133	0.147	0.110	0.120	0.095	0.102	0.05
	0.363	0.494	0.242	0.291	0.182	0.207	0.145	0.161	0.121	0.132	0.104	0.112	0.10

* α is producer's risk; Yes, designates testing with replacement of failed items; No, designates testing without replacement of failed items.

EXAMPLE 11-3   Ten units are placed on test using a life test plan (with replacement of failed items) that will accept a lot having a mean life of 2000 h with a probability of 90%. Determine whether the lot is acceptable if the terminating fifth failure occurs at 175 hours

Solution

$$\theta_0 = 2000 \text{ h}$$

$$\alpha = 1 - 0.90 = 0.10$$

$$r_F = 5$$

$$n = 10 = 2r_F$$

To find constant k enter Table 11-2 on the horizontal line for $r_F = 5$ and $\alpha = 0.10$. Read off constant $k = 0.243$ under the column for $2r_F$ and "Yes" (for "replace failed items"). From equation (93),

$$T = 0.243 \times 2000$$
$$= 486 \text{ h}$$

Since the fifth failure occurred before $T = 486$ h, the lot should be rejected.

EXAMPLE 11-4 Repeat Example 11-3, except this time without replacement of failed items and with the fifth failure occurring at 650 hours.

Solution The only change is the value of constant k. Its value is obtained from Table 11-2 on the horizontal line for $r_F = 5$ and $\alpha = 0.10$ and the column for $2r_F$ and "No," designating testing without replacement of failed items.

$$k = 0.314$$
$$T = 0.314 \times 2000$$
$$= 628 \text{ h}$$

Since the fifth failure occurs after $T = 628$ h of testing, the lot is acceptable.

DoD *Handbook 108A* also contains operating characteristic curves for each of the life-test plans. This permits the determination of the probability of acceptance of lots having different ratios of actual mean life to acceptable mean life.

11-3 SEQUENTIAL RELIABILITY TESTING

When a sequential reliability test is performed, no decision is made in advance as to the number of hours to be used for the test. Instead, the accumulated results of the test at any point serve as a criterion for making one of three possible decisions:

1. Accept the equipment.
2. Reject the equipment.
3. Continue testing.

The sequential test plans are also significantly affected by the risks both the producer and consumer are willing to accept in connection with the decision imposed by the results of the test. The risk assumed by the producer (α) is that equipment will be rejected by the test even though its true MTBF equals the specified MTBF. Similarly, the risk assumed by the consumer (β) is that equipment which is accepted by the plan just barely meets the minimum acceptable MTBF. As a result, the choice of an appropriate test plan may have significant economic implications for both producer and consumer.

In setting up a sequential test plan to determine the acceptability of the MTBF of equipment, we must first designate two specific values of MTBF: the specified MTBF, θ_0, and the minimum MTBF, θ_1, which we are willing to accept.*

---

* Note different meaning of θ_0 in this section.

The next step is to decide on the values of producer's risk (α) and consumer's risk (β). The values for α and β usually range from 0.01 to 0.30. Common values are 0.05 and 0.10. A value of $\alpha = 0.05$, for example, means that in 5 of 100 cases, the particular plan will reject equipments that should have been accepted.

Figure 11-1 is a plot for a sequential test plan. The horizontal axis shows test time; the vertical, the number of failures. The space between consists of the regions which define the decision status at any point of the test. The test begins at the origin of the graph. As the test continues, failures are plotted versus time in a continuous stairlike plot. As long as the plotted curve remains in the "Continue testing" region, the test goes on; when the curve crosses either the accept or reject lines, the test is stopped and the corresponding decision is made.

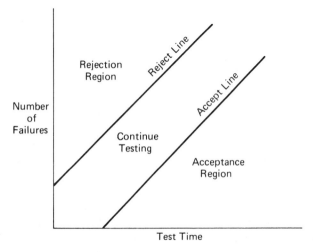

Figure 11-1 *Sequential test plan.*

Four numerical quantities must be selected to construct the graph for this test:

$\theta_0 =$ specified MTBF
$\theta_1 =$ minimum MTBF we are willing to accept
$\alpha =$ risk that the plan will reject equipment with an acceptable MTBF θ_0 (producer's risk)
$\beta =$ risk of accepting equipment with the minimum acceptable MTBF θ_1 (consumer's risk)

We construct two lines, the *accept line* and the *reject line*, after determining the constants that define these lines. The lines are defined by the following equations:

$$t_1 = \frac{\ln\left(\frac{\theta_0}{\theta_1}\right)}{\frac{1}{\theta_1} - \frac{1}{\theta_0}}(r) - \frac{\ln\left(\frac{\beta}{1 - \alpha}\right)}{\frac{1}{\theta_1} - \frac{1}{\theta_0}} \qquad \text{accept line} \qquad (94)$$

$$t_2 = \frac{\ln\left(\frac{\theta_0}{\theta_1}\right)}{\frac{1}{\theta_1} - \frac{1}{\theta_0}}(r) - \frac{\ln\left(\frac{1-\beta}{\alpha}\right)}{\frac{1}{\theta_1} - \frac{1}{\theta_0}} \qquad \text{reject line} \qquad (95)$$

where

$$\frac{\ln\left(\frac{\theta_0}{\theta_1}\right)}{\frac{1}{\theta_1} - \frac{1}{\theta_0}} = \text{slope of the accept and reject lines}$$

$$r = \text{number of failures}$$

$$\frac{\ln\left(\frac{\beta}{1-\alpha}\right)}{\frac{1}{\theta_1} - \frac{1}{\theta_0}} = \text{time-axis intercept of the accept line}$$

$$\frac{\ln\left(\frac{1-\beta}{\alpha}\right)}{\frac{1}{\theta_1} - \frac{1}{\theta_0}} = \text{time-axis intercept of the reject line}$$

See Tables A-14, A-15, and A-16, Appendix A, for values of $\ln(\theta_0/\theta_1)$, $\ln[\beta/(1-\alpha)]$, and $\ln[(1-\beta)/\alpha]$.

EXAMPLE 11-5 A piece of electronic equipment has a specified MTBF of $\theta_0 = 240$ h. The minimum MTBF the customer is willing to accept is $\theta_1 = 80$ h. Producer's and consumer's risks are both set at 10% ($\alpha = 10\%$, $\beta = 10\%$). Suppose that 10 pieces of equipment are randomly selected from each month's production and placed on a sequential reliability test. Shipment of the month's production is contingent upon passing the sequential test. Let us assume that the 10 pieces are drawn from the January production on test. The times when failures occurred were recorded as follows:

1st failure:	14 h
2nd failure:	28 h
3rd failure:	39 h
4th failure:	54 h
5th failure:	64 h

Failed pieces are immediately replaced so that 10 units remain on test until the termination of the test.

What decision may be made up to and including the fifth failure relative to product acceptance? If the result of this decision is to continue testing after the fifth failure, how many more test hours without an additional failure would permit accepting the January production run?

Solution

(a) From Tables A-15 and A-16, Appendix A, for $\alpha = \beta = 0.10$:

$$\ln\left(\frac{\beta}{1-\alpha}\right) = -2.197 \qquad \ln\left(\frac{1-\beta}{\alpha}\right) = +2.197$$

(b) From Table A-14, Appendix A, for

$$\frac{\theta_0}{\theta_1} = \frac{240}{80} = 3$$

$$\ln\left(\frac{\theta_0}{\theta_1}\right) = 1.099$$

(c) Using equation (94),

$$t_1 = \frac{1.099}{\frac{1}{80} - \frac{1}{240}}(r) - \frac{-2.197}{\frac{1}{80} - \frac{1}{240}}$$

$$= \frac{1.099}{0.00833}(r) + \frac{2.197}{0.00833}$$

$$t_1 = 131.9(r) + 263.7 \text{ h (accept line)}$$

Using equation (95),

$$t_2 = \frac{1.099}{\frac{1}{80} - \frac{1}{240}}(r) - \frac{2.197}{\frac{1}{80} - \frac{1}{240}}$$

$$= \frac{1.099}{0.00833}(r) - \frac{2.197}{0.00833}$$

$$t_2 = 131.9(r) - 263.7 \text{ h (reject line)}$$

(d) Accept and reject lines per the equations above are plotted for assumed values of r (see Figure 11-2 for this plot). The intercepts of the accept and reject lines with the time axis are $+263.7$ h and -263.7 h, respectively.

The region between the accept and reject lines is the "Continue testing" region. The area to the left of the reject line is the rejection region, and the area to the

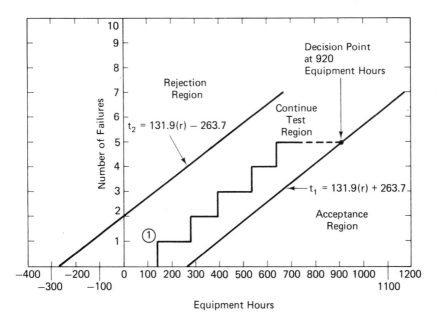

Figure 11-2 *Sequential reliability sampling test.*

right of the accept line is the acceptance region. These three regions are clearly marked in Figure 11-2.

Since the 10 pieces operate concurrently and failed units are replaced, the equipment hours given in Table 11-3 are accumulated at each of the failures.

Table 11-3 *Accumulated Equipment Hours at Each Failure Occurrence (Failed Equipment Is Replaced).*

Number of Failure	Hour of Failure	Equipment Hours
1	14	$10 \times 14 = 140$
2	28	$10 \times 28 = 280$
3	39	$10 \times 39 = 390$
4	54	$10 \times 54 = 540$
5	64	$10 \times 64 = 640$

When the accumulated equipment hours for the corresponding failures are plotted in the graph, a staircase curve results. For example, the first failure occurs on one of the 10 pieces of equipment after 14 h. The total number of equipment hours accumulated by the 10 units at that point are $10 \times 14 = 140$ h. We therefore locate the first failure at the intersection of a vertical drawn through 140 equipment hours and a horizontal representing one failure (see ① in Figure 11-2). The remaining points of the staircase curve are found in the same manner.

It is noted that until 64 h after the start of the test (640 equipment hours), when the fifth failure occurs, the staircase curve remains between the accept and reject lines; hence, no decision as to rejection or acceptance is possible and we have to continue testing. The staircase curve crosses the accept line at 920 equipment hours before the occurrence of an additional failure. The test may therefore be terminated and the lot accepted 92 h after the start of the test and 28 hours after occurrence of the fifth failure. Had the staircase curve at any time crossed the reject line, this would have meant rejection of the lot.

Variations of the described sequential life test are in use, such as sequential life test without replacement of failed equipments. Also, methods of termination of tests referred to as *truncation* are in use for instances where the staircase curve remains between the accept and reject lines for an excessive test time. Depending on contract requirements, the test is truncated at some time T if no prior accept or reject decision can be made.

11-4 RELIABILITY TEST STANDARD MIL-STD-781B

(a) General

The AGREE[3] report, issued in 1957, contained a report by Task Group 3 introducing a new approach to reliability testing of procured equipments. This involved the

---

[3] "Reliability of Military Electronic Equipment," Advisory Group on Reliability of Electronic Equipment, Office of the Assistant Secretary of Defense, June 4, 1957.

determination of the MTBF under realistic environmental conditions. The group also recommended that sequential test procedures be adopted for reliability testing as a means of obtaining economical estimates of the true value of the reliability of equipments.

Following the issuance of the AGREE report, reliability testing of equipment using sequential test methods and realistic environmental conditions gained wide acceptance, especially by the military services. In 1963 the Department of Defense issued a military standard for reliability testing, MIL-STD-781, which incorporated many of the recommendations of the AGREE report. Since November 15, 1967, Revision B to the Standard, MIL-STD-781B; has been in effect. MIL-STD-781B contains a wide selection of reliability test plans for equipment. The plans are based on the exponential distribution. Selected portions of the standard and excerpts are included in this chapter to provide the necessary understanding of the philosophy and methods used.

(b) Definitions of Terms

The definitions of the terms used in MIL-STD-781B are given in Paragraph 11.5(b) (see p. 303).

(c) Test Plans

The test plans may be used for both reliability demonstration tests (qualification) and reliability production acceptance tests (sampling). The categories of test plans are discussed in Paragraph 11.5(g) and its subparagraphs (see p. 306). There are a total of 29 test plans in the standard, with a range of values for producer's and consumer's risk, discrimination ratios, accept/reject criteria, and test duration (in the case of fixed-length test plans). The test plans are grouped in three tables (Tables 11-5, 11-6, and 11-7). Table 11-5 covers standard probability ratio sequential test plans PRST I–VI. Table 11-6 covers short-run high-risk PRST plans VII–IX. Table 11-7 covers fixed-length test plans X–XXVII, longevity test plan XXVIII, and an all equipment screening plan XXIX.

(d) Test Levels

Table 11-4 of MIL-STD-781B (p. 305) summarizes the available environmental test levels. A total of 10 test levels are given. Test levels *A–D* include no temperature cycling of the equipment but specify increasing levels of ambient temperature, vibration exposure, and equipment on/off cycling. Test level A-1 is restricted to operating the equipment at $25 \pm 5°C$ and includes no other environmental exposure. Test levels *E–J* specify an ambient temperature range from either $-54°C$ or $-65°C$ to as high as $+125°C$, depending on the level chosen. The requirements for test levels *C* and *E* are reproduced for illustrative purposes (see pp. 305 and 306).

(e)   Requirements

Paragraph 11.5(h) and its subparagraphs (pp. 308 to 310) contain specific requirements as to

1. Equipment performance.
2. Equipment quantity to be tested.
3. Length of test.
4. Reduced testing.

while Paragraph 11.5(i) contains specific requirements for tightened testing.

(f)   Qualification (Demonstration) Phase— Production

Paragraph 11.5(j) and its subparagraphs (pp. 310 and 311) contain specific details for qualification testing, including

1. Sample size.
2. Evaluation criteria.

(g)   Production Acceptance (Sampling)— Production

Paragraph 11.5(k) and its subparagraphs (pp. 311 and 312) contain details in connection with

1. Sample size.
2. Evaluation criteria.

(h)   Requirements for Test Plans XXVI, XXVII, XXVIII, and XXIX

These test plans are fixed-time plans which serve specific purposes:

1.   Test Plan XXVI

Fixed-length demonstration test (see Paragraph 11.5(l) and its subparagraphs, p. 312).

2.   Test Plan XXVII

Fixed-length production verification test (see Paragraph 11.5(m) and its subparagraphs, p. 313).

3.   Test Plan XXVIII

Longevity test (see Paragraph 11.5(n) p. 313).

4.　Test Plan XXIX

All equipment screening test (see Paragraph 11.5(p) and its subparagraphs pp. 317 and 318).

(i)　Accept/Reject Criteria

Selected accept/reject criteria for test plans I–IX are contained in Paragraph 11.5(o) and its subparagraphs. Accept/reject criteria for test plans I, III, and VII are reproduced for illustrative purposes (see pp. 313 to 316). Accept/reject criteria for test plans X–XXV are contained in Paragraph 11.5(o)-4 (p. 316). Accept/reject criteria for test plans XXVI–XXIX are covered in section (h) above.

(j)　Detail Reliability Test Procedures

Paragraph 11.5(q) (p. 319) deals with the requirement for detailed test procedures.

(k)　Inspection

Inspection of the test facility is described in Paragraph 11.5(q)-1 (p. 319).

(l)　Thermal and Vibration Surveys

Thermal and vibration survey requirements are contained in Paragraphs 11.5(q)-2 and 11.5(q)-3 (pp. 319 and 320). The thermal survey is made on equipment to be tested under this environmental stress condition. The vibration survey may be conducted to find the presence of any resonant conditions.

(m)　Burn-In (Debugging) Period

Requirements for this optional, pretest, burn-in are contained in Paragraph 11.5(q)-4 (p. 320).

(n)　Equipment Cycling

Requirements for equipment on–off cycling are contained in Paragraph 11.5(r)-1 and subparagraphs (p. 320).

(o)　Voltage Cycling

Requirement for voltage cycling of equipment input voltage is contained in Paragraph 11.5(r)-5 (p. 321).

(p)　Test Facilities

For test facilities requirements, see Paragraph 11.5(s) and subparagraphs (p. 321).

(q) Selection of Equipments for Test

Requirements for selection of equipments for test are contained in Paragraph 11.5(t) (p. 322)

(r) Determination of Compliance

Considerations in connection with determining compliance to test plan requirements are contained in Paragraph 11.5(u) and subparagraphs (pp. 322 to 324).

(s) Failure Actions

When failures occur during testing, certain failure actions, verifying repairs and corrective actions are necessary. See Paragraphs 11.5(v) and subparagraphs, 11.5(x) and 11.5(y) (pp. 324 to 327). The following is a reproduction of selected portions of the text and tables of Military Standard MIL-STD-781B, "Reliability Tests: Exponential Distribution," November 15, 1967.

11-5 TEXT OF MIL-STD-781B*

(a) Scope

1. Scope

This standard outlines test levels and test plans for reliability qualification (demonstration), reliability production acceptance (sampling) tests, and for longevity tests. The test plans are based upon the exponential, or Poisson distribution, and are intended for the testing of equipment. These tests do not replace design, performance, environmental, preproduction, individual, or other required tests (i.e., all functional and environmental tests) specified for the equipment.

2. Purpose

This standard is intended to provide uniformity in reliability testing by:

(a) Facilitating the preparation of Military Specifications and Standards through the establishment of standard test levels and test plans.
(b) Restricting the variety of reliability tests so that those conducting tests can establish facilities.
(c) Facilitating the determination of more realistic correlation factors between test reliability and operational reliability.
(d) Facilitating the direct comparison of MTBF test results through the establishment of uniform test levels and plans.

* Reproduction is partial and edited to comply with the format of this book.

(b) Definitions

1. General

Meanings of terms not defined herein are in accordance with the definitions in Standard MIL-STD-721.

2. Mean Time Between Failures (MTBF)

The mean time between failure is also the reciprocal of the failure rate. Observed Mean Time Between Failure (θ) is equal to the total operating time of the equipment divided by number of failures.

3. Minimum Acceptable Mean Time Between Failures (θ_1)

Minimum acceptable mean time between failures (θ_1) is a value so selected that an associated and specified risk of accepting equipment of this value is tolerable.

4. Specified Mean Time Between Failures (θ_0)

Specified mean time between failures (θ_0) is the MTBF value specified in the contract or equipment specification. Its value is determined by multiplying the minimum acceptable MTBF (θ_1) by the discrimination ratio (θ_0/θ_1) of the selected test plan. It is used to limit producer's decision risk (α).

When specifiying an MTBF value as a test requirement, clearly state whether the value is *minimum acceptable MTBF* (θ_1) or *specified MTBF* (θ_0).

5. Discrimination Ratio (θ_0/θ_1)

The ratio of the specified MTBF (θ_0) to the minimum acceptable MTBF (θ_1).

(c) Decision Risks:

1. Consumer's Decision Risk (β)

The probability of accepting equipment(s) with a true MTBF equal to the minimum acceptable MTBF (θ_1). [The probability of accepting equipment(s) with true MTBF less than the minimum acceptable MTBF (θ_1) will be less than β.]

2. Producer's Decision Risk (α)

The probability of rejecting equipment(s) with a true MTBF equal to the specified MTBF (θ_0). (The probability of rejecting equipment(s) with true MTBF greater than the specified MTBF will be less than α.)

(d) Failure

Details involving failure criteria should be stated in the test procedures, but in general the following shall apply:

1. Equipment Failure

The inability of a previously acceptable item to perform its required function within previously established limits.

2. Pattern Failure

The occurrence of two (2) or more failures of the same part in identical or equivalent application whose combined failure rate exceeds that predicted. When two (2) or more dependent failures of the same part occur whose combined failure rate exceeds that predicted and which are due to more than one independent failure, the dependent failures are relevant pattern failures.

3. Independent Failure

A failure which will independently cause equipment performance outside of specified limits—one which occurs without being related to the failure of associated items.

4. Dependent Failure

A failure of a part which is a direct result of an independent failure—one which is caused by the failure of an associated item(s). Dependent failures are not necessarily present when simultaneous failures occur [see Paragraph 11.5(w)-2].

(e) Longevity

The specified period of time during which it is economically feasible to repair the equipment and return it to the original operating condition. As used herein, longevity is synonymous with useful operating life.

(f) General Requirements

1. Test Levels

The test levels describe the conditions of temperature, vibration, input voltage, and the cycling of these conditions to be applied to the equipment during the test.

Table 11-4 *Summary of Test Levels.*

Test Level	Temperature (°C)	Temperature Cycling	Vibration*	Equipment on/off Cycle†
A	25 ± 5	None	Yes	Yes
A − 1	25 ± 5	None	None	None
B	40 ± 5	None	Yes	Yes
C	50 + 5 − 0	None	Yes	Yes
D	65 ± 5	None	Yes	Yes
E	−54 to +55	Yes	Yes	Yes
F	−54 to +71	Yes	Yes	Yes
G	−54 to +95	Yes	Yes	Yes
H	−65 to +71	Yes	Yes	Yes
J	−54 to +125	Yes	Yes	Yes

Input voltage shall be the nominal specified voltage $+5 - 2\%$.
Input voltage cycling may be applied to any of the test plans.

* 2.2G $\pm$ 10% at any frequency between 20 and 60 Hz. (The vibration level shall be $1g \pm 10\%$ for equipment designed for use on Navy ships.) If the equipment is designed to meet a vibration requirement less severe than the requirement contained herein, the vibration may be reduced to the level specified as a design requirement [see Paragraph 11.5(s)-2].

† See Paragraph 11.5(r)-1.

2. Test Level C

Temperature	$50° + 5° - 0°C$ (122°F to 131°F)
Temperature cycling	None
Vibration	2.2G $\pm$ 10% peak acceleration value at any non-resonant frequency between 20 and 60 cps measured at the mounting points on the equipment. The duration of vibration shall be at least 10 minutes during each hour of equipment operating time [see Paragraph 11.5(S)-2].
Equipment on–off cycling	Turn on and let temperature stabilize, hold for 3 hours, then turn off and let temperature stabilize. This cycle shall continue throughout the test.
Input voltage	Nominal specified voltage $+5 - 2\%$.
Input voltage cycling	See Paragraph 11.5(r)-5.

3. Test Level E

Temperature	$-54°C$ to $+55°C$ ($-65°F$ to 131°F)
Temperature cycling	Temperature cycling shall be timed to stabilize at low temperature followed by time to stabilize at the high temperature, plus 2 hours [see Paragraph 11.5(r)-3]

Vibration	2.2G $\pm$ 10% peak acceleration value at any non-resonant frequency between 20 and 60 cps measured at the mounting points on the equipment. The duration of vibration shall be at least 10 minutes during each hour of operating time [see Paragraph 11.5(s)-2]
Equipment on–off cycling	Equipment off during cooling cycle and on during heating cycle [see Paragraph 11.5(r)-1]
Input voltage	Nominal specified voltage $+5 - 2\%$.
Input voltage cycling	See Paragraph 11.5(r)-5.

(g) Test Plans

When reliability assurance tests in accordance with MIL-STD-781 are required, testing will consist of a reliability qualification (demonstration) test [see Paragraph 11.5(j)], and reliability production acceptance (sampling) tests [see Paragraph 11.5(k)].

1. General

The test plans of this standard consist of the statistical criteria for determining the acceptability of the reliability characteristic of the equipment submitted to test, and are used for both reliability qualification (demonstration) and reliability production acceptance (sampling) tests. The test plans are categorized as follows:

1. Standard probability ratio sequential tests (PRST) (test plans I–VI)
2. Short-run high-risk PRST plans (test plans VII–IX)
3. Fixed-length test plans (test plans X–XXVII)
4. Longevity test (test plan XXVIII)
5. All equipment screening test (test plan XXIX)

Tables 11-5, 11-6 and 11-7 present the parameters of the test plans.

Table 11-5 *Summary of Test Plans (Standard PRST).*

Test Plan	Decision α Risks β		Discrimination Ratio, θ_0/θ_1	Accept/Reject* Criteria (paragraph)	Sample Size Qual. (paragraph)	Sampling
I	10%	10%	1.5	11.5(o)-1	11.5(j)-1	11.5(k)-1, 2, 3
II	20%	20%	1.5	—	11.5(j)-1	11.5(k)-1, 2, 3
III	10%	10%	2.0	11.5(o)-2	11.5(j)-1	11.5(k)-1, 2, 3
IV	20%	20%	2.0	—	11.5(j)-1	11.5(k)-1, 2, 3
IVa	20%	20%	3.0	—	11.5(j)-1	11.5(k)-1, 2, 3
V	10%	10%	3.0	—	11.5(j)-1	11.5(k)-1, 2, 3
VI	10%	10%	5.0	—	11.5(j)-1	11.5(k)-1, 2, 3

* Accept/reject criteria for test plans I and III are reproduced on pages 313 to 315.

Table 11-6 *Summary of Test Plans (Short-Run High-Risk PRST).*

Test Plan	Decision α Risks β	Discrimination Ratio, θ_0/θ_1	Accept/Reject* Criteria (paragraph)	Sample Size Qual. (paragraph)	Sampling (paragraph)
VII	30% 30%	1.5	11.5(o)-3	11.5(j)-1	11 5(k)-1, 2, 3
VIII	30% 30%	2.0	—	11.5(j)-1	11.5(k)-1, 2, 3
IX	35% 40%	1.25	—	11.5(j)-1	11.5(k)-1, 2, 3

* Accept/reject criteria for test plan VII are reproduced on pages 315 and 316.

Table 11-7 *Summary of Test Plans (Fixed-Length, Longevity and All Equipment).*

Test Plan	Decision α Risks β	Discrimination Ratio, θ_0/θ_1	Test Duration (multiples of θ_0)	ACCEPT/REJECT FAILURES Reject (Equal or more)	Accept (Equal or Less)	Sample Size Qual. (paragraph)	Sampling (paragraph)
X	10% 10%	1.25	100.0	112	111	11.5(j)-1	11.5(k)-1, 2, 3
XI	10% 20%	1.25	72.0	83	82	11.5(j)-1	11.5(k)-1, 2, 3
XII	20% 20%	1.25	44.0	50	49	11.5(j)-1	11.5(k)-1, 2, 3
XIII	30% 30%	1.25	14.9	17	16	11.5(j)-1	11.5(k)-1, 2, 3
XIV	10% 10%	1.5	30.0	37	36	11.5(j)-1	11.5(k)-1, 2, 3
XV	10% 20%	1.5	19.9	26	25	11.5(j)-1	11.5(k)-1, 2, 3
XVI	20% 20%	1.5	14.1	18	17	11.5(j)-1	11.5(k)-1, 2, 3
XVII	30% 30%	1.5	5.3	7	6	11.5(j)-1	11.5(k)-1, 2, 3
XVIII	10% 10%	2.0	9.4	14	13	11.5(j)-1	11.5(k)-1, 2, 3
XIX	10% 20%	2.0	6.2	10	9	11.5(j)-1	11.5(k)-1, 2, 3
XX	20% 20%	2.0	3.9	6	5	11.5(j)-1	11.5(k)-1, 2, 3
XXI	30% 30%	2.0	1.84	3	2	11.5(j)-1	11.5(k)-1, 2, 3
XXII	10% 10%	3.0	3.1	6	5	11.5(j)-1	11.5(k)-1, 2, 3
XXIII	10% 20%	3.0	1.8	4	3	11.5(j)-1	11.5(k)-1, 2, 3
XXIV	20% 20%	3.0	1.46	3	2	11.5(j)-1	11.5(k)-1, 2, 3
XXV	30% 30%	3.0	.37	1	0	11.5(j)-1	11.5(k)-1, 2, 3
XXVI	N/A N/A	N/A	3.0	(See Para 11.5(l)-2)		11.5(l)-1	N/A
XXVII	N/A N/A	N/A	3.0	(See Para 11.5(m)-2)		N/A	11.5(m)-1
XXVIII	* *	*	*	*	*	N/A	N/A
XXIX	† †	†	†	†	†	N/A	ALL

* Plan XXVIII is the longevity test plan. See Paragraph 11.5(n) for details.

† Plan XXIX is the all equipment screening test. See Paragraph 11.5(p) for details.

2. Discussion of Test Plans

Standard PRST plans shall be used when a sequential test plan with normal (10 to 20%) producer and consumer risks are desired. Short-run high-risk PRST plans may be used when a sequential test plan is desired, but circumstances require the use of a short test with a low discrimination ratio and both the producer and the consumer are willing to accept relatively high decision risks. Fixed-length test plans may be used when the exact length and costs of tests must be known beforehand. However, PRST plans with similar risks and discrimination ratio will be more efficient and economical than the fixed-length test plans. The longevity test plan is designed to determine the capability of the equipment to meet performance requirements over a specified extended period of time. The all equipment screening test is designed for use when defects due to inadequate quality control, new production lines, design changes, or any incipient defects must be eliminated at the producer's facilities.

3. Use of Test Plans

The initial production lot of equipments shall be submitted to the qualification (demonstration) phase test [see Paragraph 11.5(j)]. Subsequent lots shall be submitted to production acceptance (sampling) phase tests [see Paragraph 11.5(k)]. Each phase of tests consists of selecting samples of equipments at random, placing these equipments on test in accordance with the specified test level, and comparing the time and failures occurring to the accept/reject criteria of the specified test plan. Normally, the procuring activity will select a reliability production acceptance (sampling) plan having decision risks, or a discrimination ratio, or both, equal to or greater than those of the reliability qualification (demonstration) test plan. The equipment is considered to have demonstrated compliance with the specified reliability requirements when an accept decision is reached using the designated test plan.

(h) Requirements

1. Specification of Test Plan and Test Level

The test plan and test level to be used in qualification (demonstration) and reliability production acceptance (sampling) phase tests and the requirement for longevity tests shall be as specified in the contract or equipment specification.

2. Equipment Performance

The parameters to be measured and the applicable acceptance limits shall be included in the test procedures outlined in Paragraph 11.5(q) of this standard.

3. Equipment Quantity

The quantity of equipments to be tested, not necessarily simultaneously, shall be determined as required herein, or as specified in the contract.

4. Test Length

Testing will continue until the total unit hours together with the total count of relevant equipment failures permit either an accept or reject decision in accordance with the specified test plan. Only equipment "on" time may be used in MTBF or longevity determinations. Testing shall be monitored in such a manner that the times to failure may be estimated with reasonable accuracy. The monitoring instrumentation and/or technique proposed by the contractor should be disclosed in the detail procedure of Paragraph 11.5(q). No single equipment "on" time shall be less than one half the average operating time of all equipments on test.

5. Reduced Testing

When the test history indicates significantly better results (in terms of mean time between failures) than the specified value, testing shall be reduced in accordance with the following:

1. Testing shall commence, and continue, using the test plan specified, until the rules given below indicate a change in the test plan.
2. Reduced testing shall be implemented as detailed below only upon authorization of the procuring activity.
3. The reduced testing plans shall be used only when production is continuous.
4. When eight successive lots or tests have been completed without the occurrence of a reject decision and the observed MTBF (θ) is equal to or greater than one and one-half times the specified MTBF ($1.5\theta_0$), subsequent tests shall use the test plan given for reduced testing in 11.5(h)-5.
5. At any time a reject decision occurs during testing at a reduced level, corrective action shall be taken and the test shall be repeated using the test plan specified by contract or equipment specification [see 11.5(u)-5]. Testing shall continue using the specified test plan until eligibility for reduced testing is again established. If the production line is shut down for sixty calendar days or longer, then the test plan specified by contract shall be used.
   Reduced testing may be instituted in accordance with the rules above when the contract or equipment specification requires testing using test plans I–V or test plan XXIX.

Specified Test Plan	Reduced Test Plan
I	III
II	IV
III	V
IV	IVa
V	VI
XXIX	60%*

* Testing reduced to 60% of the time specified for each equipment.

6. Conversion for Reduced Testing

The value of the specified MTBF (θ_0) shall be held constant for both normal and reduced tests. If the contract or equipment specification designates the minimum acceptable MTBF (θ_1), the value of the specified MTBF (θ_0) must be calculated in order to use the accept/reject criteria given in the test plan. This same value of specified MTBF (θ_0) shall be used in both specified and reduced testing, even though the discrimination ratio may not be the same in the reduced test plans as it was in the specified test plan. The specified MTBF (θ_0) may be calculated by multiplying the minimum acceptable MTBF (θ_1) by the discrimination ratio of the specified test plan.

(i) Tightened Testing

1. Specified to Tightened Testing

The procuring activity reserves the right to require tightened testing in accordance with 11.5(u)-5 whenever any two of five consecutive lots experience reject decisions.

2. Tightened to Specified Testing

Whenever five successive lots satisfactorily meet the acceptance criteria while on tightened testing, a return to the specified test plan shall be made on the next lot.

(j) Qualification (Demonstration) Phase of Production Reliability Tests

These tests are performed to evaluate the lot MTBF with respect to the specified MTBF. These tests may be conducted after development, after a major design change, or on initial production samples for contracts involving periodic shipments. (Normally, the reliability qualification test is accomplished on the first month's production lot.)

1. Sample Size

The sample size required for the qualification phase test plans shall be as specified in the contact, or, as agreed by the contractor and the procuring activity. Table 11-8 presents the recommended sample sizes.

2. Evaluation Criteria

The accept/reject criteria for each of the test plans in Tables 11-5 and 11-6 shall be in accordance with the paragraph referenced for the plan. For test plans X–XXV, the accept/reject criteria are given in Table 11-7.

Table 11-8 *Recommended Quantity of Equipment for Qualification (Demonstration) Phase Test.*

Lot Size	Recommended Sample Size	Maximum Sample Size
1–3	All	All
4–16	3	9
17–52	5	15
53–96	8	19
96–200	13	21
over 200	20	22

(k) Production Acceptance (Sampling) Phase of Production Reliability Tests

The production acceptance (sampling) phase, as distinguished from the qualification (demonstration) phase, is a periodic series of tests to indicate continuing production of acceptable equipments. The right to apply the production acceptance (sampling) phase is earned only by passing the qualification (demonstration) phase. A production acceptance test shall be performed on each lot produced [normally, one month's production is considered a lot. See also 11.5(t)].

1. Sample Size

The quantity of equipment required for a sampling phase test plan shall be determined by considering the production rate and test time available per month. Sufficient samples shall be selected from each lot to ensure completion of tests in one month.

2. Maximum Sample Size

The maximum number is not fixed, but depends on the production rate, test time available, and the restrictions of Paragraph 11.5(h)-4.

3. Minimum Sample Size

The minimum number of samples to be tested per lot is three equipments. When insufficient equipments are procured to provide three samples, the sample shall be made up of all available equipments. When the equipment is large, complex and expensive, the sample size may be smaller and shall be as specified by the procuring activity. (For fixed-length plans there shall be a minimum of two equipments on test.)

4. Evaluation Criteria

The accept/reject criteria for each of the test plans in Tables 11-5 and 11-6 shall be in accordance with the paragraph referenced for the plan. For test plans X–XXV, the accept/reject criteria are given in Table 11-7.

(l)   Test Plan XXVI (Fixed-Length Demonstration Test) :

1. Quantity Required

The number of equipments selected for this test shall be based on a 5000 minimum part count (electronic, electrical, and electromechanical parts which are replaceable but not repairable), i.e., 5000 divided by the individual equipment part count will determine the number of equipments to be tested by taking the quotient to the next highest integer. The minimum number of equipments shall be 2, or 0.0088 times the specified MTBF (θ_0) (to the next higher integer), whichever is greater, and the maximum number shall be 20.

2. Evaluation Procedure

The number of equipments selected shall be tested in accordance with the test level specified in the specification or contract document. There are two alternatives, based on test time of 3 MTBF's (specified) per equipment or 500 hours per equipment, whichever is less, for reaching an accept/reject decision. The procedure for each alternative is as follows:

Minimum of three (3) MTBF's (specified) but less than 500 hours: In this procedure, the equipment shall be operated for a minimum of 3 MTBF's (specified) per equipment prior to consideration of an accept decision. The accept/reject criteria of paragraph 11.5(o)-1 test plan I shall be used thereafter (until the 500-hour period is reached) with the stipulation that the accept condition shall be void of pattern failures. The test shall continue until an accept/reject decision is reached or there is an accumulation of 500 operating hours per equipment. When the 500-hour level has been reached, the following procedure shall be followed.

Five hundred (500) operating hours per equipment: The test shall be stopped when each equipment has accumulated 500 hours of operation. The accept/reject decision shall be determined by taking the quotient of total test time divided by the number of relevant failures.

If the quotient of total test time divided by the number of relevant failures is greater than the minimum acceptable MTBF value and no pattern failures are present, the equipment is acceptable. If the quotient is lower, or if pattern failures exist, the equipment is rejected.

(m) Test Plan XXVII (Fixed-Length Production Verification Test):

1. Sample Size

The sample size, while being subject to the constraints of production rate, test facilities, costs, schedules, etc., shall be as large as possible to reach an early decision. The quantity selected shall be as specified in the contract document; if it is not specified, the contractor will recommend a quantity to the procuring activity. The minimum sample size shall be 2 or $0.022\theta_0$, whichever is greater.

2. Evaluation Criteria

The accept/reject criteria of Paragraph 11.5(o)-1 test plan I shall be used for this test with the stipulation that an accept decision shall be void of pattern failures. If an accept/reject decision is not made prior to the accumulation of 200 operating hours per equipment or three times the specified MTBF, whichever is less, the test shall be terminated and an accept decision shall be made provided there is no indication of pattern failures.

(n) Test Plan XXVIII (Longevity Test):

1. General

The equipment shall be tested at the specified test level for a period of time equivalent to the life requirement in the equipment specification. If not specified, the longevity test shall be 2000 operating hours for each equipment. Test time accumulated on the equipment during a reliability test may be applied toward the longevity test when the same sample units are used.

2. Sample Size

At least two (2) equipments shall be subjected to the longevity test.

3. Evaluation of Longevity Test

No accept/reject criteria for this test is specified. The evaluation shall be limited to the analysis of all failures, identification of design and pattern failures, and the submission of recommended design changes to the procuring activity.

(o) Accept/Reject Criteria:

1. Accept/Reject Criteria for Test Plan I

Decision risks   10%
Discrimination ratio   1.5:1

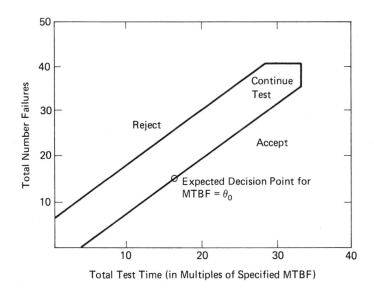

Figure 11-3 *Accept-reject criteria (test plan I).*

	TOTAL TEST TIME*			TOTAL TEST TIME*	
No. of Failures	Reject (Equal or Less)	Accept (Equal or more)	No. of Failures	Reject (Equal or Less)	Accept (Equal or More)
0	N/A	4.40	21	12.61	21.43
1	N/A	5.21	22	13.42	22.24
2	N/A	6.02	23	14.23	23.05
3	N/A	6.83	24	15.04	23.86
4	N/A	7.64	25	15.85	24.67
5	N/A	8.45	26	16.66	25.48
6	0.45	9.27	27	17.47	26.29
7	1.26	10.08	28	18.29	27.11
8	2.07	10.89	29	19.10	27.92
9	2.88	11.70	30	19.90	28.73
10	3.69	12.51	31	20.72	29.54
11	4.50	13.32	32	21.53	30.35
12	5.31	14.13	33	22.34	31.16
13	6.12	14.94	34	23.15	31.97
14	6.93	15.75	35	23.96	32.78
15	7.74	16.56	36	24.77	33.00
16	8.55	17.37	37	25.58	33.00
17	9.37	18.19	38	26.39	33.00
18	10.18	19.00	39	27.21	33.00
19	10.99	19.81	40	28.02	33.00
20	11.80	20.62	41	33.00	N/A

* Total test time is total unit hours of "equipment on" time and is expressed in multiples of the specified MTBF. Refer to Paragraph 11.5(h)-4 for minimum test time per equipment.

2. Accept/Reject Criteria for Test Plan III

Decision risks　　　　10%
Discrimination ratio　　2.0: 1

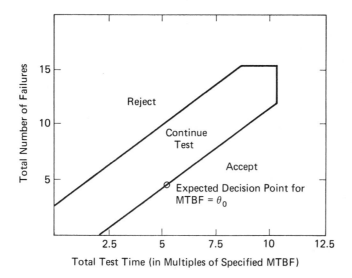

Figure 11-4 *Accept-reject criteria (test plan III).*

No. of Failures	TOTAL TEST TIME* Reject *(Equal or Less)*	Accept *(Equal or More)*	No. of Failures	TOTAL TEST TIME* Reject *(Equal or less)*	Accept *(Equal or more)*
0	N/A	2.20	9	4.51	8.44
1	N/A	2.89	10	5.20	9.13
2	N/A	3.59	11	5.90	9.83
3	0.35	4.28	12	6.59	10.30
4	1.04	4.97	13	7.28	10.30
5	1.74	5.67	14	7.97	10.30
6	2.43	6.36	15	8.67	10.30
7	3.12	7.05	16	10.30	N/A
8	3.82	7.75			

* Total test time is total unit hours of "equipment on" time and is expressed in multiples of the specified MTBF. Refer to Paragraph 11.5(h)-4 for minimum test time per equipment.

3. Accept/Reject Criteria for Test Plan VII

Decision risks　　　　30%
Discrimination ratio　　1.5: 1

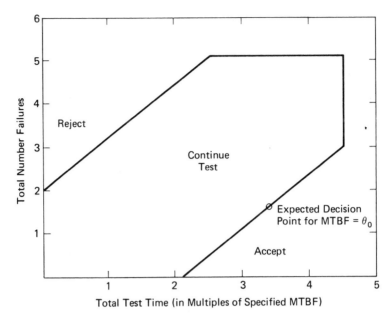

Figure 11-5 *Accept-reject criteria (test plan VII).*

No. of Failures	TOTAL TEST TIME*	
	Reject (Equal or Less)	Accept (Equal or more)
0	N/A	2.10
1	N/A	2.91
2	N/A	3.72
3	0.81	4.53
4	1.62	4.53
5	2.43	4.53
6	4.53	N/A

* Total test time is total unit hours of "equipment on" time and is expressed in multiples of the specified MTBF. Refer to Paragraph 11.5(h)-4 for minimum test time per equipment.

4. Accept/Reject Criteria for Test Plans X–XXV

The test duration and acceptable number of failures for fixed-length test plans X–XXV are given in Table 11-7. For a given plan, each lot is accepted if no more than the acceptable number of failures occurs. A lot is rejected if at any time during the test, the number of failures equals or exceeds the reject number.

(p) Test Plan XXIX (All Equipments Screening Test):

1. General

This test plan is intended for use when defects due to inadequate quality control, new production lines, design changes, or any incipient defects must be eliminated at the producer's facilities. This test plan is not intended for use in the reliability qualification (demonstration) phase.

2. Test Period

Each equipment produced, other than those submitted to the reliability qualification phase test, shall be tested for ____ hours as specified in the contract or equipment specification. If no value is specified, the test period shall be fifty hours or one-fourth the specified MTBF, whichever is less.

3. Pre-Test Burn-In

A burn-in period prior to the test period of 11.5(p)-2 may be used at the option of the contractor. If used, the burn-in must be the same for each equipment in time and severity [See 11.5(q)-4].

4. Evaluation of Test

To determine whether the MTBF is being met at any time during the contract, the operation test hours and the failures (not counting burn-in failures or burn-in operation time) shall be totaled and the results compared with the reject line of test plan II (extend the reject line as necessary to accommodate the data). These totals shall accumulate so that at any one time the experience from the beginning of the contract is included. If the line of plotted test results indicates a reject situation, that is, if the plotted point crosses the reject line, then the procuring activity shall be notified immediately and corrective action in accordance with Paragraphs 11.5(u)-5 and 11.5(y) shall be instituted.

(*Note:* As the total test time is increased, α approaches 25% and β approaches 0%.)

5. Accept/Reject Criteria for Test Plan XXIX

Decision risks $25\%:0\%$
(Derived from reject line of test plan II)

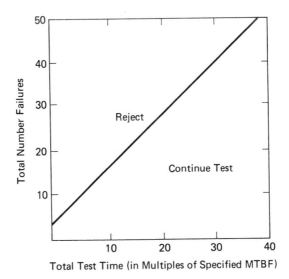

Figure 11-6 *Reject criteria (test plan XXIX).*

TOTAL TEST TIME*		TOTAL TEST TIME*		TOTAL TEST TIME*	
No. of Failures	Reject (Equal or Less)	No. of Failures	Reject (Equal or Less)	No. of Failures	Reject (Equal or Less)
0	N/A	18	12.33	36	26.92
1	N/A	19	13.14	37	27.73
2	N/A	20	13.95	38	28.54
3	0.16	21	14.76	39	29.36
4	0.97	22	15.57	40	30.17
5	1.78	23	16.38	41	30.98
6	2.60	24	17.19	42	31.79
7	3.41	25	18.00	43	32.60
8	4.22	26	18.81	44	33.41
9	5.03	27	19.62	45	34.22
10	5.84	28	20.44	46	35.03
11	6.65	29	21.25	47	35.84
12	7.46	30	22.06	48	36.65
13	8.27	31	22.87	49	37.46
14	9.08	32	23.68	50	38.28
15	9.89	33	24.49	.	.
16	10.70	34	25.30	.	.
17	11.52	35	26.11	etc.	etc.

* Total test time is total unit hours of "equipment on" time and is expressed in multiples of the specified MTBF.

(q) Detail Reliability Test Procedures

The contractor shall prepare detail test procedures (when required by the contract) and obtain approval of the procedures from the procuring activity as soon as possible. These procedures should be submitted for review a minimum of 60 days prior to the proposed commencement date of testing. These procedures shall reflect the requirements of the contract, the equipment specifications and this standard. Approval must be obtained before reliability tests begin. The test procedures should include the following details.

1. A list and brief description of all units comprising the equipment and a specific listing of those units which will be placed on test.
2. Test equipment to be used.
3. Monitoring test equipment.
4. Graphical presentation of the required thermal survey of the equipment made prior to start of testing, with narrative and analysis leading to establishment of monitored temperature stabilization points.
5. Burn-in time.
6. The time, temperature and other details of the cycle.
7. The duty cycle.
8. Vibration stress and duration.
9. Input voltage.
10. Preventive maintenance measures to be performed.
11. Performance parameters to be measured.
12. Performance limits beyond which a failure has occurred.
13. Failures classified as nonrelevant.
14. Sample of report or log forms to be used.

1. Inspection

Visitation to the contractor's facility, use of procuring activity inspection personnel, and any other appropriate means for assuring compliance with reliability requirements will be made by the procuring activity. The contractor shall provide necessary administrative support to permit such personnel to properly perform their authorized duties.

2. Thermal Survey

A thermal survey shall be made of the equipment to be tested, under the temperature cycling and duty cycle of the test level required, prior to the initiation of testing for the identification of the component of greatest thermal inertia and the establishment of the time temperature relationships between it and the chamber air. These relationships shall be used for determining equipment thermal stabilization during the test. The lower test level temperature stabilization takes place when the temperature of the point of maximum thermal inertia is within 2°C of the lower test level temperature and its rate of change is less than 2°C/hour. Upper test level temperature stabilization takes place when the rate of change of the point of maximum thermal inertia at the upper temperature limit is less than 2°C/hour. The techniques and results of the thermal survey shall be described, plotted, submitted to, and approved by the procuring activity prior to the initiation of testing. Temperatures of the heating–cooling air shall be recorded continuously during both survey and testing. The equipment thermal survey need be made only once for each identical equipment type representative of the manufacturer's normal production under the current production.

3. Vibration Survey

A vibration survey may be conducted to search for resonant conditions between 20 and 60 cycles per second in order that they may be avoided during the reliability tests.

4. Burn-In (or Debugging) Period

The contractor may elect to operate the equipment for a certain time prior to the initiation of reliability testing. Any such burn-in applied to test samples must be applied to all items of equipment submitted for acceptance. Failures occurring during this period may be considered nonrelevant but shall be recorded and reported. Burn-in time for all items of equipment submitted for acceptance shall be at the same levels of environment and for the same integral number of cycles as the burn-in conducted on the test samples.

(r) Equipment Cycling

1. Equipment On-Off Cycling:

Unless otherwise specified, the equipment shall be "on" during the heating cycle and "off" during the cooling cycle. (See Paragraph 11.5(r)-3.)

2. Duty Cycle

The duty cycle (or performance profile) is the time phase apportionment of modes of operation and functions to be performed by the equipment during the on-time portion of the environmental test cycle. It is intended that the duty cycle be representative of field operation. The equipment specification or contract referencing this standard shall specify the duty cycle to be used during the tests. If no equipment on–off cycling is required by the test level, the specified duty cycle shall continue throughout the test.

3. Temperature Cycling

The method and duration of temperature cycling may be specified in the contract or order referencing this standard. If no method of temperature cycling is specified, the standard cycling method of Paragraph 11.5(r)-4 shall be used.[4]

4. Standard Method of Temperature Cycling

Stabilize the equipment temperature at the lower test level temperature. Turn the equipment on and allow it to experience the combined effect of chamber heating and equipment operation until the point of maximum thermal inertia stabilizes at the higher temperature as determined in accordance with the thermal survey. Maintain the equipment at the upper test level temperature for a period of time specified by the procuring activity. If no such time is specified, the period shall be two (2) hours. Turn the equipment off and allow it to experience chamber cooling until the point of maximum thermal inertia stabilizes at the lower test level temperature. This cycle shall continue for the duration of the test.

5. Voltage cycling

When so directed by the procuring activity, voltage cycling shall be accomplished as follows: The input voltage shall be maintained at one hundred ten percent (110%) nominal for one-third of the equipment "on" cycle, at the nominal value for the second one-third of the equipment "on" cycle, and at ninety percent (90%) for the final one-third of the equipment "on" cycle. This cycling procedure is to be repeated continuously throughout the reliability test.

(s) Test Facilities

Test facilities shall be capable of maintaining the conditions specified for the applicable test level and of measuring equipment characteristics to the specified accuracy for the duration of the test. Test facilities shall be subject to the approval of the procuring activity. The various test levels include high and low temperature and a single vibration frequency. Consequently, it is suggested that simplified test facilities be designed and constructed for these reliability acceptance tests.

1. Test Chambers

Test chambers shall be capable of maintaining the ambient and forced air temperatures at the specified test level temperatures $\pm 2°C$ during the test. The rate of temperature change of the thermal medium in both the heating and cooling cycles when test levels E through J are used shall average not less than $5°C/minute$. Chamber and forced air temperatures shall be monitored continuously. Thermostats shall be installed to

---

[4] An alternative method is listed in MIL-STD-781B but is not included here.

interrupt the programming motor used in automatic control of environmental cycling until maximum and minimum air temperature requirements are satisfied. The time to stabilize at both temperatures shall be determined for use during the test [see Paragraph 11.5(q)-2.]

2. Vibration

The vibration shall be $2.2g \pm 10\%$ peak acceleration value at any nonresonant frequency between 20 and 60 Hz measured at the mounting points on the equipment. (Equipment designed for use on Navy ships shall be vibrated at $1g \pm 10\%$. If the equipment is designed to meet a vibration requirement less severe than the requirement contained herein, the vibration may be reduced to the level specified as a design requirement.) The duration of vibration shall be at least 10 minutes during each hour of equipment operating time. When the equipment being tested contains circuit boards or cards, vibration shall be normal to the plane of the majority of the cards. Otherwise, the direction of vibration is not critical. This vibration requirement may be met by mounting the equipment solidly, (that is, without the equipment shock vibration mounts) to a strong flat plate which is supported by vibration isolators. The desired amplitude and frequency may be obtained by a simple vibration machine with an asymmetric weight on a shaft attached to the plate and driven by a suitable motor. It is recommended that the vibration equipment be checked for proper operation each 24 hours of operation, and that the vibration transducer be "on" and monitored continuously during vibration.

(t) Selection of Equipment

If the contract or order does not define what constitutes a lot then one month's production shall constitute a lot; however, if less than three equipments are produced during a month, then the lot shall consist of two or more month's production, the actual quantity to be negotiated with the procuring activity. The equipments to be tested shall be representative of the entire lot from which they are selected; and selections shall be made at random throughout the lot, but without compromising the delivery schedule. All equipments selected shall have first passed the individual tests described in the acceptance test portion of the equipment specification. The equipments to be used for the reliability qualification plan shall begin with the first lot available after the preproduction or other test programs not required by this standard have been completed. Equipments used for special tests or life tests as outlined in the equipment specification shall not be considered for the reliability tests.

(u) Determination of Compliance:

1. General

Compliance shall be determined by the accept/reject criteria of the test plan selected unless otherwise specified in the contract or order. Compliance shall be reviewed after each equipment failure or at any other appropriate time. The basis for compliance shall be the total count of equipment failures and the test time, using the applicable accept/reject criteria. Test time is to be understood as meaning equipment "on" time, or equipment operating time for purposes of MTBF or longevity determination, and shall be the accumulated unit test hours expressed in terms of a multiple of the specified MTBF. For equipments with two or more separate components (i.e., power supply, receiver, transmitter), the test time on the equipment shall be the minimum accumulated time on any of its components. A decision to accept, continue testing, or reject shall be made and shall result in the actions delineated in the following paragraphs.

2. Accept

If the decision is to accept and requirements for lot acceptance have been met and no pattern failures are evident, all equipments in the production lot may be accepted provided those equipments subjected to reliability testing can again pass the individual tests described in the acceptance test portion of the equipment specification. When reduced testing is permitted, the data should be reviewed to see if a different test plan may be used on the next lot.

3. Continue Test

The test will continue until an accept or reject decision is reached. A review will be made of the test data each month to determine whether there will be a change in test plan (i.e., normal to reduced).

4. Reject in Reliability Qualification (Demonstration) Phase

If the qualification (demonstration) test results in a reject decision, or if pattern failures are present, corrective action shall be accomplished in accordance with Paragraph 11.5(y) and, using the same or a greater sample size, a retest shall be conducted as follows:

1. If the corrective action is determined by the procuring activity as not affecting the thermal design of the equipment, a reliability qualification (demonstration) test shall be repeated.
2. If the corrective action is determined by the procuring activity as affecting the thermal design of the equipment, then the reliability qualification (demonstration) test shall be repeated beginning with Paragraph 11.5(q)-2 of this standard.

5. Reject in Reliability Production Acceptance (Sampling) Phase

If the reliability production acceptance (sampling) phase tests result in a reject decision, or if a pattern failure is present, further shipment of units shall be discontinued until corrective action can be accomplished in accordance with Paragraph

11.5(y) and, using the same or a greater sample size, a retest shall be conducted as follows:

1. If the corrective action is determined by the procuring activity to be a design change, the lot shall be requalified in the reliability qualification (demonstration) phase beginning with that portion of the test stipulated in Paragraph 11.5(q)-1.
2. If the corrective action is determined by the procuring activity not to be a design change, the reliability production acceptance (sampling) test shall be repeated using the specified test plan. If the procuring activity determines that the corrective action does not affect the thermal design of the equipment, a reliability production acceptance (sampling) test shall be repeated beginning with the burn-in (or debugging) period as stipulated in paragraph 11.5(q)-4. If the procuring activity determines that the corrective action affects the thermal design of the equipment, the reliability production acceptance (sampling) test shall be repeated beginning with Paragraph 11.5(q)-2 of this standard.
3. If testing is being performed under the reduced testing of 11.5(h)-5, the test shall be repeated in accordance with 11.5(u)-5(1) and (2) above. Subsequent tests shall use the test plan specified by contract until eligibility for reduced testing is again established.
4. Whenever two in any sequence of five lots experience a reject decision using test plans III, IV, V, or VI, the procuring activity may require tightened testing on subsequent production acceptance (sampling) lots in accordance with the following table.

Specified Test Plan	Tightened Test Plan
III	I
IV	II
V	III
VI	V

Testing shall continue using the tightened test plan until eligibility for testing at the specified test plan is established in accordance with 11.5(i)-2.

(v) Failure

1. Failure Actions:

On the occasion of a failure (any operation outside specified limits) entries shall be made on the appropriate data logs and the failed equipment shall be removed from test and repaired with minimum interruption of test of equipments which continue to meet test performance requirements. All failed parts shall be replaced; any part which has deteriorated but does not exceed specified tolerance limits shall not be replaced. After a failed and repaired equipment has been returned to operable condition, it shall be returned to test with appropriate entries in the data logs and without interruption to the equipments continuing on test. The absence of one or more equipments for the purpose of failure repair shall not affect the ability to make decisions from log data.

(w) Failure Categories

The types of failures and required actions are as follows:

1. Relevant Failures

All failures are relevant unless determined by the procuring activity to be caused by a condition external to the equipment under test which is not a test requirement and not encountered in service. Certain parts of known limited life, such as batteries, may have a life stipulated prior to initiation of testing and approved by the procuring activity. Failures of these parts occurring prior to the end of the stipulated period are relevant. Failures of these parts occurring after the stipulated period are nonrelevant but any dependent failures caused thereby are relevant. Failures chargeable to workmanship deficiencies in the equipment shall be counted as relevant.

2. Multiple Failures

In the event simultaneous part failures occur, each failed part which will independently prevent satisfactory equipment performance shall be counted as an equipment failure except as follows: If the contractor and the procuring activity agree that the failure of one part was entirely responsible for the failure of any other parts, then each such dependent part failure shall not be counted as an equipment failure. All relevant failures shall be counted as equipment failures and included in MTBF calculations. At least one equipment failure shall be counted when a dependent failure is claimed.

3. Pattern Failures

When a definite pattern of independent or dependent failures is identified, the equipments shall be subjected to corrective action in accordance with Paragraph 11.5(y).

4. Analysis of Failures

The cause of each equipment or part failure, including government furnished equipment or parts to be included in or as a part of the test group equipment, shall be determined by investigation and analysis. The investigation and analysis shall consist of any applicable method (such as test, application study, dissection, X-ray analysis, microscopic analysis, or spectrographic analysis) which may be necessary to determine the cause of failure. No part or unit may be replaced in any equipment involved in the reliability test unless such part or unit can be unmistakably demonstrated by the repair activity to be outside of specification tolerances. When no part or unit can be demonstrated to be beyond specification limits, whether deteriorated or otherwise, the equipment failure must be classified as a relevant failure and shall be counted in the total number of equipment failures. Details of the circumstances shall be referred to the design or quality control activity for close study. An inadvertent replacement made under the above conditions shall be counted as a relevant failure and called to the attention of the procuring activity.

5. Failure Confirmation

All failures observed during the equipment reliability test should be confirmed. Failure to do so is insufficient grounds for deletion of such failures in the count of total failures. Lack of failure confirmation should instigate close review of the test facility. If the test facility can be shown to be both at fault and completely accountable for the failure to the satisfaction of the procuring activity, then and only then can the count of such failure be eliminated.

(x) Verifying Repair

Following repair or corrective action and prior to the resumption of test, it shall be permissible to operate equipment in the test facility. Test procedures should specify for all components of the system under test, the period of operation (or number of cycles) which will be used to verify the effectiveness of the repair. Post-repair "burn-in" time will in no case exceed that employed for new equipment acceptance. Failures and time during this period should be recorded and reported but not used in determining compliance with MTBF requirements. However, all failures shall be subject to analysis.

(y) Corrective Action

When any reliability test reaches a reject decision, discontinue the test, immediately notify the procuring activity, and promptly develop and propose a plan for correction of the deficiencies to the procuring activity. The specified performance and the required characteristics of the equipment shall not be changed to achieve reliability requirements. When it is determined by the contractor's analysis that a part failure was due to operation of the part beyond its rated capability, a replacement may be made with a like part having a higher rating consistent with the applied stress. Any such replacement shall be applied in like manner to all equipment under test before further testing and to all subsequent equipment being produced. Such corrective action shall be reported in detail with supporting data as part of the current monthly summary report. The procuring activity will review such corrective actions and may require handling as a design change or initiate modification based on the nature of the corrective action. Any corrective action beyond increasing the rated capability which amounts to a design change, modification, or replacement with an unlike part shall be proposed to the procuring activity as an engineering change proposal. Any necessary retrofitting of equipment already submitted shall be treated as specified in the contract.

11-6 EXAMPLES—MIL-STD-781B

The following examples will serve to illustrate the use of MIL-STD-781B.

> **EXAMPLE 11-6** A contract for 300 electronic devices requires a qualification demonstration test in accordance with MIL-STD-781B: test plan XVI, test level *E* is invoked, voltage cycling is required. The specified MTBF is 1000 h; the minimum acceptable MTBF is 667 h. The quantity of units to be placed on test is to be sufficient to achieve the necessary device operating hours in a time period not exceeding 1 month.
>
>   (a) Describe the environmental test conditions.
>   (b) What are the producer's and consumer's risks?
>   (c) How many devices are placed on test?
>   (d) Assuming that 8 failures occur, do the devices meet the MTBF requirement?

Solution
(a) Refer to Paragraph 11.5(f)-3 (p. 305), which describes the environmental conditions of test level *E*. In accordance with these requirements, the devices are placed in an environmental test chamber and the temperature of the chamber is lowered to $-54°C$. During this cooling cycle, power to the devices is turned off.

After stabilization at this temperature the power is turned on again and the temperature of the chamber is raised to $+55°C$. After stabilization at $+55°C$, the devices are operated for 2 h. Following the 2-hour operation at 55°C, power is turned off and the chamber is cooled to $-54°C$, thus resuming another cycle.

During each hour of on-time the devices are furthermore subjected to a vibration level of $2.2g \pm 10\%$ peak acceleration for at least 10 min. The requirement for input voltage to the devices is nominal value $+5 -2\%$.

Input voltage cycling is carried out as follows. During the first one third of the on-cycle, the input voltage to the devices is maintained at 110% nominal value, then, for the next one-third of the on-cycle, nominal voltage is used; and for the final one-third of the on-cycle, the input voltage is lowered to 90% of nominal.
(b) Refer to Table 11-7(p. 307). For test plan XVI, the producer's and consumer's risks are both 20%. The test duration is 14.1 multiples of the specified MTBF θ_0. Since $\theta_0 = 1000$ h, the test duration is

$$14.1 \times 1000 = 14,100 \text{ equipment hours}$$

The requirement is that the test be completed in 1 month's time (720 h). Accordingly, the number of devices to be placed on test for a period of 1 month is

$$\frac{14,100}{720} = 19.6, \text{ say 20 devices}$$

Table 11-7 refers to Paragraph 11.5(j)-1 (p. 310) for recommended sample sizes for qualification testing. A sample size of 20 for a lot size over 200 is recommended. Therefore, a 1-month test program using 20 devices on test is satisfactory.

Table 11-7 gives an accept criterion of 17 or less failures. Since only 8 failures are stated to have occurred the MTBF of the devices is acceptable.

EXAMPLE 11-7 A production acceptance test is implemented using the minimum permitted sample size of three equipments. Test plan III, test level C of MIL-STD-781B, applies. Voltage cycling is not required. The specified MTBF is $\theta_0 = 2000$ h; the minimum acceptable MTBF is $\theta_1 = 1000$ h. Assume that 5 failures occurred during 5000 equipment hours. Does the lot meet the acceptance criteria?

Solution Refer to Table 11-5 (p. 306). For the accept/reject criterion for test plan III reference is made to Paragraph 11.5(o)-2 (p. 315). To ascertain whether 5 failures still permit acceptance of the observed MTBF, we first need to convert the 5000 equipment hours to multiples of the specified MTBF θ_0:

$$\frac{5000}{2000} = 2.5 \text{ multiples}$$

Referring to the sequential test plot of Figure 11-4 it is noted that the intersection of 5 failures and 2.5 multiples of θ_0 falls inside the "Continue testing" region; hence, no decision is possible and the test is to be continued.

EXAMPLE 11-8 Refer to Example 11-7. Assume that the test is continued and no further failures occur. After how many additional equipment test hours is an accept decision possible?

Solution Refer to Paragraph 11.5(o)-2. For 5 failures an accept decision can be made after completion of 5.67 multiples of θ_0. Hence,

$$5.67 \times 2000 = 11,340 \text{ equipment hours}$$

The additional test time, following the earlier 5000-h testing, is therefore

$$11,340 - 5000 = \underline{6340 \text{ equipment hours}}$$

EXAMPLE 11-9 Production of a certain device is continuous. The qualification (demonstration) test per MIL-STD-781B has been successfully completed, and 8 successive lots have passed production acceptance sampling per test plan I. The specified MTBF is $\theta_0 = 2000$ h. The observed MTBF (total test hours divided by the number of failures encountered in the 8 lots) was found to be 3000 h. Reduced testing is started beginning with the ninth production lot in accordance with the provisions of MIL-STD-781B.

If the next test were to produce any of the following results:

(a) 10 failures in 11,340 equipment hours
(b) 3 failures in 650 equipment hours
(c) 6 failures in 13,600 equipment hours
(d) 5 failures in 5000 equipment hours

what would the accept/reject decision be in each of these instances?

Solution
(a) Per Paragraph 11.5(h)-5 (p. 309), the reduced test plan for an initial test plan I is test plan III.

(b) Determine the equipment hours in multiples of θ_0:

$$\frac{11,340}{2000} = 5.67$$

$$\frac{650}{2000} = 0.325$$

$$\frac{13,600}{2000} = 6.8$$

$$\frac{5000}{2000} = 2.5$$

(c) Refer to Paragraph 11.5(o)-2 (p. 315) and ascertain the accept/reject criterion for the respective number of failures and the test time expressed in multiples of θ_0:

No. Failures	Computed Test Time in Multiples of θ_0	Reject, if Computed Test Time in Multiples of θ_0 is Equal or Less Than:	Accept if Computed Test Time in Multiples of θ_0 Is Equal or More Than:
10	5.67	5.20	9.13
3	0.325	0.35	4.28
6	6.8	2.43	6.36
5	2.5	1.74	5.67

Accordingly, the following decisions would be made for the above cases:

10 failures: continue testing
3 failures: reject the lot
6 failures: accept the lot
5 failures: continue testing

PROBLEMS

11-1 What is the purpose of reliability testing?

11-2 What is the criterion for acceptability in a reliability test?

11-3 Name the three types of reliability tests discussed in this chapter and give a brief explanation for each.

11-4 State the definitions for producer's risk and consumer's risk as used for the sequential test plan.

11-5 A life test is conducted on a random sample of 5 units drawn from a production lot. Failed units are replaced. The producer's risk is 10%. The acceptable mean life is 3000 h. The test is to be terminated upon occurrence of the fourth failure. This point is reached after 1100 hours operation. Determine whether the lot is acceptable.

11-6 A life test is conducted on 6 units, but failed units are not replaced. The specified θ_0 is 2500 h. The test is to be stopped upon reaching the fifth failure. The producer's risk is 5%. Assume that the five failures occurred after the following hours of testing:

1st failure: 240
2nd failure: 325
3rd failure: 630
4th failure: 925
5th failure: 1350

Have the units passed the life test?

11-7 A life test is conducted on 8 units and any failed unit is replaced. The acceptable mean life is 3000 h. The producer's risk is 10%. The test is to be terminated upon occurrence of the fourth failure. If the fourth failure occurs upon reaching 1450 h of operation, have the units passed the test successfully?

11-8 Refer to Problem 11-7. Repeat the problem for the case where failed units are not replaced. The first three failures occurred at 1012, 1150, and 1330 h.

11-9 (a) What are the three possible decisions that apply at any point of a sequential reliability test prior to truncation?
(b) What is the advantage of using a sequential test plan?

11-10 (a) State the formulas for the slope and the time-axis intercepts of the accept and reject lines.
(b) What are the meanings for θ_1 and θ_0 in connection with sequential testing?

11-11 A group of 8 units is subjected to a sequential test and failed units are replaced. The specified MTBF is 1200 h. The minimum acceptable MTBF is 600 h. Producer's risk and consumer's risk are both 10%. Failures occurred at the following times:

1st failure: 100 h
2nd failure: 400 h
3rd failure: 500 h
4th failure: 700 h

Plot the sequential test curve and indicate after what test time an accept or reject decision is possible. Which failure immediately precedes the decision point?

11-12 What were the recommendations of the AGREE report with respect to reliability testing?

11-13 (a) What are the three major groups of test plans offered by MIL-STD-781B?
(b) What is the discrimination ratio?

11-14 Discuss briefly the environmental test levels offered by MIL-STD-781B.

11-15 (a) One of the two major categories of tests in MIL-STD-781B is the qualification (demonstration) test. What is the other category?
(b) What are the thermal and vibration surveys?

11-16 (a) Discuss the longevity and the all equipment screening test plans in MIL-STD-781B.
(b) Under what conditions may tightened inspection be used?
(c) What is a relevant failure?

11-17 A radar system is to be subjected to a qualification demonstration test per MIL-STD-781B, test plan VII, test level *C*. Voltage cycling is required. The specified MTBF is

1500 h. Describe the environmental test conditions and identify producer's and consumer's risks. Assuming that a total of 5 failures occurred in the equipment during 3000 h of testing, has the equipment passed the reliability test? Use Figure 11-5.

11-18 A production acceptance test is performed on 6 units using test plan I, test level E of MIL-STD-781B. The specified MTBF is 800 h. Do the units pass the test if 4 of the 6 units failed during 4000 equipment test hours?

11-19 Reduced Production Acceptance Sampling is introduced per MIL-STD-781B after 8 successive lots passed successfully. Initially, test plan I was used. The specified MTBF is 2500 h. Identify the reduced test plan. If the test on units from the ninth production lot produces 3 failures in 2800 equipment hours, what accept/reject decision may be made? (See paragraph 11.5(h)-5).

11-20 Discuss the required action per MIL-STD-781B when a reject decision is reached during the production acceptance (sampling) phase.

11-21 Discuss the three failure categories defined in MIL-STD-781B.

11-22 Discuss the requirements in MIL-STD-781B for the analysis of failures.

12

MAINTAINABILITY

12-1 MAINTAINABILITY ENGINEERING

Failures will occur no matter how reliable equipment is made. These failures have to be diagnosed and repaired. Depending on the severity of the failure, the ability to diagnose the problem, the ease of access to the failed item, the availability of spare parts, etc., the repair of the failure may take place immediately or it may have to be deferred. Repairs requiring the replacement of failed parts usually require equipment shutdown and loss of operational time. The harder the diagnosis of the problem and the accessibility to the affected area, the longer the equipment downtime that may be expected.

The prediction of equipment downtime is quite important to the ultimate user. He requires a quantitative prediction of the maintainability of the equipment in order to assess whether the downtime is consistent with operational requirements and to plan corrective and preventive maintenance procedures and logistic support. A suggested definition of maintainability is: *Qualitatively, maintainability is the aggregate of all design features which promote equipment serviceability. Quantitatively, it is commonly expressed as the average time required to restore the equipment to service.*

Since the MTBF of equipment defines the average time between failures, we can easily estimate the probable number of failures in a given time period. For example, suppose the MTBF of a piece of electronic equipment is 50 h; we may, on the average, expect approximately $744/50 = 15$ failures each month, assuming the equipment is operating continuously. If the mean time to repair (MTTR) runs about 1 h, a 15-h equipment downtime per month may then, on the average, be expected.

The objectives of a maintainability program should be to influence equipment design to assure that maintenance of the equipment can be accomplished efficiently and safely. The *downtime*, which is the time the equipment is inoperative because of maintenance activities, is obviously of vital interest to the user. He is also interested in an estimate of the *availability* of the equipment, that is, the probability that the equipment is not down due to required maintenance actions. The concept of availability as a function of MTBF and MTTR will be introduced later in this chapter.

12-2 DESIGNING FOR MAINTAINABILITY

Complex electrical and mechanical systems should be designed so that there is quick access to all points in the equipment. Where possible, electronic circuits should be standardized on plug-in modules and assemblies that are replaced with spare units when they fail. The plug-in concept allows rapid restoration of service, minimizing downtime; failed units removed from the equipment are repaired while the equipment continues to operate.

Maintainability design features (a comprehensive checklist is given in Table 12-2) include access to replaceable assemblies, clear markings for convenience and safety, rapid release/captive fasteners, maximum use of interchangeability, elimination of unnecessary adjustments, clearly written maintenance instructions, and adequate use of test points for fault isolation in electronic equipment.

The more complex the equipment, the more time-consuming will be fault location and testing. Whereas less sophisticated machines and devices still depend on the trouble shooting capability of the maintenance technician, complex electronic systems use diagnostic computer programs which store coded instructions for automatic fault isolation and detection. Automated fault diagnosis has even found its way into the automotive world for rapid analysis of engine condition.

In densely packaged digital electronics it often is not economical to isolate faults to a single plug-in module out of thousands. An approach of isolating faults only down to a group of, say, 10 modules may prove more cost-effective in system design trade-offs. To minimize downtime and sparing requirements, the 10-module group may be reduced further by *probabilistic replacement*—replacing a subgroup (1 or more modules) at a time (starting with the subgroup most likely to contain the failed unit), retesting the system after each subgroup replacement, and ending the sequence when the fault has been isolated and corrected. Like so many other aspects of reliability, maintainability and logistics, the *probabilistic replacement* strategy is suited to computer optimization.

The concern by the military that designers should consider equipment supportability (and hence, maintainability) is illustrated by the following excerpt from an army pamphlet[1]:

> Some of our large complex weapon systems are being phased out of the defense inventory, not because of inability to do the job, but rather because of excessive support costs. For the most part, these costs resulted from inade-

---

[1] Department of the Army Pamphlet 705-1. Research and Development of Material, Maintainability Engineering, Headquarters, Department of the Army, June 1966.

quate support planning. Contracts were awarded based on initial hardware procurement costs, rather than on total system life-cycle costs—hardware costs plus support costs over the life of the system. All this meant that little consideration was given, in the design stage, to the costs of supporting a system after it became operational. The end result has inevitably been abandonment of effective systems because of the costs of maintaining them.

Judging from the high incidence of consumer complaints about products from automobiles to washing machines, some manufacturers could profit from the lessons learned by the military.

To summarize, equipment maintainability is vital both to military and commercial equipment. The equipment must be maintainable at reasonable cost and with acceptable downtimes.

Sections 12.3 through 12.6 are based on and contain selected material from a current Navy Maintainability Design Criteria Handbook.[2]

12-3 MAINTAINABILITY ASSURANCE

(a) Maintainability Program

Maintainability programs rank in importance with other equipment program elements, such as performance, reliability, quality, and safety. An effective maintainability program must provide the designers with the necessary maintainability design criteria and requirements. These criteria must be both qualitative and quantitative and must of course agree with contractual equipment maintainability requirements.

The development of a maintainability program should begin when the equipment's performance requirements are being formulated, or at the latest, during that phase of development in which the preliminary design and engineering effort are verified and accomplished. During the early design phase, part of the maintainability information gained by the designer will necessarily be limited to a qualitative nature because of the limited amount of detailed equipment design data. As the design progresses, increased maintainability information becomes available to the designer to quantitatively measure the effects of his maintainability design in achieving the maintainability requirements. The depth and scope of the maintainability program should at all times be economically consistent with the type and complexity of the equipment design and the equipment procurement cycle.

(b) Maintainability Program Development

The designer should develop a maintainability program at the earliest time in equipment design consistent with the type and complexity of the equipment design. The maintainability program should describe in sufficient detail how the designer intends to conduct the maintainability program in order to achieve the qualitative and quantitative maintainability requirements for the equipment design. To assist the designer

[2] NAVSHIPS 0967-312-8010 Maintainability Design Criteria Handbook for Designers of Shipboard Electronic Equipment (Change 4), Department of the Navy, July 1972.

in identifying essential tasks of an effective maintainability program, Figure 12-1 may be used. This figure illustrates not only the essential tasks of an effective main-

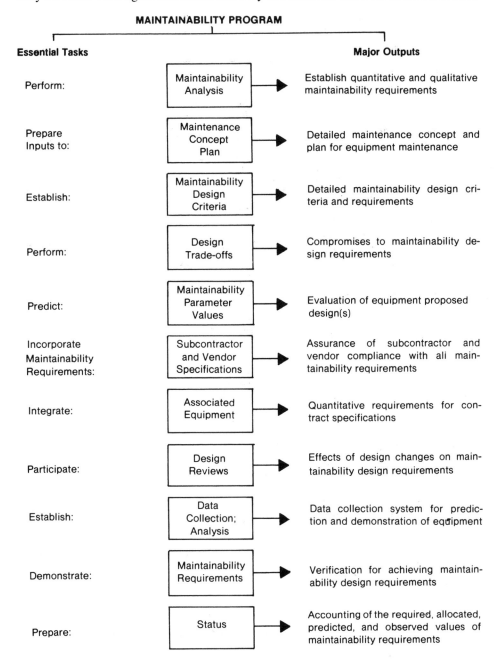

Figure 12-1 *Essential tasks and major outputs of an equipment maintainability design NAVSHIPS 0967-312-8010.*

tainability program but also contains a brief description of the major outputs in accomplishing the essential tasks. In addition to the essential tasks, the maintainability program should identify the work to be accomplished for each essential task, its time phasing, and organizational element responsible for its implementation. Also, interfaces with other program activities should be defined.

(c) Maintainability Analysis

A maintainability analysis of the equipment design is an essential step in assuring design effectiveness. Inputs to a maintainability analysis usually consist of (1) preliminary equipment design data, (2) information from the designer's other program activities affecting the equipment design, and (3) information and constraints contained in the procurement contract. Such constraints could involve higher-level maintenance concepts, cost, skill or tool limitations, and so on.

In essence, a maintainability analysis is the process that translates the above information into detailed qualitative and quantitative maintainability design criteria and requirements for the equipment design. A major task in the analysis is the allocation of quantitative maintainability requirements (from the overall maintainability requirements) to lower levels of the equipment design.

(d) Maintenance Concept and Maintenance Plan

As a result of the maintainability analysis effort, the designer should have sufficient information to formulate an effective and economical, broad, maintenance concept for the equipment design.

The resultant data from the maintenance concept effort are used to formulate the detailed maintenance plan for the equipment. Therefore, the designer should include in the equipment's maintenance plan all pertinent decisions that were made during the development of the maintenance concept for the equipment. Decisions regarding such factors as depth and frequency of maintenance requirements at each equipment functional level, facilities required for equipment support, support equipment and tools required, and maintenance personnel skill levels all impact on the equipment maintenance concept.

Included in the detailed maintenance plan, therefore, should be sufficient information on each type of scheduled maintenance action that is required by maintenance personnel. Refer to chapter 13 for a detailed discussion of the maintenance engineering analysis.

(e) Maintainability Design Criteria

In developing maintainability design criteria, the design guidelines provided in Table 12-1 as well as the comprehensive maintainability design check list shown in Table 12-2 may be used.

Table 12-1 *General Maintainability Design Guidelines (NAVSHIPS 0967-312-8010).*

Maintainability Design Criteria	Maintainability Design Criteria
1. Reduce complexity of maintenance by:	a. Providing adequate accessibility, work clearance b. Proving for interchangeability of like components, materials and parts within the equipment c. Utilizing MIL-Standard parts and items within government inventories d. Limiting the number and variety of tools, accessories and support equipment e. Insuring compatibility among equipment and facilities
2. Reduce the need for and frequency of design-dictated maintenance activities by using:	a. Fail-safe features b. Components which require little or no preventive maintenance c. Tolerances which allow for use and wear throughout life d. Adequate corrosion prevention and/or control features
3. Reduce maintenance down-time by designing for:	a. Rapid and positive detection of equipment malfunction or perform-mance degradation b. Rapid and complete preparation to begin maintenance tasks c. Rapid and positive localization of malfunctions to the repair level for which skills, spares, and test equipment are planned d. Ease of fault correction e. Rapid and positive adjustment and calibration f. Rapid and positive verification of correction
4. Reduce design-dictated maintenance support costs	a. The need for specialized maintenance tools, support equipment and facilities b. The requirements for depot or factory maintenance, consistent with system/cost effectiveness c. The need for extensive maintenance technical data
5. Limit maintenance personnel requirements by applying	a. Identification and accessibility of parts, test points, adjustments and corrections b. Easy handling, mobility, transportability and storeability c. Logically sequenced maintenance tasks d. A feasible range of relevant personnel physiological parameters
6. Reduce the potential for maintenance error by designing to eliminate:	a. The possibility of incorrect connection, assembly, and installation b. Awkward and tedious maintenance tasks c. Ambiguity in maintenance labeling, coding, and required technical documentation

Table 12-2 *Maintainability Design Checklist (NAVSHIPS 0907-312-8010).*

DESCRIPTION	RECOMMENDATIONS
Accessibility	1. Optimum accessibility should be provided in all equipment and components requiring maintenance, inspection, removal or replacement.
	2. A transparent window or quick-opening metal cover should be used for visual inspection accesses.
	3. A hinged door should be used where physical access is required (instead of a cover plate held in place by screws or other fasteners).
	4. If lack of available space for opening the access prevents use of a hinged opening, a cover plate with captive quick-opening fasteners should be used.
	5. When a screw-fastened access plate is used, no more than four screws should be used.
	6. On hinged access doors, the hinge should be placed on the bottom or a prop provided so that the door will stay open without being held if unfastened in a normal installation.
	7. Parts should be located so that other large parts which are difficult to remove do not prevent access to them.
	8. Components should be placed, so that there is sufficient space to use test probes, soldering irons, and other required tools without difficulty.
	9. Units should be placed so that structural members do not prevent access to them.
	10. Components should be placed so that all throwaway assemblies or parts are accessible without removal of other components.
	11. Screwdriver operated controls should be located so that when controls are adjusted, the handle will clear any obstruction.
	12. Enough access room should be provided for tasks which necessitate the insertion of two hands and two arms through the access.
	13. The access opening should provide enough room for personnel's hands or arms and still provide for an adequate view of work area.
	14. Irregular extensions, such as bolts, tables, waveguides, and hoses should be easy to remove before the unit is handled.
	15. Access doors should be made in whatever shape it is necessary to permit passage of components and implements.
	16. Units should be removable from the installation along a straight or moderately curved line.
	17. Heavy units (more than about 25 lbs.) should be installed within normal reach of personnel for purposes of replacement.
	18. Provisions should be made for support of units while they are being removed or installed.
	19. Rests or stands should be provided on which units can be set to prevent damage to delicate parts.

Table 12-2 (Continued) *Maintainability Design Checklist*

DESCRIPTION	RECOMMENDATIONS
Accessibility (Cont'd)	20. Access points should be individually labeled so they can be easily identified with nomenclature in the Maintenance Requirement Cards (MRC) and maintenance manuals.
	21. Accesses should be labeled to indicate what can be reached through this point (label on cover or close thereto).
	22. Access openings should be free of sharp edges or projections which could injure the personnel or snag clothing.
	23. Parts which require access from two or more openings should be marked in order to avoid and/or damage by trying to repair or remove through only one access. Double openings of this type should be avoided wherever possible.
	24. Human strength limits should be considered in designing all devices which must be carried, lifted, pulled, pushed, and turned.
	25. Environmental factors (cold weather, darkness, etc.) should be considered in design and location of all manipulatible equipment items.
	26. The same type fasteners should be used for all access openings on a given equipment.
	27. Ventilation holes with screening of small enough mesh should be provided to prevent entry of probes or conductors that could inadvertently contact high voltages.
	28. A combination RFI and moisture proof gasket should be provided for each chassis or access opening in the enclosure.
	29. Accessibility of external test points should be assured under use conditions.
	30. Test points should be grouped for accessibility and convenient sequential arrangement of testing.
	31. Each test point should be labeled with name or symbol appropriate to that point.
	32. Test points should be located close to controls and displays with which they are associated.
	33. Test points should be located so as to reduce hunting time (near main access openings, in groups, properly labeled, near primary surface to be observed from working position).
	34. Test points which retain test probe should be provided so that personnel will not have to hold the probe.
Identification	1. All units should be labeled and, if possible, with full identifying data.
	2. Each terminal should be labeled with the same code symbol as the wire attached to it.
	3. On equipment utilizing color coding, meaning of colors should be given in manuals and on an equipment panel.

Table 12-2 (Continued) *Maintainability Design Checklist*

DESCRIPTION	RECOMMENDATIONS
Identification (Cont'd)	4. Color coding should be consistent throughout system, equipment and maintenance supports. 5. Display and control labels should clearly indicate their functional relationship. Displays should be labeled by functional quantity rather than operational characteristics (i.e., volts, ohms, etc.). 6. Schematics and instructions should be attached directly to, or adjacent to chassis for all units which may require troubleshooting. 7. When selector switches may have to be used with a cover panel off, a duplicate switch position label should be provided on the internal unit so that maintenance personnel does not have to refer to label on the case or cover panel. 8. Color codes for identifying test points or tracing wire or lines should be easily identifiable under all conditions of illumination and should be resistant to damage or wear. 9. Functional organization of displays and controls should be emphasized by use of such techniques as color coding, marked outline, symmetry of grouping, and/or differential plane of mounting.
Standardization	1. Functional interchangeability should exist where physical interchangeability is possible. 2. Sufficient information should be provided on identification plates and within related Maintenance Requirements Cards (MRC's) so the user can adequately judge whether two similar parts are interchangeable. 3. Parts, fasteners, connectors, lines and cables, etc., should be standardized throughout the system and, particularly, from unit to unit within the system. 4. Interchangeability should be provided for components having high mortality wherever practical.
Safety	1. Edges of components and maintenance access openings should be rounded or protected by rubber, fiber, or plastic protectors to prevent personnel injury. 2. Audible warning signals should be distinctive and unlikely to be obscured by other noises. 3. Critical warning lights should be isolated from other less important lights for best effectiveness. 4. When selecting warning lights, they should be compatible with ambient illumination levels expected. (A dim light will not be seen in bright sunlight, and a bright light may be detrimental to dark adaptation). Are dimmer controls utilized if necessary? 5. On-off or fail-safe circuits should be utilized wherever possible to minimize failures without operator knowledge. 6. Bleeding devices should be provided for high-energy capacitors which must be removed during maintenance operations.

340

Table 12-2 (Continued) *Maintainability Design Checklist*

DESCRIPTION	RECOMMENDATIONS
Safety (Cont'd)	7. Struts and latches should be provided to secure hinged and sliding components against accidental movement which could cause injury to personnel during maintenance operations.
	8. Limit stops should be provided on drawers or fold-out assemblies which could cause personnel injury if not restrained.
	9. Safety interlocks should be used wherever necessary.
	10. Components located and mounted so that access may be achieved without danger to personnel from electrical charge, heat, sharp edges and points, moving parts, and chemical contamination.
Cables and Wiring	1. Cables should be of sufficient length to provide movement of each functioning unit to convenient positions for checking and/or to allow free access to cable connectors during installation or removal.
	2. All cables should be color coded and both ends tagged. Colors should be selected which cannot be confused because they are too nearly alike or may not be recognized because of poor illumination.
	3. Cables and lines should be directly accessible to the personnel wherever possible (not under panels which are difficult to remove).
	4. Cables should be routed so they need not be bent or unbent sharply when being connected or disconnected.
	5. Cables should be routed so they cannot be pinched by doors, lids, etc., or so they will not be stepped on or used as hand holds by maintenance personnel.
	6. Cables or lines should be attached to units which can be partially removed (chassis on slide racks) and attached so unit can be replaced conveniently without damaging cable or interfering with securing of unit.
	7. The necessity for removal of connectors or splicing of lines should be avoided.
	8. Cable entrances on the front of cabinets should be avoided where it is apparent they could be "bumped" by passing equipment or personnel.
Connectors	1. Adjacent solder connections should be far enough apart so work on one connection does not compromise integrity of adjacent connections.
	2. Connector plugs should be designed so that pins cannot be damaged (aligning pins extended beyond electrical pins).
	3. Self-locking safety catches should be provided on connector plugs rather than safety wire.
	4. Connectors should be designed so that it is physically impossible to reverse connections or terminals in the same or adjacent circuits.
	5. Quick-disconnect devices should be used wherever possible to save time and minimize human error which could occur in soldering, etc. (fractional-turn, quick-snap action, press-fit, etc.)

Table 12-2 (Continued) *Maintainability Design Checklist*

DESCRIPTION	RECOMMENDATIONS
Connectors (Cont'd)	6. Markings on plugs, connectors, and receptacles should show proper position of keys for aligning pins for proper insertion position.
	7. Connectors should be located for easy accessibility for replacement or repair.
Controls	1. All adjustments should be located on a single panel.
	2. Controls should be located where they can be seen and operated without disassembly or removal of any part of installation.
	3. Controls should be placed on panel in the order in which they will be normally used.
	4. Controls when used in a fixed procedure should be numbered in operational sequence.
	5. For concentric shaft vernier controls, larger diameter control should be used for the fine adjustment.
	6. Knobs, for precision settings should have a 2-inch minimum diameter.
	7. Controls should be labeled with functional statement.
	8. Control position markings should be descriptive rather than coded or numbered.
	9. Control scales should be only fine enough to permit accurate setting.
	10. Except for detents or selector switches, controls should be used which have smooth, even resistance to movement.
	11. Selector switches should be used which have sufficient spring loading to prevent indexing between detents.
	12. Spring-loaded pushbuttons should be used which require no excessive finger pressure.
	13. Tool-operated controls operable by screwdriver or other medium size standard tool should be used.
	14. Coaxial knobs should be adequately coded to avoid confusion.
	15. Calibration instructions should be placed as close to the calibrating controls as possible.
	16. Adjustment controls should be easy to set and lock.
	17. Visual, auditory, or tactual feedback should be provided for all physical adjustment procedures.
	18. Some type of indexing should be provided for adjustment controls. (It is difficult to remember positional settings without some marking: indexing can also show gradual deterioration in performance as settings change from time to time.)

Table 12-2 (Continued) *Maintainability Design Checklist*

DESCRIPTION	RECOMMENDATIONS
Displays	1. All displays used in system checkout should be located so they can be observed from one position.
	2. On units having operator displays, maintenance displays should be located behind access doors on operator's panel.
	3. On units without operator panel, maintenance displays should be located on one face accessible in normal installation.
	4. Display scales should be limited to only that information needed to make decisions or to take some action.
	5. When center-null displays are used, circuits should be designed so that if power fails indicator will not rest in the in-tolerance position.
	6. Numerical scales should be used only when quantitative data is required.
	7. Scales should be provided with only enough graduation for required accuracy without interpolation.
	8. Special calibration point should be provided on scale or on a separate overlay if edges and midpoint of tolerance range are not sufficient for accurate calibration.
	9. Displays which provide tolerance ranges should be coded so both the correct reading and tolerance limits are identified.
	10. Critical warning lights should be isolated from other less important lights for best effectiveness.
	11. Counters should be used in which numbers snap rather than shift into place.
	12. Counters should be mounted close to the panel so that numbers are not obscured by bezel opening.
Handles	1. Handles should be used on units weighing over 10 pounds.
	2. Handles should be provided on smaller units which are difficult to grasp, remove, or hold without using components or controls as hand holds.
	3. Handles should be placed above the center of gravity and positioned for balanced loads.
	4. Handles requiring firm grip, bale openings, should be provided at least 4.5 inches wide and 2 inches deep.
	5. Handles should be located to prevent accidental activation of controls.
Mounting	1. Wrong installation of unit should be prevented by virtue of size, shape, and configuration, etc.
	2. Modules and mounting plates should be labeled.
	3. Guides should be used for module installation.

343

Table 12-2 (Continued) *Maintainability Design Checklist*

DESCRIPTION	RECOMMENDATIONS
Mounting (Cont'd)	4. Means should be provided for pulling out drawers and slide-out racks without breaking electrical connections when internal in-service adjustments are required.
	5. Parts should be mounted on one side of a surface with associated wiring on the other side.
	6. Easily damaged components should be mounted or guarded so they will be protected.
	7. When fold-out construction is employed, parts and wiring should be positioned so as to prevent damage by opening and closing.
	8. When screwdriver adjustments must be made by touch, screws should be mounted vertically so that the screwdriver will not fall out of the slot.
	9. Internal controls (switches, adjustment screws) should be located away from dangerous voltages.
	10. Screwdriver guides should be provided on adjustments which must be located near high voltage.
	11. Parts should be located so that other large parts (such as indicator and magnetron tubes) that are difficult to remove do not block access to them.
	12. Parts, assemblies, and components should be placed so there is sufficient space to use test probes, soldering irons, and other tools.
	13. All throwaway items should be accessible without removal of other items.
	14. All fuses should be located so that they can be seen and replaced without removal of any other item.
	15. Fuse assemblies should be designed and placed so that tools are not required to replace fuses.
	16. Fasteners for assemblies and subassemblies should be designed to operate with maximum of one complete turn.
	17. If bolts are used, they should require a minimal number of turns (less than 10).
	18. When tool operated fasteners are required, only those operable with standard tools should be used.
	19. Guide pins on units and assemblies should be provided for alignment during mounting.
	20. When tool-driven screws must be used, types should be used which can be driven be several tools (screwdriver, wrench, or pliers, where possible, i.e., a hex head with screwdriver slot).
	21. Access cover fasteners of the captive type should be used.

(f) Design Trade-Offs

In optimizing system design effectiveness for the equipment, the designer should ensure that the maintainability requirements and maintainability design criteria receive appropriate consideration. In the event of any changes in the equipment's maintainability design which are deemed necessary, the designer should see to it that these changes are documented and reflected in the maintainability analysis of the equipment. In many instances improvements in system design effectiveness may be obtained more economically through the judicious application of established maintainability design features (reductions in equipment repair time) than through reliability improvements (increases in the equipment mean time between failure).

(g) Design Reviews

Informal and formal design reviews should be conducted to ensure compliance with overall maintainability requirements of the equipment and to measure the effects design changes may have in meeting the overall maintainability requirements. The depth and scope of the informal and formal design reviews will depend to a large extent on the type and complexity of the equipment design and the design phase of the equipment procurement cycle.

(h) Data Collection and Analysis

In developing the equipment maintainability program, the designer should include provisions for establishing data-collection systems for (1) maintainability predictions during equipment design, and for (2) the evaluation of the equipment demonstration test results. Wherever possible, these data-collection systems should be integrated with similar data-collection systems for other equipment program activities.

The data-collection system for equipment maintainability predictions should be formulated as early as possible during equipment design but not later than equipment contract definition. The data-collection system should also be used during the equipment design and development. The data-collection system for the equipment demonstration test evaluation should be planned as early as possible but should be finalized in the maintainability demonstration test plan prior to equipment demonstration testing.

(i) Demonstration of Maintainability
Requirements

Once equipment has been designed to meet a specified degree of maintainability, a maintainability demonstration test of the actual equipment will usually be necessary for a final verification of compliance with the overall maintainability requirements. Moreover, the maintainability demonstration test plan should be integrated in the equipment program plan with other equipment testing requirements such as proof of design, prototype, and production and acceptance testing. The maintainability demonstration test plan should follow the procedures outlined in MIL-STD-471,[3] "Maintainability Demonstration." Very often, Method 3 of MIL-STD-471 will be used for demonstration testing of equipment unless the equipment's specifications provide requirements for other methods. In general, the maintainability demonstration test plan should emphasize the following type of information:

1. Demonstration conditions:
   (a) Quantitative maintainability requirements.
   (b) Maintenance concept.
   (c) Maintenance demonstration environment.
   (d) Level(s) of maintenance to be demonstrated.
   (e) Demonstration sites.
   (f) Facility requirements.
   (g) Participating agencies.
   (h) Mode of operation for the demonstration test, including configurations and mission requirements.
   (i) Items to be subjected to demonstration tests.
2. Simulation of faults or failures. In general, fault simulation for corrective maintenance tasks should be performed by such methods as introduction of faulty parts and deliberate misalignment. Fault simulation tasks should be planned and performed in such a manner that no damage to the equipment will occur. The failure modes and actual random selection of the tasks to be demonstrated should be jointly decided by the designer and the procuring activity.

The maintainability demonstration test data are used to verify the achievement of the maintainability parameter values determined from the maintainability analysis and maintainability predictions. Therefore, the data from the demonstration test should include the following:

1. All data collected.
2. Factors that influence the data.
3. Analysis of data.
4. Results of the demonstration.
5. Assessment of qualitative factors.
6. Deficiencies.
7. Recommendations for
   (a) Correcting deficiencies.
   (b) Suggesting improvements.
8. Results of retest, if applicable.

---

[3] MIL-STD-471A Military Standard, "Maintainability Verification/Demonstration/Evaluation," March 27, 1973, Department of Defense, Washington, D.C.

The formal maintainability demonstration test that is performed to determine the designer's compliance with the overall maintainability requirements outlined in the equipment specification should be conducted in an operational or simulated operational environment.

12-4 QUALITATIVE AND QUANTITATIVE MAINTAINABILITY REQUIREMENTS

Both qualitative and quantitative maintainability requirements enter into the establishment of maintainability design objectives. *Qualitative maintainability requirements* are general nonquantitative statements of a desired feature or function to be incorporated in an equipment design. For example, minimizing part complexity, designing for minimal use of external test equipment, providing a fault location device, optimizing accessibility of high-failure-rate parts, and so on, are all nonquantitative maintainability requirements which may be included in an equipment specification.

A *quantitative maintainability requirement*, however, is a definite statement of the allowable resources or time provided to perform equipment support. For example, an equipment specification may include an availability (A) requirement, a dependability (D) requirement, and a mean-time-to-repair (MTTR) requirement which must be achieved by the designer. In other instances, a designer may be required by the equipment specification to ascertain achieved levels of equipment availability (A), dependability (D), and mean time to repair (MTTR).

Designers should have a firm understanding of the impact quantitative maintainability requirements have on their maintainability design objectives. To an extent, all qualitative maintainability requirements are reflected in the quantitative maintainability assessment of an equipment design. Accessibility of parts, for example, contributes to the equipment repair time. Similarly, fault location and isolation also contribute to the equipment repair time. Therefore, early quantitative evaluations of designs, including alternative designs, are important to designers, since they provide measurements of effectiveness in maintainability design.

12-5 EQUIPMENT AVAILABILITY

Equipment effectiveness can be measured by designers in terms of (a) operational availability and (b) inherent availability.

(a) Operational Availability

Operational availability (A_o) can be defined as the probability that equipment, when used under stated conditions in an actual support environment, will operate satisfactorily at any time. It may be expressed as

$$A_o = \frac{\text{MTBM}}{\text{MTBM} + \text{MDT}} \tag{96}$$

where MTBM = mean of the distribution of the time intervals between maintenance actions (either preventive, corrective, or both)

MDT = sum of the mean corrective and preventive maintenance time intervals, including supply downtime and administrative downtime

MTBM becomes MTBF when preventive maintenance downtime is zero or is not considered.

Operational availability (A_o) is a significant measurement of a developed equipment. Its greatest value lies in planning operations in which the equipment will be used. Operational availability (A_o) has limited application for designers during the early phases of equipment development when changes to the design are still possible. For this reason, inherent availability (A_i) is more important to designers since it can provide an early measurement of the equipment effectiveness.

(b) Inherent Availability

Inherent availability (A_i) can be defined as the probability that equipment, when used under stated conditions in an ideal support environment, will operate satisfactorily at any given time. Scheduled or preventive maintenance downtime, ready time, supply downtime, and waiting or administrative downtime are excluded from the computation. Also, an ideal support environment is assumed with all necessary tools, parts, manpower, manuals, and so on. Inherent availability may be expressed as

$$A_i = \frac{\text{MTBF}}{\text{MTBF} + \text{MTTR}} \tag{97}$$

where

MTBF = mean time between failure
MTTR = mean time to repair

Inherent availability (A_i) can be used to determine the extent to which a designer has achieved his maintainability (and reliability) objectives. As such, equipment specifications may very well include availability statements which are based on equation (97) as a maintainability design requirement. The availability nomograph reproduced in Figure 12-2 provides the designer with a convenient means of solving expression (97) for most practical values of MTTR and MTBF. A straight line drawn through the appropriate values of MTTR and MTBF will intersect the A scale at the desired value. For example, the dashed line of Figure 12-2 shows that the inherent availability is 0.9995 when the MTTR is 1.5 h and MTBF is 3000 h.

In equation (97), inherent availability (A_i) will become larger as the MTBF is increased while the MTTR is held constant. Since MTBF is a function of the equipment failure rate, improving equipment in this area should be the responsibility of the reliability engineers. Owing to the complexity of equipment and limitations in the state of the art it is, however, often very difficult to obtain an MTBF of sufficiently high value to produce the desired availability. Therefore, it is necessary to reduce the MTTR to meet a rigorous availability requirement.

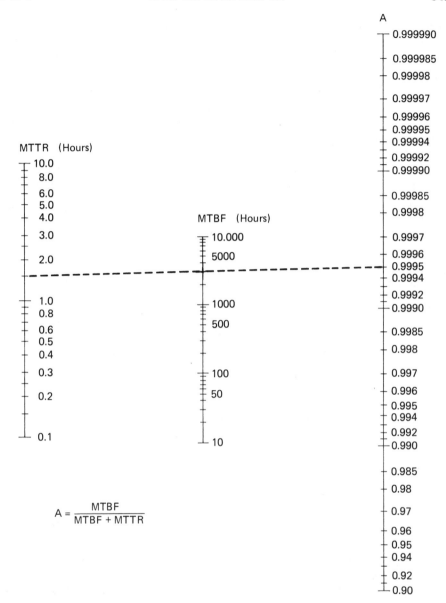

Figure 12-2 *Availability nomograph (source: NAVSHIPS 0967-312-8010).*

12-6 MTBF AND MTTR TRADE-OFF

When equipment is being designed to meet a specified availability figure, and when specific values for MTBF and MTTR have not been given, the designer often has some degree of freedom for a trade-off between reliability and maintainability. Since electronic parts, in general, are made as reliable as economically possible within the present state of the art, usually any improvement in equipment reliability can be obtained only by utilizing redundancy techniques at the expense of increased equipment complexity, size, and weight. On the other hand, improvement in equipment maintainability can often be readily accomplished through judicious application of established principles presented earlier in the chapter. Thus, improvements in availability may, in many cases, be obtained more economically through maintainability improvements (reductions in MTTR) than through reliability improvements (increases in MTBF).

In making a trade-off between reliability and maintainability, however, the designer may often be limited to a minimum value of MTBF and a maximum value of MTTR determined by the operational requirements for the equipment. Therefore, he may not have freedom to compromise one in favor of the other in obtaining a given availability. For example, assume that a minimum availability of 0.999 is specified. Also assume that the MTTR cannot exceed 4 h, and that the MTBF cannot be less than 700 h. Figure 12-3 shows the nomograph of Figure 12-2 on which these limiting conditions have been plotted. The equipment can be designed for the most economical com-

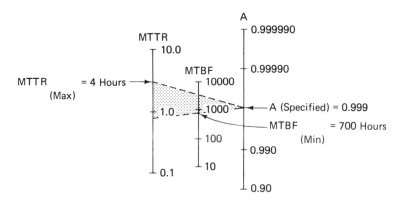

Figure 12-3 *MTTR-MTBF Trade-off limits, (source: NAVSHIPS 0967-312-8010).*

bination of MTTR and MTBF that will give the specified availability provided the limits indicated by the shaded portion of Figure 12-3 are not exceeded. In making such trade-offs, it must be kept in mind that if the MTBF is allowed to fall to the lower limit of 700 h, the MTTR is limited to a maximum of 0.7 h in order to maintain the specified availability. Conversely, if the MTTR is allowed to be 4 h, the MTBF cannot be less than 4000 h.

EXAMPLE 12-1 Equipment is required to meet an inherent availability requirement of $A_i = 0.985$ and a mean time between failures of MTBF = 100 h. What is the permissible mean time to repair (MTTR)?

Solution   Using equation (97),

$$A_i = \frac{MTBF}{MTBF + MTTR}$$

Solving for MTTR,

$$MTTR = \frac{MTBF\,(1 - A_i)}{A_i}$$

Therefore,

$$MTTR = \frac{100(1 - 0.985)}{0.985} = \underline{1.52\ h}$$

EXAMPLE 12-2   Determine the operational availability of a system with a mean of the time intervals between corrective maintenance actions, MTBM = 900 h. The sum of the mean corrective maintenance time intervals including supply downtime and administrative downtime, MDT = 10 h.

Solution   Using equation (96),

$$A_o = \frac{MTBM}{MTBM + MDT}$$

$$= \frac{900}{900 + 10} = \underline{0.989}$$

12-7   MTTR PREDICTION

(a)   General Background

MIL-HDBK-472[4] was developed by the Department of Defense and released in 1966. It contains, in addition to analytic foundations, several methods for maintainability prediction. Both corrective and preventive maintenance actions are considered. As the names imply, corrective maintenance is carried out as the result of a prior failure. Preventive maintenance is designed to prevent the occurrence of a failure by appropriate and timely replacement of parts and other needed maintenance actions (periodic lubrication of bearings, for example).

Procedure II, Part A, of MIL-HDBK-472 is concerned with corrective maintenance during the final design stages of the product development cycle. This procedure is frequently employed and will be presented here. It applies principally to electronic equipment but can be adapted to other types of hardware. As applied in the procedure, corrective maintenance time includes only active repair time. Active repair time does not include administrative delays or logistic time (time spent obtaining maintenance personnel, spares, tools and test equipment).

The procedure measures maintainability in terms of mean time to repair (MTTR), expressed in hours. It uses tabulated maintenance task times, also recorded in hours, which have been established from past experience. These task times are not included

[4] MIL-HDBK-472 Military Standardization Handbook, "Maintainability Prediction," Department of Defense, May 24, 1966.

here. Rather, it is suggested that time values for the corrective maintenance tasks (fault localization and isolation, disassembly, interchange of failed unit, reassembly, alignment, and checkout) may best be estimated from more recent experience with similar equipment. The magnitude of the total repair time is the sum of the individual maintenance tasks.

The MTTR may be computed with the aid of equation (98):

$$\text{MTTR} = \frac{\sum_{i=1}^{m} n_i \lambda_i R_{p_i}}{\sum_{i=1}^{m} n_i \lambda_i} \tag{98}$$

where

$\text{MTTR} = $ mean time to repair
$n_i = $ number of units of type i
$\lambda_i = $ failure rate, failures/10^6 h of unit type i
$R_{p_i} = $ average repair time in hours for unit type i
$m = $ number of types of units

(b) Corrective Maintenance Prediction

The corrective maintenance action is divided into the following corrective maintenance tasks:

1. *Localization:* Determining the location of a failure to the extent possible, without using accessory test equipment.
2. *Isolation:* Determining the location of a failure to the extent possible, by the use of accessory test equipment.
3. *Disassembly:* Equipment disassembly to the extent necessary, to gain access to the item that is to be replaced.
4. *Interchange:* Removing the defective item and installing the replacement.
5. *Reassembly:* Closing and reassembly of the equipment after the replacement has been made.
6. *Alignment:* Performing any alignment, testing, and/or adjustment made necessary by the repair action.
7. *Checkout:* Performing the minimum checks or tests required to verify that the equipment has been restored to satisfactory performance.

(c) Maintenance Worksheet Preparation

The form used to prepare the corrective maintenance prediction is the worksheet shown in Table 12-3. The step-by-step procedure is as follows:

1. *Column A, Part Identification:* List identifying part numbers, names, and type designation of each part.
2. *Column B, Quantity of Parts:* List the quantity of parts for each of the listed part types.

3. *Column C, Failure Rate (λ):* The failure rate of each part is determined in accordance with MIL-STD-756[5] or other suitable sources. The failure rates in failures per 10^6 h are listed in Column C.

4. *Column D, Total Failure Rate ($n\lambda$):* The total failure rate for each part category is the product of columns B and C.

5. *Column E, Localization; F, Isolation; G, Disassembly; H, Interchange; I, Reassembly; J, Alignment; and K, Checkout:* List the estimated average time required to perform each of the listed corrective maintenance tasks corresponding to the respective part identification.

6. *Column L, Average Repair Time/Failure, R_p:* The sum of the values recorded in columns E, F, G, H, I, J, and K is determined and recorded in column L.

7. *Column M, Repair Time/10^6 Hours, $n\lambda R_p$:* The value of each R_p multiplied by the respective value of total failure rate ($n\lambda$) is recorded in column M opposite the respective part identification. The sum of all $n\lambda R_p$ values on each sheet is determined and recorded in the space indicated ($\sum n\lambda R_p$) at the bottom of column M.

The mean time to repair per equation (98) is computed by dividing the summation of column M by the summation of column D.

EXAMPLE 12-3   A piece of communications equipment consists of the assemblies listed below. Also included in this table is the failure rate, λ, for each unit.

Listing of Assemblies Contained in Communication Equipment, Including Failure Rates and Quantities.

Assembly Identification	Failure Rate, λ, per Unit (Failures/10^6 h)	Number of Units
A. Plug-in circuit card	70	40
B. Signal generator	235	1
C. Power supply	500	2
D. Bandpass filter	35	4
E. Amplifier	200	6
F. Mixer	150	4
G. Attenuation network	30	20
H. Connector assembly	10	16
I. Fan	100	10

Determine the MTTR of the equipment. Assumptions:

(a) Preventive maintenance is not required.

(b) Alignment (column J of Table 12-3) is not required.

---

[5] MIL-STD-756A, Military Standard, "Reliability Prediction," May 15, 1963, Defense Supply Agency, Washington, D.C.

Table 12-3 Worksheet for Corrective Maintenance Prediction.

| A Part Identification | B Quantity n | C Failure Rate, λ (Failure/10^6 h) | D Total Failure Rate, $n\lambda$ | AVERAGE TIME TO PERFORM CORRECTIVE MAINTENANCE TASKS | | | | | | | L Average Repair Time per Failure, R_p | M Average Repair Time/10^6 h $n\lambda R_p$ |
				E Localization	F Isolation	G Disassembly	H Interchange	I Reassembly	J Alignment	K Checkout		
A	40	70	2800	0.02	0.01	0.10	0.10	0.15	N/A	0.04	0.42	1176.00
B	1	235	235	0.02	0.02	0.15	0.20	0.16	N/A	0.04	0.59	138.65
C	2	500	1000	0.02	0.01	0.12	0.30	0.15	N/A	0.04	0.64	640.00
D	4	35	140	0.02	0.01	0.10	0.15	0.15	N/A	0.04	0.47	65.80
E	6	200	1200	0.01	0.01	0.13	0.13	0.20	N/A	0.04	0.52	624.00
F	4	150	600	0.01	0.02	0.10	0.12	0.14	N/A	0.04	0.43	258.00
G	20	30	600	0.02	0.02	0.11	0.20	0.11	N/A	0.04	0.50	300.00
H	16	10	160	0.01	0.01	0.10	0.10	0.10	N/A	0.04	0.36	57.60
I	10	100	1000	0.01	0.01	0.10	0.11	0.10	N/A	0.04	0.37	370.00
			$\sum n\lambda = 7735$									$\sum n\lambda R_p$ 3630.05

Per equation (98), $\text{MTTR} = \dfrac{\sum n\lambda R_p}{\sum n\lambda} = \dfrac{3630.05}{7735} = 0.469 \text{ h}$

(c) Estimate the time to perform the corrective maintenance tasks E, F, G, H, I, and K in Table 12-3.

Note: The selection of realistic times to perform the several corrective maintenance tasks is based on experience with similar tasks. For the purpose of this example, exact values for the corrective maintenance tasks are not necessary. Refer to Table 12-3 for all computations.

Maintainability specifications sometimes include an upper limit on repair duration expressed as a time (M_{max}) which 90% or 95% of all corrective maintenance actions must not exceed. To calculate M_{max} it is necessary to assume a statistical distribution of repair times. The log normal distribution[6] is frequently used. A suggested distribution is given on page 2-4 of MIL-HDBK-472; a method of calculating M_{max} is presented on pages 3-12 and 3-13 of the same publication.

PROBLEMS

12-1 Why is downtime of vital importance to the equipment user?

12-2 Define maintainability.

12-3 What is the objective of a maintainability program?

12-4 What is the definition for "availability" of an equipment?

12-5 Certain conclusions reached in Department of the Army Pamphlet 705-1 (June 1966) illustrate the Army's position relative to the operational availability of military equipment. Elaborate.

12-6 How is fault isolation and detection accomplished on highly complex systems?

12-7 Give three practical recommendations in each of the following areas covered by Table 12-2:
(a) Accessibility.
(b) Identification.
(c) Standardization.
(d) Safety.
(e) Cables and wiring.
(f) Connectors.
(g) Controls.

12-8 Differentiate between qualitative and quantitative maintainability requirements.

12-9 Give examples of quantitative maintainability measures.

12-10 State the difference between operational and inherent availability and give the formulas for each.

12-11 Using the nomograph on availability (Figure 12-2), determine the availability of an equipment having a MTTR = 2.0 h and an MTBF = 100 h.

[6] In the log normal probability distribution the logarithms of observed repair times are normally distributed.

12-12 An equipment availability $A = 0.999$ is required. Using the nomograph (Figure 12-2), list four combinations of MTTR and MTBF that would satisfy this requirement.

12-13 The MTTR of a system cannot exceed 3 h. An availability of 0.9995 is specified, and the minimum MTBF required is 1000 h. Sketch the availability nomograph showing the MTTR-MTBF trade-off limits.

12-14 A system has an MTBM $= 70$ h and an operational availability of $A_0 = 0.95$. What is its mean downtime, MDT?

12-15 What is the difference between corrective and preventive maintenance?

12-16 Give examples of preventive maintenance actions.

12-17 Give a brief description of Procedure II, Part A, of MIL-HDBK-472 as used in this text.

12-18 State the formula for the determination of MTTR as used in the procedure above and explain the terms used.

12-19 An electronic system consists of the seven assemblies listed below. One of each of these assemblies is used. Failure rates and maintenance times are shown below. Determine the MTTR of the equipment by using a worksheet as shown in Table 12-3.

Assembly	Failure Rate (Failures/10^6 h)
RF amplifier	30
Oscillator	50
Mixer	40
IF amplifier	80
Detector	60
Power amplifier	150
Power supply	200

The following average times in hours to perform corrective maintenance tasks will apply for all assemblies, except as noted

Localization	0.025
Isolation	0.015
Disassembly	0.15
Interchange	0.20
Reassembly	0.17
Alignment	0.20 (not required for power amplifier and power supply)
Checkout	0.08

12-20 Discuss the essential tasks of a maintainability program and the major output from these tasks.

12-21 Discuss maintainability demonstration tests. What military specification is applicable?

13

INTEGRATED LOGISTIC SUPPORT

Maintenance and support problems can seriously limit the operational availability of equipment. Methods of maintenance, training of maintenance personnel, spare-parts policies and provisioning, documentation, choice of test and servicing equipment— all are essential elements in a relatively recent equipment support concept introduced by the Department of Defense, known as *integrated logistic support* (ILS). This chapter serves as a brief introduction to this new discipline.

13-1 THE ILS CONCEPT

(a) What Is ILS?

ILS is a composite of all the elements necessary to assure the effective and economical support of a system, or equipment, at all levels of maintenance for its programmed life cycle. While full scale ILS programs have been used primarily on larger government procurement contracts, ILS has important applications in commerce and industry.

During the last two decades the Department of Defense has been actively involved in applying system management techniques to the logistic and support problems associated with complex and sophisticated electronic and weapons systems. Initial efforts eventually led in June 1964 to the issuance of DoD Directive 4100.35, "Development of Integrated Logistic Support for Systems and Equipments." Logistic support elements as delineated by this document include all the resources necessary to maintain and operate equipment. These resources include the following major elements:

1. Maintenance.
2. Personnel.
3. Logistic information and data.
4. Spares and repair parts.
5. Facilities.

Two well-known U.S. Navy-generated documents on ILS were issued in 1969: ILS Requirements OR-30[1] and military specification MIL-M-24365.[2] The material that follows is based on these publications.

(b) Objectives

ILS is concerned with system support requirements and management throughout the life cycle of the equipment. The life cycle encompasses concept, design, fabrication, test, operation, and eventual disposal. It is vital that decisions made early in the conceptual phase of an equipment design consider logistic supportability.

ILS as invoked on many government equipment contracts requires the interfacing of management, design engineering, reliability, maintainability, and other associated functions. Elements of ILS can be equally valuable to many non-government activities such as airlines, trucking, communications and widely distributed consumer products.

The essential objectives of an ILS program[3] are as follows:

1. Organize the necessary management effort to assure effective and economical logistic support.
2. Delineate the management organization, disciplines, and controls in the ILS program plan.
3. Influence hardware design decisions, including engineering changes, to simplify and standardize the logistic resource requirements necessary to support the design.
4. Assure appropriate interface among design engineering, data management, standardization, and other related concurrent functions.
5. Validate the effectiveness of the total logistic support plan.

13-2 LIFE-CYCLE COST

Life-cycle cost (LCC) refers to the total cost of acquisition and ownership incurred by the purchaser of an equipment or system throughout its full life cycle. For newly developed equipments, LCC includes

---

[1] "Integrated Logistic Support Program Requirements," OR-30, Naval Ordnance Systems Command, Department of the Navy, March 3, 1969.

[2] Military Specification MIL-M-24365, "Maintenance Engineering Analysis General Specification," Department of Defense, May 26, 1969.

[3] Based on OR-30.

1. Research and development costs.
2. Production costs.
3. System operating costs.
4. Replacement costs of failed components.
5. Disposal costs at end of life.

One of the main decisions to be made concerning the replacement cost of failed components or assemblies is whether to repair them (and at what level), or discard and replace them. In other words, a *level of repair* (LOR) must be identified for each component or assembly. Some of the cost elements which enter into an LOR decision are:

1. Number and geographical distribution of using locations
2. Replacement parts
     (a) Procurement
     (b) Storage
     (c) Transportation
     (d) Handling
3. Repair facilities
4. Tools and test equipment
5. Labor
6. Records and administration.

Computerized techniques for optimizing LOR decisions as a function of life cycle cost are frequently used.

Life-cycle costing also offers advantages when applied to *make* versus *buy* decisions as in evaluating bids from potential suppliers. Some equipment may have a lower acquisition cost but will be considerably more costly to maintain in operation than initially more expensive types. Conversely, whereas an increase in reliability and maintainability of equipment tends to reduce the maintenance costs, it may be more than offset by the high cost of initial purchase. The only way a true picture can be obtained is by a life-cycle-cost trade-off analysis encompassing all the cost elements.

13-3 MAINTENANCE ENGINEERING ANALYSIS

Maintenance engineering analysis (MEA) is an integral part of most ILS programs. The MEA is a process by which engineers with specialized experience in the area of maintenance critically examine the proposed or actual design of equipment to establish the most efficient corrective and preventive maintenance procedures and to identify the logistic resources required to support the end use of the equipment. Frequently, feedback from the MEA constructively influences the maintainability and performance test features of the design. Activities using MEA data as inputs include *spares provisioning, technical manuals, training and facilities planning.*

The MEA describes the necessary corrective and preventive maintenance tasks and identifies the spares, tools, personnel, and facilities needed to support each maintenance task. Generally, a MEA is performed for each functional system, equipment, unit, assembly or subassembly which requires corrective or preventive maintenance.

The MEA is performed at the following broad equipment levels:[4]

1. Major systems.
2. Subsystems.
3. Repairable units and assemblies, when their complexity or essentiality warrants separate identification of logistic resources rather than including them in the next-higher-level MEA.

Special consideration must be given to:

1. Items with reliability characteristics that require frequent maintenance (high failure rate items).
2. Items that require special handling, initial adjustment, or periodic checks.
3. High-value items.
4. High usage items.
5. Identical items that have significantly different applications and environments.

The MEA interacts directly with the level of repair (LOR) analysis. While the initial MEA reflects baseline LOR decisions made early in a program it may in turn cause updating of the LOR concept by highlighting impractical or uneconomical support requirements.

Each maintenance requirement (task) is uniquely identified in the MEA. Types of maintenance requirements and their applicability are as follows:[5]

Inspection: Inspection will apply when the basic intent is the determination that a requisite condition exists. The requirement statement shall be amplified to fully describe what the inspection will prove.

Servicing: Servicing will apply when the basic intent is the replenishment of consumables, that is, fuel, oil, air, liquid oxygen, and so forth. Servicing is also applicable when consumables must be drained and filled, or replenished at specified intervals, or when a ship, for example, must be prepared to perform a specific mission. Servicing is not applicable when it is the result of or a part of other maintenance requirements, such as servicing of a hydraulic system subsequent to the replacement of a hydraulic line.

Lubrication: Lubrication will apply when the basic intent is the application of lubricants to any device on a scheduled basis. Lubrication is not applicable when it is a part of or the result of other maintenance requirements, that is, lubricating

---

[4] Based on OR-30
[5] Based on MIL-M-24365

a hinge pin prior to its installation, or repacking a wheel bearing during wheel replacement.

Remove: Remove will apply whenever the removal is a necessary removal. Remove is not applicable when the removal is required to gain access, to facilitate maintenance, or for similar reasons.

Install: Install requirements will be established to couple with each remove requirement.

Adjust, Calibrate, Align, Rig, and So Forth: Adjust, calibrate, align, rig, and so forth will apply when such actions represent the basic intent or objective, or the verification of these actions is the basic intent or objective.

Functional Test: Functional test will apply when the basic intent is to ensure proper system, subsystem, assembly, or module operation. At the assembly or module level, functional test includes a bench check to verify the need for repair, maintenance level, or to determine serviceability after repair is completed.

Preserve or Depreserve: Preserve or depreserve will apply when the basic intent is to apply preservatives or to remove preservatives.

Repair: A repair requirement will exist for each malfunction symptom. Repair will apply when the basic intent or objective is to return the MEA item to a serviceable status. The kind or type of work that the word "repair" implies will not be the same in the system MEA as it is in the assembly MEA. The basic difference is that system repair normally involves isolation of a fault to an item, replacement of the item, and verification of repair. Assembly repair normally involves isolation of a fault to the replaceable module or part, replacement or repair of the latter, followed by verification of the repair.

Fault Isolation: Fault isolation will apply when the basic intent is to identify the specific item that caused the failure symptom. The fault isolation procedure may ·include operation of built-in test equipment, self-check, or special support equipment testers.

Handling: Handling will apply when positioning of support equipment, mooring or tie-down, installing covers, moving to new positions, decontamination, or special handling equipment is required.

Cleaning: Cleaning will apply when washing, acid bath, buffing, sand blasting, degreasing, and so forth, are required to facilitate inspection and to control corrosion. Cleaning accomplished during repair will be accounted for in the repair requirement.

The need to perform maintenance tasks and the time interval between them must relate to the failure and wear characteristics of the particular item. This assessment must also take into account the effect of a failure on equipment operation, ease of accomplishing an inspection, and the economy of the maintenance action.

The *maintenance engineering analysis record* (MEAR) is the set of worksheets on which the MEA is performed and documented. It serves as an input to the various activities which analyze total sparing, facilities, and manpower requirements and which write technical manuals. Quantitative data provided in the MEAR for each maintenance task includes: task frequency (derived from reliability predictions in the case of corrective maintenance tasks), elapsed subtask and task times, number and types of maintenance technicians required, types and quantities of spare parts, test equipment, tools, fixtures and facilities required.

For greater detail on the MEA and the MEAR, with sample worksheets, see MIL-M-24365.

13-4 SUPPORT REQUIREMENTS

(a) General Support Areas

The support categories listed below are initially identified in the MEAR on a task by task basis. Total (system wide) requirements are then developed using a variety of mathematical methods.

1. *Tools and test equipment:* a maximum degree of standardization must be sought.
2. *Personnel:* the necessary skill levels must be defined and training programs developed.
3. *Maintenance resources:* queueing theory may be employed to determine the optimum arrangement of repair facilities as a function of demand rate and repair times.[6]
4. *Fixed facilities:* mathematical techniques such as linear programming determine optimum locations of repair shops and warehouses to minimize transportation time and costs.[7]
5. *Spare parts:* quantities of spares to be allocated depend on equipment failure rate, desired spares protection level (probability of having a spare available when needed), and spares cost. This topic will be treated in more detail in paragraph 13.4(b) because of its dependence on equipment reliability and its analogy to the redundancy models discussed in Chapter 10.

(b) Spare-Parts Provisioning

1. General Considerations

It is necessary that a sufficient number of spares be provided for replaceable parts, units, modules, or assemblies in order to maintain equipment in operational status

---

[6] C. West Churchman, Russel L. Ackoff, E. L. Arnoff, *Introduction to Operations Research* John Wiley & Sons, Inc., New York, 1957 Chapter 14.

[7] James L. Riggs, *Economic Decision Models for Engineers and Managers* Chapter 3 McGraw Hill Book Co., New York, 1968.

throughout its required operating time, t. Spare-parts provisioning is a vital ILS support function; it is concerned with providing a logistically feasible and economical spares policy. Factors entering into the making of this policy include:

1. Reliability of the unit to be spared, expressed as failure rate (λ).
2. The number of units used.
3. The required probability that a spare will be available when needed during time t.
4. Criticality of application.
5. Physical and geographical location of units.
6. Storage, shipping, and handling considerations.
7. Cost considerations.

Time t is the operating time between spares stock replenishments. When all the necessary spares for a total mission are provided at the start of the mission, t is simply operating time. Quantities were once determined on the basis of an arbitrary percentage of the parts used, say 10 % or 20 %. This frequently resulted in an insufficient number of spares of one type and a wasteful surplus of others. It was difficult to gage the optimum number of spares by this method.

Today's complex military systems and products depend on more scientific spares provisioning procedures to minimize life-cycle cost and enhance operational equipment availability. This is frequently accomplished by using mathematical models and computer programs to analyze and to optimize the variables involved.

2. Probability That a Spare Will Be Available When Needed

Of the factors listed earlier, which are considered in the formulation of a spares policy, the probability that a spare will be available, when needed, during time t will be examined in more detail. As we shall see, another factor, the reliability of the spared unit, also enters into this probability determination.

If we assume that failures occur in accordance with the exponential distribution (constant failure rate), which is applicable during the useful life of the equipment, we may postulate our spares problem as follows. Assuming that we have a single unit with a certain reliability working in a piece of equipment and we provide one spare, what is the probability that we will have a spare available when needed during operating time t? Failures are, of course, presumed to be corrected as they occur. The single spare provided does not give us positive assurance of causing no further problems. A failure of the initially operating unit may occur, which would necessitate the deployment of the spare; but it, too, could fail prior to completion of the required operating time t. This would create the situation in which no spare is available when needed. There is an inherent chance of this situation developing whenever failures occur randomly. In determining the probability that spares will be available when needed, we shall assume that the spares have 100 % reliability prior to being used.

3. Analysis of a Single Unit Backed By One Spare

Let us now examine the sparing problem numerically; assume, for example, a basic part reliability of $P = 0.705$.

$$P = 0.705 = e^{-\lambda t}$$

where λt represents the expected number of failures. From Table A10, Appendix A, we determine that for $P = 0.705$, the value of $\lambda t = 0.35$. Upon a little reflection it is recognized that the case of an operating part backed by a nonoperating spare is analogous to the parallel-redundant system on standby which was discussed in Chapter 10. Equation (84) (p. 282) gave the probability that at least one of the two parts will be operative during time t as

$$P = e^{-\lambda t} + \lambda t \cdot e^{-\lambda t}$$

where λt is the expected number of failures. Applied to the case of a spare backing an operating unit, this is, likewise, the probability that a spare will be available throughout the time period t should the basic unit fail.

Substituting $\lambda t = 0.35$ into the probability equation, P_{SP}, the probability of having a spare becomes

$$P_{SP} = e^{-0.35} + 0.35 \times e^{-0.35}$$
$$= 0.705 + 0.35 \times 0.705$$
$$= \underline{0.951}$$

Different values of part reliability will furnish different values for λt and hence will result in different probability values when substituted in the equation above.

Table 13-1 lists computed values of the probability of having a spare when needed for a range of part reliabilities P_P. In each of these cases it is assumed that an operating part is backed by one spare, ready to take the place of a failed operating part anytime during the time period t.

Table 13-1 *Probability of Having a Spare When Needed During Time t. (One Spare Provided)*

Part Reliability, P_P Over Time t	Expected Number of Failures, λt	Probability of Having a Spare When Needed During Time t, P_{SP}
0.95	0.051	0.9985
0.90	0.105	0.9945
0.85	0.163	0.9879
0.80	0.223	0.9780
0.75	0.288	0.9656
0.70	0.357	0.9500
0.60	0.511	0.9070

4. Analysis of a Number of Unique Units
Each Backed by Its Own Unique Spare

So far our considerations have been restricted to the simplest case of one operating unit backed by a single spare. Consider the case of a small chassis carrying four different replaceable modules. For example, the modules could all be bandpass filter circuits, each in its own molded package. The difference between the modules is a slightly different midband frequency, involving only slightly different tuning capacitor values. The design of the four modules is identical, and hence their reliability is the same, but they, as well as their spares, are not interchangeable; each unit can only be replaced by its respective spare. Assuming that the reliability of each module is 0.80, what is the probability of having a spare, when needed, throughout the operating time t? This overall probability encompasses the probability that unit 1 has its spare available when needed, that unit 2 has its spare available when needed, and so on. Since the failure of unit 1 is independent of that of any other unit, we may use the expression for the probability of independent events:

$$P = P_1 \times P_2 \times P_3 \times \ldots$$

Accordingly, from Table 13-1 for $P_P = 0.80$ we determine $P_{SP} = 0.9780$.

The probability of having a spare during time t for any of the four modules, when needed, is therefore:

$$P = P_{SP_1} \times P_{SP_2} \times P_{SP_3} \times P_{SP_4}$$

Since

$$P_{SP_1} = P_{SP_2} = P_{SP_3} = P_{SP_4} = 0.9780,$$

$$P = (0.9780)^4 = \underline{0.91486}$$

Hence, when the reliability of each module is 0.8, there is a 91.49% chance of being able to substitute a spare for any of the four modules, when required, during operating time t.

Slightly more complex is the problem involving two or more different groups of units with each unit having its own spare. The units are unique; that is, they are not interchangeable within their own group or with any of the units from other groups. The units within each group have, however, the same reliability. An example will illustrate this case.

EXAMPLE 13-1   Three unique modules have a reliability of 0.6 each and four unique parts have a reliability of 0.8 each. Each of the seven items is backed by its own unique spare. What is the probability of having a spare when needed for any of the seven items during operating time t?

Solution

Modules:

$$P_P = 0.6$$

P_{SP} from Table 13-1 is 0.907.

$$P = (P_{SP})^3 = (0.907)^3 = 0.7461$$

Parts:

$$P_P = 0.8$$

P_{SP} from Table 13-1 is 0.9780.

$$P = (P_{SP})^4 = (0.9780)^4 = 0.9149$$

$$P = P_{\text{Modules}} \times P_{\text{Parts}} = 0.7461 \times 0.9149$$

$$= \underline{0.6826}$$

Table 13-2 gives the probability of having a spare, when needed, during operating time t for a range of 1 to 100 unique parts. Each part is backed by a corresponding spare. None of the parts nor their spares are interchangeable. The use of Table 13-2 is illustrated in the following two examples.

Table 13-2 *Probability of Having a Spare When Needed During Time t.*

Number of Unique Parts	INDIVIDUAL PART RELIABILITY						
	0.60	0.70	0.75	0.80	0.85	0.90	0.95
1	0.9070	0.9500	0.9656	0.9780	0.9879	0.9945	0.9985
2	0.8226	0.9025	0.9324	0.9565	0.9759	0.9890	0.9970
3	0.7461	0.8574	0.9003	0.9354	0.9641	0.9836	0.9955
4	0.6768	0.8145	0.8693	0.9149	0.9525	0.9782	0.9940
5	0.6138	0.7738	0.8394	0.8947	0.9409	0.9728	0.9925
6	0.5567	0.7351	0.8106	0.8751	0.9296	0.9675	0.9910
7	0.5050	0.6983	0.7827	0.8558	0.9183	0.9621	0.9895
8	0.4580	0.6634	0.7557	0.8370	0.9072	0.9568	0.9880
9	0.4154	0.6302	0.7298	0.8186	0.8962	0.9516	0.9866
10	0.3768	0.5987	0.7046	0.8006	0.8854	0.9463	0.9851
25	0.0871	0.2774	0.4168	0.5734	0.7376	0.8712	0.9632
50	0.0076	0.0769	0.1737	0.3288	0.5440	0.7590	0.9277
100	0.0001	0.0059	0.0300	0.1081	0.2960	0.5761	0.8606

Note: Each unique operating unit is backed by its own spare.

EXAMPLE 13-2 Nine unique parts, each having a reliability of $P_P = 0.85$, are backed by one spare each. What is the probability of having a spare when needed during operating time t?

Solution The intersection of row 9 and column 0.85 in Table 13-2 renders a probability of 0.8962.

EXAMPLE 13-3 What is the required part reliability to assure a probability of at least 94% of having a spare, when required, during operating time t, when 5 unique parts are backed by one corresponding spare each. None of the parts nor their spares are interchangeable.

Solution Scanning the row for 5 units from left to right shows a probability of 0.9409 under column heading 0.85. Hence, the individual part reliability should be 85%.

5. Analysis of Single or Multiple Interchangeable Units Backed by One or More Interchangeable Spares

Up to now we have discussed spares provisioning for cases involving unique, basic units and their unique spares. We have also extended these considerations to groups of unique basic units backed by their corresponding unique spares. The only common factor of all units within a group was that they were assumed to have the same reliability.

Let us now increase the number of spares per basic unit. To start, we shall add a second spare to a single basic unit so that any one of the three units is usable in the basic unit location. To calculate the probability of having a spare available, when required, we merely add one additional term of the Poisson expansion to the probability equation (see Table 2-7 for the Poisson terms). For convenience, several terms of the Poisson expansion are shown below with λt replacing a and with the common factor $e^{-\lambda t}$ extracted:

$$P = e^{-\lambda t}\left[1 + \lambda t + \frac{(\lambda t)^2}{2} + \frac{(\lambda t)^3}{6} + \frac{(\lambda t)^4}{24} + \frac{(\lambda t)^5}{120} + \cdots\right] \tag{99}$$

Hence, for the above case of one basic unit and two spares, where all three units are interchangeable, the probability of having a spare, when needed, during time t is

$$P = e^{-\lambda t}\left[1 + \lambda t + \frac{(\lambda t)^2}{2}\right] \tag{100}$$

Adding one more spare yields

$$P = e^{-\lambda t}\left[1 + \lambda t + \frac{(\lambda t)^2}{2} + \frac{(\lambda t)^3}{6}\right] \tag{101}$$

Further additions of spare units are treated likewise by adding one more Poisson term from equation (99). An inspection of equations (100) and (101) reveals that following the number 1 inside the brackets, there are as many Poisson terms as there are spares. If there were five spares for a single basic unit, we would include the Poisson terms up to $(\lambda t)^5/120$.

So far, the configurations discussed dealt with noninterchangeable units. This lack of standardization of parts was reflected in a lower probability that a spare would be available, when needed, than for an arrangement where one spare is able to replace any unit in a group. With standardized parts the potential deployment range of a spare is greater, a fact that permits the use of fewer spares for the same probability of spares availability. The advantage of parts standardization is therefore the ability to reduce the logistic supply burden for spares and thereby the cost of the parts-provisioning program, without decreasing spares availability. The net result is a lower life cycle cost for the equipment which contains these parts.

As long as we dealt with a single basic unit, the expected number of failures, for a failure rate λ, and time t, was λt. When two basic parts are used, the expected number of failures is obviously $2\lambda t$. For three basic parts we have $3\lambda t$, and so on. The Poisson expansion for the case of two interchangeable basic units then becomes

$$P = e^{-2\lambda t}\left[1 + 2\lambda t + \frac{(2\lambda t)^2}{2} + \frac{(2\lambda t)^3}{6} + \frac{(2\lambda t)^4}{24} + \frac{(2\lambda t)^5}{120} + \cdots\right] \qquad (102)$$

and for 3 interchangeable units:

$$P = e^{-3\lambda t}\left[1 + 3\lambda t + \frac{(3\lambda t)^2}{2} + \frac{(3\lambda t)^3}{6} + \frac{(3\lambda t)^4}{24} + \frac{(3\lambda t)^5}{120} \cdots\right] \qquad (103)$$

and so forth. The number of terms following the number 1 inside the brackets is again equal to the total number of spares.

Table 13-3 gives a listing of a variety of basic parts and spares configurations with the corresponding probability expression. The first column represents the configuration (white squares are operating units, black squares are standby spares). The second column lists the expected number of failures and the third column represents the probability of having a spare available when needed during time t. All the units in each configuration are fully interchangeable.

EXAMPLE 13-4 Using Table 13-3, determine the probability of having a spare available for a configuration of two operating units backed by two spares. All units are interchangeable and their reliability is 85%.

Solution From Table 13-1 we determine

$$\lambda t = 0.163$$

The applicable equation from Table 13-3 is

$$P = e^{-2\lambda t}\left[1 + 2\lambda t + \frac{(2\lambda t)^2}{2}\right]$$

Substituting,

$$P = e^{-0.326}\left[1 + 0.326 + \frac{(0.326)^2}{2}\right]$$

From Table A10, Appendix A, we determine $e^{-0.326}$ for $\lambda t = 0.326$. The value is 0.72181. Hence,

$$P = 0.72181\left(1 + 0.326 + \frac{0.1063}{2}\right)$$

$$= \underline{0.9954}$$

EXAMPLE 13-5 The spares for an equipment are replenished every 6 months. Among them is one spare for a single module with a failure rate of 0.05091 failure/ 1000 h. What is the probability that this spare will suffice until the next spare-replenishment date?

Table 13-3 *Probability of Having a Spare When Needed During Time t.*

Configuration	Expected No. Failures	Probability of Having a Spare Over Time t
(1 spare, 1 active)	λt	$P = e^{-\lambda t} + \lambda t \cdot e^{-\lambda t}$
(1 spare, 2 active)	λt	$P = e^{-\lambda t}\left(1 + \lambda t + \dfrac{(\lambda t)^2}{2}\right)$
(1 spare, 3 active)	λt	$P = e^{-\lambda t}\left(1 + \lambda t + \dfrac{(\lambda t)^2}{2} + \dfrac{(\lambda t)^3}{6}\right)$
(2 spare, 1 active)	$2\lambda t$	$P = e^{-2\lambda t}(1 + 2\lambda t)$
(2 spare, 2 active)	$2\lambda t$	$P = e^{-2\lambda t}\left(1 + 2\lambda t + \dfrac{(2\lambda t)^2}{2}\right)$
(2 spare, 3 active)	$2\lambda t$	$P = e^{-2\lambda t}\left(1 + 2\lambda t + \dfrac{(2\lambda t)^2}{2} + \dfrac{(2\lambda t)^3}{6}\right)$
(3 spare, 1 active)	$3\lambda t$	$P = e^{-3\lambda t}(1 + 3\lambda t)$
(3 spare, 2 active)	$3\lambda t$	$P = e^{-3\lambda t}\left(1 + 3\lambda t + \dfrac{(3\lambda t)^2}{2}\right)$
(3 spare, 3 active)	$3\lambda t$	$P = e^{-3\lambda t}\left(1 + 3\lambda t + \dfrac{(3\lambda t)^2}{2} + \dfrac{(3\lambda t)^3}{6}\right)$
(3 spare, 4 active)	$3\lambda t$	$P = e^{-3\lambda t}\left(1 + 3\lambda t + \dfrac{(3\lambda t)^2}{2} + \dfrac{(3\lambda t)^3}{6} + \dfrac{(3\lambda t)^4}{24}\right)$

Solution   The operational time t, between spare replenishment dates, is

$$t = 6 \text{ months} = 4380 \text{ h}$$

$$\lambda = \frac{0.05091}{1000} = 0.05091 \times 10^{-3} \text{ failure/hour}$$

$$\lambda t = 0.05091 \times 10^{-3} \times 4380 = 0.223$$

From Table 13-3,

$$P = e^{-\lambda t} + \lambda t \cdot e^{-\lambda t}$$

$$= e^{-0.223} + 0.223 \cdot e^{-0.223}$$

From Table A-10, Appendix A,

$$e^{-0.223} = 0.8$$

Hence,

$$0.8 + 0.223 \times 0.8 = \underline{0.978}$$

EXAMPLE 13-6   Repeat Example 13-1 but assume that the three modules of the first group are interchangeable among themselves as well as with any of the three spares. Determine the increase in the earlier calculated probability of having a spare when needed.

Solution   For each module P_P was 0.60, hence, per Table 13-1

$$\lambda t = 0.511$$

From Table 13-3,

$$P = e^{-3\lambda t}\left[1 + 3\lambda t + \frac{(3\lambda t)^2}{2} + \frac{(3\lambda t)^3}{6}\right]$$

$$= e^{-1.533}\left[1 + 1.533 + \frac{(1.533)^2}{2} + \frac{(1.533)^3}{6}\right] = \underline{0.9302}$$

The corresponding result in Example 13-1 was a probability of 0.7461 for the nonstandardized modules; hence, the increase is from 0.7461 to 0.9302.

Computerized techniques have been developed to allocate spares to equipment categories as a function of spares protection bought per dollar of spares cost.

PROBLEMS

13-1   Define integrated logistic support.

13-2   List the resources as delineated by DoD Directive 4100.35 necessary to maintain and operate equipments.

13-3   What are the five objectives of ILS given in the chapter?

13-4   What is a maintenance engineering analysis and what is its purpose?

13-5   Give examples of equipment levels on which a MEA is performed. What is the MEAR?

13-6 A number of maintenance requirements used in maintenance engineering analysis worksheets have been listed in the chapter. State and discuss four of these.

13-7 Name two activities that use the MEA as an input.

13-8 What is LOR?

13-9 Define life-cycle costs.

13-10 For newly developed equipment, what costs are included by LCC?

13-11 Name some of the cost elements that enter into component repair versus discard and replace decisions.

13-12 What are some of the factors that influence a spare-parts provisioning policy?

13-13 A module has a reliability of 0.85 over operating time t. It is backed by one spare. What is the probability of having a spare when needed during time t?

13-14 Six unique devices are each backed by their own spare. Each device has a reliability of 0.70 over operating time t. What is the probability of having a spare when required during time t?

13-15 Five modules, each with a reliability of 0.90 over operating time t, and four modules, each with a reliability of 0.60 over operating time t are used in an assembly. Each of the nine units is backed by its own spare. What is the probability of having a spare when required during time t?

13-16 Ten unique parts are backed by one corresponding spare over operating time t each. Using Table 13-2, determine the part reliability to assure a probability of at least 80% of having a spare, when required, during time t.

13-17 Give the Poisson formula for the probability of having a spare when required during operating time t, when three interchangeable modules are backed by four spares. If the module reliability is $P = 0.90484$, what is the probability of having a spare when required?

13-18 Explain why standardization aids in spare-parts provisioning.

13-19 With the aid of Table 13-3, determine the probability of having a spare available when required during operating time t for a configuration of three operating units backed by two spares. The reliability of all units is 0.70 over operating time t and they are all interchangeable.

13-20 A piece of equipment contains two interchangeable modules, backed by three spares. The failure rate of each module is 0.000021 failure/hour. The spares are replenished every 10 months (7300 hours). What is the probability that the three spares will suffice until the next spare replenishment date?

APPENDIX A

*Tables A1 Through A16 and
Figures A1 and A2*

TABLE A1 NUMERICAL INDEX OF TEST METHODS OF MIL-STD-202E.

Method number	Date	Title
		Environmental tests (100 Class)
101D	16 April 1973	Salt spray (corrosion)
102A	Cancel effective 31 December 1973 (See note on method 102.)	
103B	12 September 1963	Humidity (steady state)
104A	24 October 1956	Immersion
105C	12 September 1963	Barometric pressure (reduced)
106D	16 April 1973	Moisture resistance
107D	16 April 1973	Thermal shock
108A	12 September 1963	Life (at elevated ambient temperature)
109B	16 April 1973	Explosion
110A	16 April 1973	Sand and dust
111A	16 April 1973	Flammability (external flame)
112B	16 April 1973	Seal
		Physical-characteristics tests (200 Class)
201A	24 October 1956	Vibration
202D	16 April 1973	Shock (specimens weighing not more than 4 pounds) (Superseded by method 213.)
203B	16 April 1973	Random drop
204C	16 April 1973	Vibration, high frequency
205E	16 April 1973	Shock, medium impact (Superseded by method 213)
206	12 September 1963	Life (rotational)
207A	12 September 1963	High-impact shock
208C	16 April 1973	Solderability
209	18 May 1962	Radiographic inspection
210A	16 April 1973	Resistance to soldering heat
211A	14 April 1969	Terminal strength
212A	16 April 1973	Acceleration
213B	16 April 1973	Shock (specified pulse)
214	9 November 1966	Random vibration
215	9 November 1966	Resistance to solvents
216	Cancel effective 16 APRIL 1973. (See note on method 216.)	
		Electrical-characteristics tests (300 Class)
301	6 February 1956	Dielectric withstanding voltage
302	6 February 1956	Insulation resistance
303	6 February 1956	DC resistance
304	24 October 1956	Resistance-temperature characteristic
305	24 October 1956	Capacitance
306	24 October 1956	Quality factor (Q)
307	24 October 1956	Contact resistance
308	29 November 1961	Current-noise test for fixed resistors
309	27 May 1965	Voltage coefficient of resistance determination procedure
310	20 January 1967	Contact-chatter monitoring
311	14 April 1969	Life, low level switching
312	16 April 1973	Intermediate current switching

TABLE A2 METHOD 102A TEMPERATURE CYCLING

1. PURPOSE. This test is conducted for the purpose of determining the resistance of a part to the shock of repeated surface exposures to extremes of high and low temperatures for comparatively short periods of time, such as would be experienced when equipment or parts are transferred to and from heated shelters in arctic areas. These conditions may also be encountered in equipment operated intermittently in low-temperature areas. It is not required that the specimen reach thermal stability at the temperature of the test chamber during the short exposure time specified. Permanent changes in operating characteristics and physical damage produced during temperature cycling result principally from variations in dimensions and other physical properties, and from alternate condensation and freezing of atmospheric moisture. Effects of temperature cycling include cracking and de-

lamination of finishes, embedding compounds, and other materials, opening of terminal seals and case seams, and changes in electrical characteristics due to moisture effects or to mechanical displacement of conductors or of insulating materials.

2. APPARATUS. Separate chambers shall be used for the extreme temperature conditions of steps 1 and 3. The air temperature of the two chambers shall be held at each of the extreme temperatures by means of circulation and sufficient hot- or cold-chamber thermal capacity so that the ambient temperature shall reach the specified temperature within 2 minutes after the specimens have been transferred to the appropriate chamber.

3. PROCEDURE. Specimens shall be placed in such a position with respect to the air stream that there is substantially no obstruction to the flow of air across and around the specimen. When special mounting is required, it shall be specified. The specimen shall then be subjected to the specified test condition of table 102-1, for a total of five cycles performed continuously. Specimens shall not be subjected to forced circulating air while being transferred from one chamber to another. Direct heat conduction to the specimen should be minimized.

TABLE 102-1. Temperature-cycling test conditions.

Step	Test condition A [1]/ (not preferred)		Test condition B [1]/ (not preferred)	
	Temperature	Time	Temperature	Time
	°C	Minutes	°C	Minutes
1	85^{+3}_{-0}	30	65^{+3}_{-0}	30
2	25^{+10}_{-5}	10 to 15	25^{+10}_{-5}	10 to 15
3	-55^{+0}_{-3}	30	-55^{+0}_{-3}	30
4	25^{+0}_{-5}	10 to 15	25^{+10}_{-5}	10 to 15
	Test condition C (preferred)		Test condition D (preferred)	
1	-65^{+0}_{-5}	30	-55^{+0}_{-3}	30
2	25^{+10}_{-5}	10 to 15	25^{+10}_{-5}	10 to 15
3	125^{+3}_{-0}	30	85^{+3}_{-0}	30
4	25^{+10}_{-5}	10 to 15	25^{+10}_{-5}	10 to 15

4. MEASUREMENTS. Specified measurements shall be made prior to the first cycle and upon completion of the final cycle, except that failures shall be based on measurements made after the specimen has returned to thermal stability at room ambient temperature following the final cycle.

5. SUMMARY. The following details must be specified in the individual specification:

    (a) Special mounting, if applicable (see 3).

    (b) Test condition letter (see 3).

    (c) Measurements before and after cycling (see 4).

TABLE A3 METHOD 103B HUMIDITY (STEADY STATE)

1. PURPOSE. This test is performed to evaluate the properties of materials used in components as they are influenced by the absorption and diffusion of moisture and moisture vapor. This is an accelerated environmental test, accomplished by the continuous exposure of the specimen to high relative humidity at an elevated temperature. These conditions impose a vapor pressure on the material under test which constitutes the force behind the moisture migration and penetration. Hy-

groscopic materials are sensitive to moisture, and deteriorate rapidly under humid conditions. Absorption of moisture by many materials results in swelling, which destroys their functional utility, and causes loss of physical strength and changes in other important mechanical properties. Insulating materials which absorb moisture may suffer degradation of their electrical properties. This method, while not necessarily intended as a simulated tropical test, is of use in determining moisture absorption of insulating materials.

2.   PROCEDURE.

2.1   Conditioning.   The specimens shall be conditioned in a dry oven at a temperature of 40°
±5°C for a period of 24 hours.   At the end of this period, measurements shall be made as specified.

2.2   Chamber.   The chamber and accessories shall be constructed and arranged in such a manner as to avoid condensate dripping on the specimens under test, and such that the specimens shall be exposed to circulating air.

2.3   Exposure.   The specimens shall be placed in a chamber and subjected to a relative humidity of 90 to 95 percent and a temperature of 40° ±2°C for the period of time indicated in one of the following test conditions, as specified:

Test condition	Length of test
A - - - - - - - - - - -	240 hours.
B - - - - - - - - - - -	96 hours.
C - - - - - - - - - - -	504 hours.
D - - - - - - - - - - -	1,344 hours.

When specified, a direct-current potential of 100 volts or as specified shall be applied to the specimens during the exposure period.   The length of time for the application of voltage and the points of application shall be as specified.

3.   FINAL MEASUREMENTS.

3.1   At high humidity.   Upon completion of the exposure period, and while the specimens are still in the chamber, the specified measurements shall be performed.   These measurements may be compared to the initial measurements (see 2.1), when applicable.

3.2   After drying period.   Upon completion of the exposure period or following measurements at high humidity if applicable, the specimens shall be conditioned at room ambient conditions for not less than 1 hour, nor more than 2 hours unless otherwise specified, after which the specified measurements shall be performed at room ambient conditions.

4.   SUMMARY.   The following details are to be specified in the individual specification:

(a)   Measurements after conditioning (see 2.1).

(b)   Test condition letter (see 2.3).

(c)   The length of time and points of application of polarizing voltage, if
       applicable (see 2.3).

(d)   Final measurements:
       (1)   At high humidity, if applicable (see 3.1).
       (2)   After drying period (see 3.2).

TABLE A4 METHOD 106D MOISTURE RESISTANCE

1.   PURPOSE.   The moisture-resistance test is performed for the purpose of evaluating, in an accelerated manner, the resistance of component parts and constituent materials to the deteriorative effects of the high-humidity and heat conditions typical of tropical environments.   Most tropical degradation results directly or indirectly from absorption of moisture vapor and films by vulnerable insulating materials, and form surface wetting of metals and insulation.   These phenomena produce many types of deterioration, including corrosion of metals, physical distortion and decomposition of organic materials, leaching out and spending of constituents of materials; and detrimental changes in

electrical properties. This test differs from the steady-state humidity test (method 103 of this standard) and derives its added effectiveness in its employment of temperature cycling, which provides alternate periods of condensation and drying essential to the development of the corrosion processes and, in addition, produces a "breathing" action of moisture into partially sealed containers. Increased effectiveness is also obtained by use of a higher temperature, which intensifies the effects of humidity. The test includes low-temperature and vibration subcycles that act as accelerants to reveal otherwise undiscernible evidences of deterioration since stresses caused by freezing moisture and accentuated by vibration tend to widen cracks and fissures. As a result the deterioration can be detected by the measurement of electrical characteristics (including such tests as dielectric withstanding voltage and insulation resistance) or by performance of a test for sealing. Provision is made for the application of a polarizing voltage across insulation to investigate the possibility of electrolysis, which can promote eventual dielectric breakdown. This test also provides for electrical loading of certain components, if desired, in order to determine the resistance of current-carrying components, especially fine wires and contacts, to electro-chemical corrosion. Results obtained with this test are reproducible and have been confirmed by investigations of field failures. This test has proved reliable for indicating those parts which are unsuited for tropical field use.

2. APPARATUS.

2.1 Chamber. A test chamber shall be used which can meet the temperature and humidty cycling specified on figure 106-1. The material used to fabricate the platforms and standoffs, which support the specimens, shall be nonreactive in high humidty. Wood or plywood shall not be used because they are resiniferous. Materials shall not be used if they contain formaldehyde or phenol in their composition.

2.1.1 Opening of the chamber door. During the periods when the humidity is ascending or descending, the chamber door should not be opened. If the chamber door must be opened, it should be opened during the 16th cycle through the 24th cycle of the test. While the chamber is at 25°C, and the relative humidity tolerance must be maintained, the chamber door should only be allowed to be opened for a short period of time.

2.1.2 Water. Steam, or distilled and demineralized, or deionized water, having a pH value between 6.0 and 7.2 at 23°C (73°F) shall be used to obtain the specified humidity. No rust or corrosive contaminants shall be imposed on the test specimens by the test facility.

2.1.3 Dripping. Provision shall be made to prevent the dripping of condensate on the test specimens from the test chamber ceiling.

3. PROCEDURE.

3.1 Mounting. Specimens shall be mounted by their normal mounting means, in their normal mounting position, but shall be positioned so that they do not contact each other, and so that each specimen receives essentially the same degree of humidity.

3.2 Initial measurements. Prior to step 1 of the first cycle, the specified initial measurements shall be made at room ambient conditions, or as specified.

3.3 Number of cycles. Specimens shall be subjected to 10 continuous cycles, each as shown on figure 106-1.

3.4 Subcycle. During step 7, at least 1 hour but not more than 4 hours after step 7 begins, the specimens shall be either removed from the humidity chamber, or the temperature of the chamber shall be reduced, for performance of step 7a. After step 7b, the specimens shall be returned to 25°C at 90 to 98 percent relative humidity (RH) and kept there until the next cycle begins. This subcycle shall be performed during any five of the first nine cycles.

3.4.1 Step 7a. At least 1 hour but not more than 4 hours after the beginning of step 7, the specimens shall be either removed from the humidity chamber, or the temperature of the chamber shall be reduced. Specimens shall then be conditioned at -10° ±2°C with humidity not controlled, for 3 hours as indicated on figure 106-1. When a separate cold chamber is not used, care should be taken to assure that the specimens are held at -10° ±2°C for the full 3-hour period.

3.4.2 Step 7b. Within 15 minutes after completion of step 7a and with humidity not controlled, specimens shall be vibrated for 15 minutes, at room ambient temperature, using a simple harmonic motion having an amplitude of 0.03 inch (0.76 mm) (0.06 inch (1.52 mm) maximum total excursion), the frequency being varied uniformly between the approximate limits of 10 and 55 hertz (Hz). The entire frequency range, from 10 to 55 Hz and return to 10 Hz, shall be traversed in approximately 1 minute.

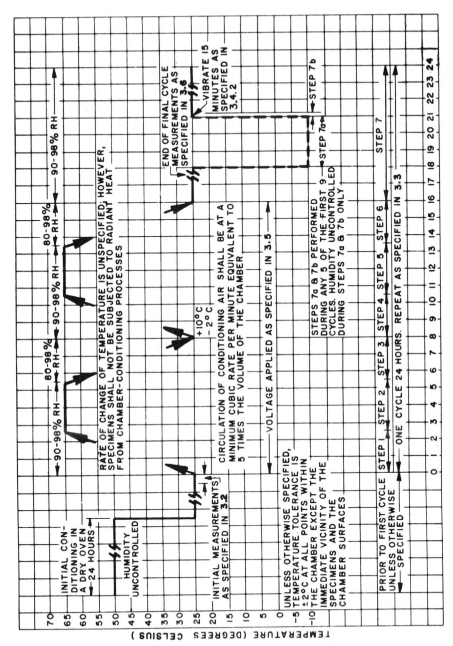

Figure 106-1 *Graphical representation of moisture-resistant test.*

378

3.5 Polarization and load. When applicable, polarization voltage shall be 100 volts, dc, or as specified. The loading voltage shall be as specified.

3.6 Final measurements.

3.6.1 At high humidity. Upon completion of step 6 of the final cycle, when measurements at high humidity are specified, the specimens shall be maintained at a temperature of 25° ±2°C, and an RH of 90 to 98 percent for a period of 1-1/2 to 3-1/2 hours, after which the specified measurements shall be made. Due to the difficulty in making measurements under high-humidity conditions, the individual specification shall specify the particular precautions to be followed in making measurements under such conditions.

3.6.2 After high humidity. Upon removal from humidity chamber, final measurements shall be made within a period of 1 - 2 hours after the final cycle. During final measurements, specimens shall not be subjected to any means of artificial drying.

3.6.3 After drying period. Following step 6 of the final cycle, or following measurements at high humidity, if applicable, specimens shall be conditioned for 24 hours at the ambient conditions specified for the initial measurements (see 3.2) after which the specified measurements shall be made. Measurements may be made during the 24-hour conditioning period; however, failures shall be based on the 24-hour period only.

4. SUMMARY. The following details are to be specified in the individual specification:

   (a) Initial measurements, and conditions if other than room ambient (see 3.2).

   (b) When applicable, the polarization voltage if other than 100 volts (see 3.5).

   (c) Loading voltage (see 3.5).

   (d) Final measurements (see 3.6).

TABLE A5 METHOD 107D THERMAL SHOCK

1. PURPOSE. This test is conducted for the purpose of determining the resistance of a part to exposures at extremes of high and low temperatures, and to the shock of alternate exposures to these extremes, such as would be experienced when equipment or parts are transferred to and from heated shelters in arctic areas. These conditions may also be encountered in equipment operated noncontinuously in low-temperature areas or during transportation. Although it is preferred that the specimen reach thermal stability at the temperature of the test chamber during the exposure specified, in the interest of saving test time, parts may be tested at the minimum exposure durations specified, which will not insure thermal stability, but only an approach thereto. Permanent changes in operating characteristics and physical damage produced during thermal shock result principally from variations in dimensions and other physical properties. Effects of thermal shock include cracking and delamination of finishes, cracking and crazing of embedding and encapsulating compounds, opening of thermal seals and case seams, leakage of filling materials, rupturing or cracking of hermetic seals and vacuum glass to metal seals, and changes in electrical characteristics due to mechanical displacement or rupture of conductors or of insulating materials.

2. APPARATUS. Separate chambers shall be used for the extreme temperature conditions of steps 1 and 3. The air temperature of the two chambers shall be held at each of the extreme temperatures by means of circulation and sufficient hot- or cold-chamber thermal capacity so that the ambient temperature shall reach the specified temperature within 2 minutes after the specimens have been transferred to the appropriate chamber.

3. PROCEDURE. Specimens shall be placed in such a position with respect to the airstream that there is substantially no obstruction to the flow of air across and around the specimen. When special mounting is required, it shall be specified. The specimen shall then be subjected to the specified test condition of table 107-1. The first five cycles shall be run continuously. After five cycles, the test may be interrupted after the completion of any full cycle, and the specimens allowed to return to room ambient temperature before testing is resumed. One cycle consists of steps 1 through 4 of the applicable test condition. Specimens shall not be subjected to forced circulating air while being transferred from one chamber to another. Direct heat conduction to the specimen should be minimized.

TABLE 107-1. Thermal-shock test conditions.

Step	Test condition	Number of cycles	Test condition	Number of cycles	Test condition	Number of cycles
	A	5	B	5	C	5
	A-1	25	B-1	25	C-1	25
	A-2	50	B-2	50	C-2	50
	A-3	100	B-3	100	C-3	100
	Temperature	Time	Temperature	Time	Temperature	Time
	°C	Minutes	°C	Minutes	°C	Minutes
1	-55^{+0}_{-3}	(See table 107-2)	-65^{+0}_{-5}	(See table 107-2)	-65^{+0}_{-5}	(See table 107-2)
2	25^{+10}_{-5}	5 max	25^{+10}_{-5}	5 max	25^{+10}_{-5}	5 max
3	85^{+3}_{-0}	(See table 107-2)	125^{+3}_{-0}	(See table 107-2)	200^{+5}_{-0}	(See table 107-2)
4	25^{+10}_{-5}	5 max	25^{+10}_{-5}	5 max	25^{+10}_{-5}	5 max

TABLE 107-1. Thermal-shock test conditions. - Continued

Step	Test condition	Number of cycles	Test condition	Number of cycles	Test condition	Number of cycles
	D	5	E	5	F	5
	D-1	25	E-1	25	F-1	25
	D-2	50	E-2	50	F-2	50
	D-3	100	E-3	100	F-3	100
	Temperature	Time	Temperature	Time	Temperature	Time
	°C	Minutes	°C	Minutes	°C	Minutes
1	-65^{+0}_{-5}	(See table 107-2)	-65^{+0}_{-5}	(See table 107-2)	-65^{+0}_{-5}	(See table 107-2)
2	25^{+10}_{-5}	5 max	25^{+10}_{-5}	5 max	25^{+10}_{-5}	5 max.
3	350^{+5}_{-0}	(See table 107-2)	500^{+5}_{-0}	(See table 107-2)	150^{+3}_{-0}	(See table 107-2
4	25^{+10}_{-5}	5 max	25^{+10}_{-5}	5 max	25^{+10}_{-5}	5 max.

TABLE 107-2. Exposure time at temperature extremes.

Weight of specimen	Minimum time (for steps 1 and 3)
	Hours
1 ounce (28 grams) and below -	1/2; or (1/4 when specified)
Above 1 ounce (28 grams) to .3 pound (136 grams) incl. - - - - - - - - - -	1/2
Above .3 pound (136 grams) to 3 pounds (1.36 kilograms) incl. - - - - -	1
Above 3 pounds (1.36 kilograms) to 30 pounds (13.6 kilograms) incl. - - -	2
Above 30 pounds (13.6 kilograms) to 300 pounds (136 kilograms) incl. - -	4
Above 300 pounds (136 kilograms) -	8

4. MEASUREMENTS. Specified measurements shall be made prior to the first cycle and upon completion of the final cycle, except that failures shall be based on measurements made after the specimen has returned to thermal stability at room ambient temperature following the final cycle.

5. SUMMARY. The following details are to be specified in the individual specification:

(a) Special mounting, if applicable (see 3).

(b) Test condition letter (see 3).

(c) Measurements before and after cycling (see 4).

TABLE A6 METHOD 108A LIFE (AT ELEVATED AMBIENT TEMPERATURE)

1. PURPOSE. This test is conducted for the purpose of determining the effects on electrical and mechanical characteristics of a part, resulting from exposure of the part to an elevated ambient temperature for a specified length of time, while the part is performing its operational function. This test method is not intended for testing parts whose life is expressed in the number of operations. Evidence of deterioration resulting from this test can at times be determined by visual examination; however, the effects may be more readily ascertained by measurements or tests prior to, during, or after exposure. Surge current, total resistance, dielectric strength, insulation resistance, and capacitance are types of measurements that would show the deleterious effects due to exposure to elevated ambient temperatures.

2. APPARATUS. A suitable chamber shall be used which will maintain the temperature at the required test temperature and tolerance (see 3.2) to which the parts will be subjected. Temperature measurements shall be made within a specified number of unobstructed inches from any one part or group of like parts under test. In addition, the temperature measurement shall be made at a position where the effects of heat generated by the parts have the least effect on the recorded temperature. Chamber construction shall minimize the influence of radiant heat on the parts being tested. Chambers which utilize circulating liquid as a heat exchanger, free-convection (gravity-type) chambers, and circulating-air chambers may be used providing that the other requirements of this test method are met. When specified, this test shall be made in still air. (Still air is defined as surrounding air with no circulation other than that created by the heat of the part being operated.) The employment of baffling devices and the coating of their surfaces with a heat-absorbing finish are permitted. When a test is conducted on parts that do not have the still-air requirement, there shall be no direct impingement of the forced-air supply upon the parts.

3. PROCEDURE.

3.1 Mounting. Specimens shall be mounted as specified by their normal mounting means. When groups of specimens are to be subjected to test simultaneously, the mounting distance between specimens shall be as specified for the individual groups. When the distance is not specified, the mounting distance shall be sufficient to minimize the temperature of one specimen affecting the temperature of another. Specimens fabricated of different materials, which may have a detrimental effect on each other and alter the results of this test, shall not be tested simultaneously.

3.2 Test temperature. Specimens shall be subjected to one of the following test temperatures with accompanying tolerances, as specified:

Temperature and tolerance 1/		Temperature and tolerance 1/	
°C	°F	°C	°F
70 ±2	(158 ±3.6)	150 ±3	(302 ±5.4)
85 ±2	(185 ±3.6)	200 ±5	(392 ±9)
100 ±2	(212 ±3.6)	350, tolerance as specified.	(662, tolerance as specified.)
125 ±3	(257 ±5.4)	500, tolerance as specified.	(932, tolerance as specified.)

1/ For tests on resistors only, in a still-air environment, the maximum temperature tolerance shall be ±5° C (±9° F).

3.3 Operating conditions. The test potential, duty cycle, load, and other operating conditions, as applicable, applied to the specimen during exposure shall be as specified.

3.4 Length of test. Specimens shall be subjected to one of the following test conditions, as specified:

Test condition	Length of test, hours
A - - - - - - - - - - - - - -	96
B - - - - - - - - - - - - - -	250
C - - - - - - - - - - - - - -	500
D - - - - - - - - - - - - - -	1,000
F - - - - - - - - - - - - - -	2,000
G - - - - - - - - - - - - - -	3,000
H - - - - - - - - - - - - - -	5,000
I - - - - - - - - - - - - - -	10,000
J - - - - - - - - - - - - - -	30,000
K - - - - - - - - - - - - - -	50,000

NOTE. Test condition E (1,500-hour test) has been deleted from this test method.

4. MEASUREMENTS. Specified measurements shall be made prior to, during, or after exposure, as specified. If applicable, frequency of measurements, and portion of the duty cycle in which measurements are to be made, while the specimen is subjected to test, shall be as specified.

5. SUMMARY. The following details are to be specified in the individual specification:

(a) Distance of temperature measurements from specimens, in inches. (see 2).

(b) Still-air requirement, when applicable (see 2).

(c) Method of mounting and distance between specimens, if required (see 3.1).

(d) Test temperature and tolerance (see 3.2).

(e) Operating conditions (see 3.3).

(f) Test condition letter (see 3.4).

(g) Measurements (see 4).
(1) Prior to, during, or after exposure (see 4).
(2) Frequency of measurements, and portion of duty cycle during test, if applicable (see 4).

TABLE A7 METHOD 204C VIBRATION, HIGH FREQUENCY

1. PURPOSE. The high-frequency vibration test is performed for the purpose of determining the effect on component parts of vibration in the frequency ranges of 10 to 500 hertz (Hz), 10 to 2,000 Hz or 10 to 3,000 Hz, as may be encountered in aircraft, missiles, and tanks. The choice of test condition A, B, C, D, E, F, or G should be based on the frequency range and the vibration amplitude dictated by the applications of the component under consideration, and the state of the component art in relation to resistance-to-vibration damage.

2. PROCEDURE.

2.1 Mounting. The specimens shall be mounted as specified. For specimens with attached brackets, one of the vibration-test directions shall be parallel to the mounting surface of the bracket. Vibration input shall be monitored on the mounting fixture in the proximity of the support points of the specimen.

2.2 Test condition A (10g peak). The specimens, while deenergized or operating under the load conditions specified, shall be subjected to the vibration amplitude, frequency range, and duration specified in 2.2.1, 2.2.2, and 2.2.3, respectively (see figure 204-1).

2.2.1 Amplitude. The specimens shall be subjected to a simple harmonic motion having an amplitude of either 0.06-inch double amplitude (maximum total excursion) or 10 gravity units (g peak), whichever is less. The tolerance on vibration amplitude shall be ±10 percent.

2.2.2 Frequency range. The vibration frequency shall be varied logarithmically between the approximate limits of 10 and 500 Hz (see 2.9), except that the procedure of method 201 of this standard may be applied during the 10 to 55 Hz band of the vibration frequency range.

2.2.3 Sweep time and duration. The entire frequency range of 10 to 500 Hz and return to 10 Hz shall be traversed in 15 minutes. This cycle shall be performed 12 times in each of three mutually perpendicular directions (total of 36 times), so that the motion shall be applied for a total period of approximately 9 hours. Interruptions are permitted provided the requirements for rate of change and test duration are met. Completion of cycling within any separate band is permissible before going to the next band. When the procedure of method 201 of this standard is used for the 10 to 55 Hz band, the duration of this portion shall be the same as the duration for this band using logarithmic cycling (approximately 1-1/3 hours in each of three mutually perpendicular directions).

2.3 Test condition B (15g peak). The specimens, while deenergized or operating under the load conditions specified, shall be subjected to the vibration amplitude, frequency range, and duration specified in 2.3.1, 2.3.2, and 2.3.3, respectively (see figure 204-1).

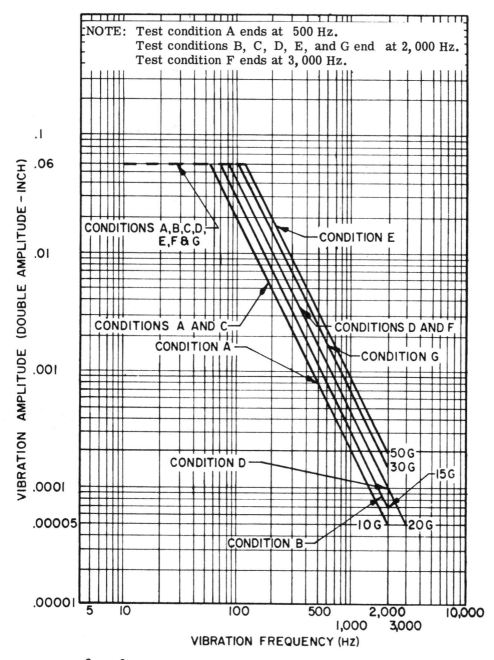

$G = .0512f^2 DA (f^2$ = frequency in hertz, DA = double amplitude in inches.$)$

Figure 204-1 *Vibration-test curves.*

2.3.1 Amplitude. The specimens shall be subjected to a simple harmonic motion having an amplitude of either 0.06-inch double amplitude (maximum total excursion) or 15g(peak), whichever is less. The tolerance on vibration amplitude shall be ±10 percent.

2.3.2 Frequency range. The vibration frequency shall be varied logarithmically between the approximate limits of 10 to 2,000 Hz (see 2.9), except that the procedure of method 201 of this standard may be applied during the 10 to 55 Hz band of the vibration frequency range.

2.3.3 Sweep time and duration. The entire frequency range of 10 to 2,000 Hz and return to 10 Hz shall be traversed in 20 minutes. This cycle shall be performed 12 times in each of three mutually perpendicular directions (total of 36 times), so that the motion shall be applied for a total period of approximately 12 hours. Interruptions are permitted provided the requirements for rate of change and test duration are met. Completion of cycling within any separate band is permissible before going to the next band. When the procedure of method 201 of this standard is used for the 10 to 55 Hz band, the duration of this portion shall be the same as the duration for this band using logarithmic cycling (approximately) 1-1/3 hours in each of three mutually perpendicular directions).

2.4 Test condition C (10g peak). The specimens, while deenergized or operating under the load conditions specified, shall be subjected to the vibration amplitude and frequency range shown on figure 204-1. The tolerance on vibration amplitude shall be ±10 percent.

2.4.1 Part 1. The specimens shall be tested in accordance with method 201 of this standard for 6 hours; 2 hours in each of three mutually perpendicular directions.

2.4.2 Part 2. The specimens shall be subjected to a simple harmonic motion having an amplitude varied to maintain a constant peak acceleration of 10g (peak), the frequency being varied logarithmically between the approximate limits of 55 and 2,000 Hz (see 2.9). The entire frequency range of 55 to 2,000 Hz (no return sweep) shall be traversed in 35 ±5 minutes, except that in the vicinity of what appears to be resonance, and in order to facilitate the establishment of a resonant frequency, the above rate may be decreased. If resonance is detected, specimens shall be vibrated for 5 minutes at each critical resonant frequency observed. This procedure shall be performed in each of three mutually perpendicular directions. Interruptions are permitted provided the requirements for rate of change and test duration are met.

2.4.3 Resonance. A critical resonant frequency is that frequency at which any point on the specimen is observed to have a maximum amplitude more than twice that of the support points. When specified, resonant frequencies shall be determined either by monitoring parameters such as contact opening, or by use of resonance-detecting instrumentation.

2.5 Test condition D (20g peak). The specimens, while deenergized or operating under the load conditions specified, shall be subjected to the vibration amplitude, frequency, range, and duration specified in 2.5.1, 2.5.2, and 2.5.3, respectively (see fig. 204-1).

2.5.1 Amplitude. The specimens shall be subjected to a simple harmonic motion having an amplitude of either 0.06-inch double amplitude (maximum total excursion) or 20g (peak), whichever is less. The tolerance on vibration amplitude shall be ±10 percent.

2.5.2 Frequency range. The vibration frequency shall be varied logarithmically between the approximate limits of 10 to 2,000 Hz (see 2.9), except that the procedure of method 201 of this standard may be applied during the 10 to 55 Hz band of the vibration frequency range.

2.5.3 Sweep time and duration. The entire frequency range of 10 to 2,000 Hz and return to 10 Hz shall be traversed in 20 minutes. This cycle shall be performed 12 times in each of three mutually perpendicular directions (total of 36 times), so that the motion shall be applied for a total period of approximately 12 hours. Interruptions are permitted provided the requirements for rate of change and test duration are met. Completion of cycling within any separate band is permissible before going to the next band. When the procedure of method 201 of this standard is used for the 10 to 55 Hz band, the duration of this portion shall be the same as the duration for this band using logarithmic cycling (approximately 1-1/3 hours in each of three mutually perpendicular directions).

2.6 Test condition E (50g peak). The specimens, while deenergized or operating under the load conditions specified, shall be subjected to the vibration amplitude, frequency, range, and duration specified in 2.6.1, 2.6.2, and 2.6.3, respectively (see fig. 204-1).

2.6.1 Amplitude. The specimens shall be subjected to a simple harmonic motion having an amplitude of either 0.06-inch double amplitude (maximum total excursion) or 50g (peak), whichever is less. The tolerance on vibration amplitude shall be ±10 percent.

2.6.2 <u>Frequency range.</u> The vibration frequency shall be varied logarithmically between the approximate limits of 10 and 2,000 Hz (see 2.9), except that the procedure of method 201 of this standard may be applied during the 10 to 55 Hz band of the vibration frequency range.

2.6.3 <u>Sweep time and duration.</u> The entire frequency range of 10 to 2,000 Hz and return to 10 Hz shall be traversed in 20 minutes. This cycle shall be performed 12 times in each of three mutually perpendicular directions (total of 36 times), so that the motion shall be applied for a total period of approximately 12 hours. Interruptions are permitted provided the requirements for rate of change and test duration are met. Completion of cycling within any separate band is permissible before going to the next band. When the procedure of method 201 of this standard is used for the 10 to 55 Hz band, the duration of this portion shall be the same as the duration for this band using logarithmic cycling (approximately 1-1/3 hours in each of three mutually perpendicular directions).

2.7 <u>Test condition F (20g peak).</u> The specimens, while deenergized or operating under the load conditions specified, shall be subjected to the vibration amplitude, frequency range, and duration specified in 2.7.1, 2.7.2, and 2.7.3, respectively (see figure 204-1).

2.7.1 <u>Amplitude.</u> The specimens shall be subjected to a simple harmonic motion having an amplitude of either 0.06-inch double amplitude (maximum total excursion) or 20g (peak), whichever is less. The tolerance on vibration amplitude shall be ±10 percent.

2.7.2 <u>Frequency range.</u> The vibration frequency shall be varied logarithmically between the limits of 10 and 3,000 Hz (see 2.9), except that the procedure of method 201 of this standard may be applied during the 10 to 55 Hz band of the vibration frequency range.

2.7.3 <u>Sweep time and duration.</u> The entire frequency range of 10 to 3,000 and return to 10 Hz shall be traversed in 20 minutes. This cycle shall be performed 12 times in each of three mutually perpendicular directions (total of 36 times), so that the motion shall be applied for a total period of approximately 12 hours. Interruptions are permitted provided the requirements for rate of change and test duration are met. Completion of cycling within any separate band is permissible before going to the next band. When the procedure of method 201 of this standard is used for the 10 to 55 Hz band, the duration of this portion shall be the same as the duration for this band using logarithmic cycling (approximately) 1-1/3 hours in each of three mutually perpendicular directions.

2.8 <u>Test condition G (30g peak).</u> The specimens, while deenergized or operating under the load conditions specified, shall be subjected to the vibration amplitude, frequency range, and duration specified in 2.8.1, 2.8.2, and 2.8.3, respectively (see figure 204-1).

2.8.1 <u>Amplitude.</u> The specimens shall be subjected to a simple harmonic motion having an amplitude of either 0.06-inch double amplitude (maximum total excursion) or 30g (peak), whichever is less. The tolerance on vibration amplitude shall be ±10 percent.

2.8.2 <u>Frequency range.</u> The vibration frequency shall be varied logarithmically between the limits of 10 and 2,000 Hz (see 2.9), except that the procedure of method 201 of this standard may be applied during the 10 to 55 Hz band of the vibration frequency range.

2.8.3 <u>Sweep time and duration.</u> The entire frequency range of 10 to 2,000 and return to 10 Hz shall be traversed in 20 minutes. This cycle shall be performed 12 times in each of three mutually perpendicular directions (total of 36 times), so that the motion shall be applied for a total period of approximately 12 hours. Interruptions are permitted provided the requirements for rate of change and test duration are met. Completion of cycling within any separate band is permissible before going to the next band. When the procedure of method 201 of this standard is used for the 10 to 55 Hz band, the duration of this portion shall be the same as the duration for this band using logarithmic cycling (approximately) 1-1/3 hours in each of three mutually perpendicular directions.

2.9 <u>Alternate procedure for use of linear in place of logarithmic change of frequency.</u> Linear rate of change of frequency is permissible under the following conditions:

(a) The frequency range above 55 Hz shall be subdivided into not less than three bands. The ratio of the maximum frequency to the minimum frequency in each band shall be not less than two.

(b) The rate of change of frequency in hertz per minute (Hz/min) shall be constant for any one band.

(c) The ratios of the rate of change of frequency of each band to the maximum frequency of that band shall be approximately equal.

2.9.1 <u>Example of alternate procedure.</u> As an example of the computation of rates of change, assume that the frequency spectrum has been divided into three bands, 55 to 125 Hz, 125 to 500 Hz, and 500 to 2,000 Hz, in accordance with 2.9(a). Let the (constant) ratio of rate of frequency change in Hz/min, to maximum frequency in Hz be k for each band. Then the rates of change for the three bands will be 125k, 500k, and 2,000k, respectively. The times, in minutes, to traverse the three frequency bands will therefore be respectively:

$$\frac{125 - 55}{125k}, \quad \frac{500 - 125}{500k}, \quad \text{and} \quad \frac{2,000 - 500}{2,000k}$$

Since the minimum total sweep time is 30 minutes,

$$30 = \frac{70}{125k} + \frac{375}{500k} + \frac{1,500}{2,000k}$$

from which: k = 0.0687

The required maximum constant rates of frequency change for the three bands are therefore 8.55, 34.2, and 137 Hz/min, respectively. The minimum times of traverse of the bands are 8.2, 10.9, and 10.9 minutes, respectively.

3. MEASUREMENTS. Measurements shall be made as specified.

4. SUMMARY. The following details are to be specified in the individual specification.

(a) Mounting of specimens (see 2.1).

(b) Electrical-load conditions, if applicable (see 2.2, 2.3, 2.4, 2.5, 2.6, 2.7, and 2.8).

(c) Test condition letter (see figure 204-1).

(d) Method of determining resonance, if applicable (see 2.4.3).

(e) Measurements (see 3).

TABLE A8 METHOD 207A HIGH-IMPACT SHOCK

1. PURPOSE. This test is performed for the purpose of determining the ability of various parts to withstand shock of the same severity as that produced by underwater explosions, collision impacts, near-miss gunfire, blasts caused by air explosions, and field conditions. Exact simulation of some of the severe shock motions experienced in the field is difficult to reproduce; however, parts that successfully complete the test of this method have been found to possess the necessary ruggedness for this use. The test apparatus utilized in this method is the same as that designated as Shock Testing Machine for Lightweight Equipment in MIL-S-901, Shock Tests, HI (High-Impact), Shipboard Machinery, Equipment and Systems, Requirements for. The purpose of this apparatus is to determine the ability of equipment installed aboard naval ships to withstand shock and still continue to perform its operational function. This test method is limited to testing of parts weighing not more than 300 pounds.

2. PRECAUTIONS. The apparatus shall be examined periodically for damage. Any hardware which has become defective by being deformed or cracked shall be replaced. Particular attention shall be given to the anvil plate which shall not be bowed more than 1 inch at the center. Proper safeguards shall be taken to protect personnel from objects that may become loosened and act as projectiles as a result of this test. A sound-warning arrangement shall be made, for use in alerting personnel in the vicinity of the test of the impending drop of the hammer.

3. APPARATUS. The apparatus used in this test method shall be as shown on figure 207-1 and the associated detail drawings (see 3.4). The parts shall be installed on a mounting fixture which is attached to the anvil plate of the shock-testing apparatus. A 400-pound hammer shall be dropped from a specified height (see 4.4) onto a shock pad located on the anvil plate. The shock motion is then transmitted by the anvil plate to the parts attached on the mounting fixture.

3.1 <u>Anvil plate.</u> The test apparatus of this method is so constructed that the anvil plate (see figure 207-3) can be installed, in sequence, in two positions. By utilizing these two installation positions and separately employing both hammers of the apparatus, shock is applied through the three principal mutually perpendicular axes of the part being subjected to test. One position is to locate the anvil in such a manner that it will receive blows through the back of the anvil plate by contact from the horizontal hammer, and blows on the top shock pad of the anvil plate by a drop of the vertical hammer as shown on figure 207-2. The other position is as shown on figure 207-1, whereby the end shock pad is contacted by the horizontal hammer.

3.2 <u>Hammers.</u> The test apparatus is equipped with two 400-pound hammers. One hammer renders a blow by a vertical drop. The other hammer applies a force in a horizontal direction. In this manner, and by changing the orientation of the anvil plate, blows may be delivered to the anvil and the parts in three directions.

3.3 <u>Mounting fixtures.</u> Figures 207-4A, 207-4B, 207-5, and 207-6 show standard mounting fixtures that shall be used when testing parts with this test apparatus. These mounting fixtures simulate platform and panel mountings. The applicable mounting fixture shall be as specified. When one of the standard mounting fixtures shown on figures 207-4A, 207-4B, 207-5, and 207-6 cannot be used, the individual specification shall specify a mounting fixture or adapter which approximates the actual rigidity encountered in service.

3.4 <u>Detail drawings.</u> Copies of detail drawings may be obtained from the Defense Electronics Supply Center, Directorate of Engineering Standardization (DESC-E), 1507 Wilmington Pike, Dayton, Ohio 45401, by those who require this information. When requesting copies of these drawings, both the identifying symbol number and the title should be stipulated, as follows:

Drawing 207 - Class HI (High-Impact) Shock Testing Machine for Lightweight
Equipment, Department of the Navy (BuShips) 10-T-2145-L.

4. PROCEDURE.

4.1 <u>Mounting method.</u> The specimens shall be installed by their normal mounting means on the mounting fixture in their normal operating position. Bolts for mounting the parts shall conform to type I, II, or III, grade 2, of MIL-B-857, Bolts, Nuts, and Studs. Mounting bolts shall be checked for tightness before each blow. Care shall be taken in the mounting procedure to prevent initial stresses being applied to the specimens prior to shock.

4.2 <u>Anvil-plate bolts and positioning springs.</u> Due to the severity of the shock applied to the anvil plate by a series of three blows, the anvil-plate bolts shall be checked for tightness before each series of blows. The spacing between stops of the positioning springs (1.5 inches) shall also be corrected before each succession of blows.

4.3 <u>Direction of shock.</u> A total of nine blows, three through each of the three principal mutually perpendicular axes for the heights indicated in 4.4, shall be delivered to the anvil plate supporting the specimens under test. Direction of the shock shall be, in order, to the back, top, and side. Back and top blows shall be applied with the anvil plate located to receive blows from the horizontal and vertical hammers. Side blows are delivered by the horizontal hammer contacting the end shock pad of the anvil plate (see 3.1).

4.4 <u>Height of hammer drops.</u> The hammer shall strike the shock pad on the anvil plate, in sequence, from heights of 1 foot, 3 feet, and 5 feet.

4.5 <u>Hammer supports.</u> During the test, the hammer not in use shall be disengaged from the lifting cable and supported so that the hammer and its support are not in contact with the anvil plate.

4.6 <u>Electrical load and operating conditions.</u> The electrical load and operating conditions applied to the specimens shall be as specified.

4.7 <u>External resilient mountings.</u> Unless otherwise specified, no external resilient mountings associated with the specimen being tested shall be used. Integral mounting devices and external resilient mountings (if specified) associated with the specimen shall remain unblocked during tests.

5. MEASUREMENTS. Monitoring of the specimens during test (e.g., delayed contact opening of relays, momentary stopping of dynamotors, calibration errors in meters) shall be as specified. Upon completion of the required number of blows, electrical and physical measurements shall be made as specified. Allowable tolerances shall be as specified.

6. SUMMARY. The following details are to be specified in the individual specification:

    (a) Mounting fixtures (see 3.3).

    (b) Electrical load on operating conditions, if applicable (see 4.6).

    (c) External resilient mountings, if required (see 4.7).

    (d) Monitoring during test, measurements after test, and allowable tolerances, as applicable (see 5).

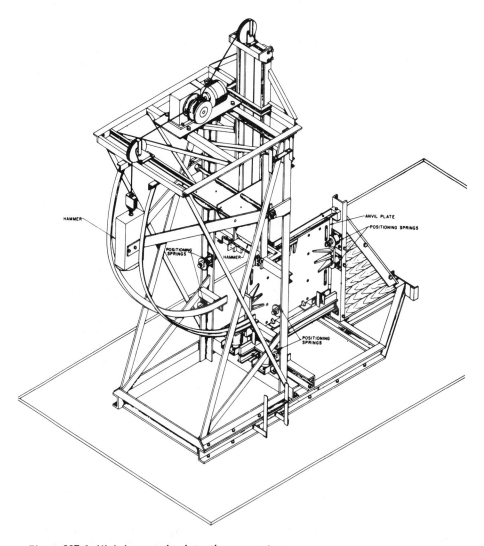

Figure 207-1 *High-impact shock-testing apparatus.*

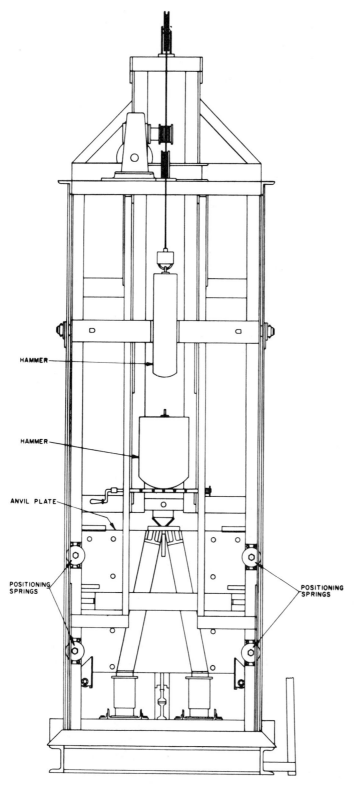

Figure 207-2 *High-impact shock-testing apparatus (backview) with anvil plate located for back and top blows.*

389

Appendix A

Table A9(a) *Table of Individual Binomial Probability Terms.* *

n	r	p				
		0.10	0.20	0.30	0.40	0.50
2	0	0.8100	0.6400	0.4900	0.3600	0.2500
2	1	0.1800	0.3200	0.4200	0.4800	0.5000
2	2	0.0100	0.0400	0.0900	0.1600	0.2500
3	0	0.7290	0.5120	0.3430	0.2160	0.1250
3	1	0.2430	0.3840	0.4410	0.4320	0.3750
3	2	0.0270	0.0960	0.1890	0.2880	0.3750
3	3	0.0010	0.0080	0.0270	0.0640	0.1250
4	0	0.6561	0.4096	0.2401	0.1296	0.0625
4	1	0.2916	0.4096	0.4116	0.3456	0.2500
4	2	0.0486	0.1536	0.2646	0.3456	0.3750
4	3	0.0036	0.0256	0.0756	0.1536	0.2500
4	4	0.0001	0.0016	0.0081	0.0254	0.0625
5	0	0.5905	0.3277	0.1681	0.0778	0.0312
5	1	0.3280	0.4096	0.3602	0.2592	0.1562
5	2	0.0729	0.2049	0.3087	0.3456	0.3125
5	3	0.0081	0.0512	0.1323	0.2304	0.3125
5	4	0.0004	0.0064	0.0284	0.0768	0.1562
5	5	0.0001	0.0003	0.0023	0.0102	0.0312
6	0	0.5314	0.2621	0.1176	0.0467	0.0156
6	1	0.3543	0.3932	0.3025	0.1866	0.0938
6	2	0.0984	0.2458	0.3241	0.3110	0.2344
6	3	0.0146	0.0819	0.1852	0.2765	0.3125
6	4	0.0012	0.0154	0.0595	0.1382	0.2344
6	5	0.0001	0.0015	0.0102	0.0369	0.0938
6	6		0.0001	0.0009	0.0041	0.0156
7	0	0.4783	0.2097	0.0824	0.0280	0.0078
7	1	0.3720	0.3670	0.2471	0.1306	0.0547
7	2	0.1240	0.2753	0.3177	0.2613	0.1641
7	3	0.0230	0.1147	0.2269	0.2903	0.2734
7	4	0.0026	0.0287	0.0972	0.1935	0.2734
7	5	0.0002	0.0043	0.0250	0.0774	0.1641
7	6		0.0004	0.0036	0.0172	0.0547
7	7			0.0001	0.0017	0.0078
8	0	0.4305	0.1678	0.0576	0.0168	0.0039
8	1	0.3826	0.3355	0.1977	0.0896	0.0312
8	2	0.1488	0.2936	0.2965	0.2090	0.1094
8	3	0.0331	0.1468	0.2541	0.2787	0.2188
8	4	0.0046	0.0459	0.1361	0.2323	0.2734
8	5	0.0005	0.0092	0.0467	0.1239	0.2188
8	6		0.0011	0.0100	0.0413	0.1094
8	7		0.0001	0.0012	0.0079	0.0312
8	8			0.0001	0.0006	0.0039

* This table is adapted from "Tables of Binomial Probability Distribution," U.S. Department of Commerce, Applied Mathematics Series 6, issued January 1950.

Table A9(a) (Continued)

				p		
n	r	0.10	0.20	0.30	0.40	0.50
9	0	0.3874	0.1342	0.0404	0.0101	0.0020
9	1	0.3874	0.3020	0.1556	0.0606	0.0176
9	2	0.1722	0.3020	0.2668	0.1612	0.0703
9	3	0.0446	0.1762	0.2668	0.2508	0.1641
9	4	0.0074	0.0661	0.1715	0.2508	0.2461
9	5	0.0008	0.0165	0.0735	0.1672	0.2461
9	6	0.0001	0.0028	0.0210	0.0743	0.1641
9	7		0.0003	0.0039	0.0212	0.0703
9	8			0.0004	0.0035	0.0176
9	9			0.0001	0.0004	0.0020
10	0	0.3487	0.1074	0.0282	0.0060	0.0010
10	1	0.3874	0.2684	0.1211	0.0403	0.0098
10	2	0.1937	0.3020	0.2335	0.1209	0.0439
10	3	0.0574	0.2013	0.2668	0.2150	0.1172
10	4	0.0112	0.0881	0.2001	0.2508	0.2051
10	5	0.0015	0.0264	0.1029	0.2007	0.2461

Table A9(b) *Table of Cumulative Binomial Probabilities.* *

				p		
n	r	0.10	0.20	0.30	0.40	0.50
2	0	1.000	1.000	1.000	1.000	1.000
2	1	0.190	0.360	0.510	0.640	0.750
2	2	0.010	0.040	0.090	0.160	0.250
3	0	1.000	1.000	1.000	1.000	1.000
3	1	0.271	0.488	0.657	0.784	0.875
3	2	0.028	0.104	0.216	0.352	0.500
3	3	0.001	0.008	0.027	0.064	0.125
4	0	1.000	1.000	1.000	1.000	1.000
4	1	0.344	0.590	0.760	0.870	0.938
4	2	0.052	0.181	0.348	0.525	0.688
4	3	0.004	0.027	0.084	0.179	0.312
4	4		0.002	0.008	0.026	0.062
5	0	1.000	1.000	1.000	1.000	1.000
5	1	0.410	0.672	0.832	0.922	0.969
5	2	0.081	0.263	0.472	0.663	0.812
5	3	0.009	0.058	0.163	0.317	0.500
5	4		0.007	0.031	0.087	0.188
5	5			0.002	0.010	0.031

* This table is adapted from Ordnance Corps Pamphlet ORDP 20-1, Tables of the Cumulative Binomial Probabilities, Sept. 1952.

Table A9(b) (Continued)

n	r	0.10	0.20	0.30	0.40	0.50
				p		
6	0	1.000	1.000	1.000	1.000	1.000
6	1	0.469	0.738	0.882	0.953	0.984
6	2	0.114	0.345	0.580	0.767	0.891
6	3	0.016	0.099	0.257	0.456	0.656
6	4	0.001	0.017	0.070	0.179	0.344
6	5		0.002	0.011	0.041	0.109
6	6			0.001	0.004	0.016
7	0	1.000	1.000	1.000	1.000	1.000
7	1	0.522	0.790	0.918	0.972	0.992
7	2	0.150	0.423	0.672	0.841	0.938
7	3	0.026	0.148	0.353	0.580	0.773
7	4	0.003	0.033	0.126	0.290	0.500
7	5		0.005	0.029	0.096	0.227
7	6			0.004	0.019	0.062
7	7				0.002	0.008
8	0	1.000	1.000	1.000	1.000	1.000
8	1	0.570	0.832	0.942	0.983	0.996
8	2	0.187	0.497	0.745	0.894	0.965
8	3	0.038	0.203	0.448	0.685	0.855
8	4	0.005	0.056	0.194	0.406	0.637
8	5		0.010	0.058	0.174	0.363
8	6		0.001	0.011	0.050	0.145
8	7			0.001	0.009	0.035
8	8				0.001	0.004
9	0	1.000	1.000	1.000	1.000	1.000
9	1	0.613	0.866	0.960	0.990	0.998
9	2	0.225	0.564	0.804	0.928	0.980
9	3	0.053	0.262	0.537	0.768	0.910
9	4	0.008	0.086	0.270	0.517	0.746
9	5	0.001	0.020	0.099	0.267	0.500
9	6		0.003	0.025	0.099	0.254
9	7			0.004	0.025	0.090
9	8				0.004	0.020
9	9					0.002
10	0	1.000	1.000	1.000	1.000	1.000
10	1	0.651	0.893	0.972	0.994	0.999
10	2	0.264	0.624	0.851	0.954	0.989
10	3	0.070	0.322	0.617	0.833	0.945
10	4	0.013	0.121	0.350	0.618	0.828
10	5	0.002	0.033	0.150	0.367	0.623

Table A10 *Values of the Descending Exponential* $e^{-\lambda t}$ *(Source: MIL-HDBK-217A).*

$a = \lambda t = 0.000$ $\qquad\qquad\qquad\qquad\qquad\qquad\qquad\qquad\qquad\qquad\qquad\qquad a = \lambda t = 0.249$

λt	$e^{-\lambda t}$	λt	$e^{-\lambda t}$	λt	$e^{-\lambda t}$	λt	$e^{-\lambda t}$	λt	$e^{-\lambda t}$
0.000	1.00000	0.050	0.95123	0.100	0.90484	0.150	0.86071	0.200	0.81873
0.001	0.99900	0.051	0.95028	0.101	0.90393	0.151	0.85985	0.201	0.81791
0.002	0.99800	0.052	0.94933	0.102	0.90303	0.152	0.85899	0.202	0.81709
0.003	0.99700	0.053	0.94838	0.103	0.90213	0.153	0.85813	0.203	0.81628
0.004	0.99601	0.054	0.94743	0.104	0.90123	0.154	0.85727	0.204	0.81546
0.005	0.99501	0.055	0.94649	0.105	0.90032	0.155	0.85642	0.205	0.81465
0.006	0.99402	0.056	0.94554	0.106	0.89942	0.156	0.85556	0.206	0.81383
0.007	0.99302	0.057	0.94459	0.107	0.89853	0.157	0.85470	0.207	0.81302
0.008	0.99203	0.058	0.94365	0.108	0.89763	0.158	0.85385	0.208	0.81221
0.009	0.99104	0.059	0.94271	0.109	0.89673	0.159	0.85300	0.209	0.81140
0.010	0.99005	0.060	0.94176	0.110	0.89583	0.160	0.85214	0.210	0.81058
0.011	0.98906	0.061	0.94082	0.111	0.89494	0.161	0.85129	0.211	0.80977
0.012	0.98807	0.062	0.93988	0.112	0.89404	0.162	0.85044	0.212	0.80896
0.013	0.98708	0.063	0.93894	0.113	0.89315	0.163	0.84959	0.213	0.80816
0.014	0.98610	0.064	0.93800	0.114	0.89226	0.164	0.84874	0.214	0.80735
0.015	0.98511	0.065	0.93707	0.115	0.89137	0.165	0.84789	0.215	0.80654
0.016	0.98413	0.066	0.93613	0.116	0.89047	0.166	0.84705	0.216	0.80574
0.017	0.98314	0.067	0.93520	0.117	0.88959	0.167	0.84620	0.217	0.80493
0.018	0.98216	0.068	0.93426	0.118	0.88870	0.168	0.84535	0.218	0.80413
0.019	0.98118	0.069	0.93333	0.119	0.88781	0.169	0.84451	0.219	0.80332
0.020	0.98020	0.070	0.93239	0.120	0.88692	0.170	0.84366	0.220	0.80252
0.021	0.97922	0.071	0.93146	0.121	0.88603	0.171	0.84272	0.221	0.80172
0.022	0.97824	0.072	0.93053	0.122	0.88515	0.172	0.84198	0.222	0.80082
0.023	0.97726	0.073	0.92960	0.123	0.88426	0.173	0.84114	0.223	0.80011
0.024	0.97629	0.074	0.92867	0.124	0.88338	0.174	0.84030	0.224	0.79932
0.025	0.97531	0.075	0.92774	0.125	0.88250	0.175	0.83946	0.225	0.79852
0.026	0.97434	0.076	0.92682	0.126	0.88161	0.176	0.83862	0.226	0.79772
0.027	0.97336	0.077	0.92589	0.127	0.88073	0.177	0.83778	0.227	0.79692
0.028	0.97239	0.078	0.92496	0.128	0.87985	0.178	0.83694	0.228	0.79612
0.029	0.97142	0.079	0.92404	0.129	0.87897	0.179	0.83611	0.229	0.79533
0.030	0.97045	0.080	0.92312	0.130	0.87810	0.180	0.83527	0.230	0.79453
0.031	0.96948	0.081	0.92219	0.131	0.87722	0.181	0.83444	0.231	0.79374
0.032	0.96851	0.082	0.92127	0.132	0.87634	0.182	0.83360	0.232	0.79295
0.033	0.96754	0.083	0.92035	0.133	0.87547	0.183	0.83277	0.233	0.79215
0.034	0.96657	0.084	0.91943	0.134	0.87459	0.184	0.83194	0.234	0.79136
0.035	0.96561	0.085	0.91851	0.135	0.87371	0.185	0.83110	0.235	0.79057
0.036	0.96464	0.086	0.91759	0.136	0.87284	0.186	0.83027	0.236	0.78978
0.037	0.96368	0.087	0.91666	0.137	0.87197	0.187	0.82944	0.237	0.78899
0.038	0.96271	0.088	0.91576	0.138	0.87110	0.188	0.82861	0.238	0.78820
0.039	0.96175	0.089	0.91485	0.139	0.87023	0.189	0.82779	0.239	0.78741
0.040	0.96079	0.090	0.91393	0.140	0.86936	0.190	0.82696	0.240	0.78663
0.041	0.95983	0.091	0.91302	0.141	0.86805	0.191	0.82613	0.241	0.78584
0.042	0.95887	0.092	0.91211	0.142	0.86762	0.192	0.82531	0.242	0.78506
0.043	0.95791	0.093	0.91119	0.143	0.86675	0.193	0.82448	0.243	0.78427
0.044	0.95695	0.094	0.91028	0.144	0.86589	0.194	0.82366	0.244	0.78349
0.045	0.95600	0.095	0.90937	0.145	0.86502	0.195	0.82283	0.245	0.78270
0.046	0.95504	0.096	0.90846	0.146	0.86416	0.196	0.82201	0.246	0.78192
0.047	0.95409	0.097	0.90710	0.147	0.86330	0.197	0.82119	0.247	0.78114
0.048	0.95313	0.098	0.90665	0.148	0.86243	0.198	0.82037	0.248	0.78036
0.049	0.95218	0.099	0.90574	0.149	0.86157	0.199	0.81955	0.249	0.77958

$a = \lambda t = 0.250$ $\qquad\qquad\qquad\qquad\qquad\qquad\qquad\qquad\qquad\qquad\qquad$ $a = \lambda t = 0.499$

λt	$e^{-\lambda t}$	λt	$e^{-\lambda t}$	λt	$e^{-\lambda t}$	λt	$e^{-\lambda t}$	λt	$e^{-\lambda t}$
0.250	0.77880	0.300	0.74082	0.350	0.70469	0.400	0.67032	0.450	0.63763
0.251	0.77802	0.301	0.74008	0.351	0.70398	0.401	0.66965	0.451	0.63699
0.252	0.77724	0.302	0.73934	0.352	0.70328	0.402	0.66898	0.452	0.63635
0.253	0.77647	0.303	0.73860	0.353	0.70258	0.403	0.66831	0.453	0.63572
0.254	0.77569	0.304	0.73786	0.354	0.70187	0.404	0.66764	0.454	0.63508
0.255	0.77492	0.305	0.73712	0.355	0.70117	0.405	0.66698	0.455	0.63445
0.256	0.77414	0.306	0.73639	0.356	0.70047	0.406	0.66631	0.456	0.63381
0.257	0.77337	0.307	0.73565	0.357	0.69977	0.407	0.66564	0.457	0.63318
0.258	0.77260	0.308	0.73492	0.358	0.69907	0.408	0.66498	0.458	0.63255
0.259	0.77182	0.309	0.73418	0.359	0.69837	0.409	0.66431	0.459	0.63192
0.260	0.77105	0.310	0.73345	0.360	0.69768	0.410	0.66365	0.460	0.63128
0.261	0.77028	0.311	0.73271	0.361	0.69696	0.411	0.66299	0.461	0.63065
0.262	0.76951	0.312	0.73198	0.362	0.69628	0.412	0.66232	0.462	0.63002
0.263	0.76874	0.313	0.73125	0.363	0.69559	0.413	0.66166	0.463	0.62939
0.264	0.76797	0.314	0.73052	0.364	0.69489	0.414	0.66100	0.464	0.62876
0.265	0.76721	0.315	0.72979	0.365	0.69420	0.415	0.66034	0.465	0.62814
0.266	0.76644	0.316	0.72906	0.366	0.69350	0.416	0.65968	0.466	0.62751
0.267	0.76567	0.317	0.72833	0.367	0.69281	0.417	0.65902	0.467	0.62688
0.268	0.76491	0.318	0.72760	0.368	0.69212	0.418	0.65836	0.468	0.62625
0.269	0.76414	0.319	0.72688	0.369	0.69143	0.419	0.65770	0.469	0.62563
0.270	0.76338	0.320	0.72615	0.370	0.69073	0.420	0.65704	0.470	0.62500
0.271	0.76262	0.321	0.72542	0.371	0.69004	0.421	0.65639	0.471	0.62438
0.272	0.76185	0.322	0.72433	0.372	0.68935	0.422	0.65573	0.472	0.62375
0.273	0.76109	0.323	0.72397	0.373	0.68867	0.423	0.65508	0.473	0.62313
0.274	0.76033	0.324	0.72325	0.374	0.68798	0.424	0.65442	0.474	0.62251
0.275	0.75957	0.325	0.72253	0.375	0.68729	0.425	0.65377	0.475	0.62189
0.276	0.75881	0.326	0.72181	0.376	0.68660	0.426	0.65312	0.476	0.62126
0.277	0.75805	0.327	0.72108	0.377	0.68592	0.427	0.65246	0.477	0.62064
0.278	0.75730	0.328	0.72036	0.378	0.68523	0.428	0.65181	0.478	0.62002
0.279	0.75654	0.329	0.71964	0.379	0.68454	0.429	0.65116	0.479	0.61940
0.280	0.75578	0.330	0.71892	0.380	0.68386	0.430	0.65051	0.480	0.61878
0.281	0.75503	0.331	0.71821	0.381	0.68318	0.431	0.64986	0.481	0.61816
0.282	0.75427	0.332	0.71749	0.382	0.68250	0.432	0.64921	0.482	0.61755
0.283	0.75352	0.333	0.71677	0.383	0.68181	0.433	0.64856	0.483	0.61693
0.284	0.75277	0.334	0.71605	0.384	0.68113	0.434	0.64791	0.484	0.61631
0.285	0.75201	0.335	0.71534	0.385	0.68045	0.435	0.64726	0.485	0.61570
0.286	0.75126	0.336	0.71462	0.386	0.67977	0.436	0.64662	0.486	0.61508
0.287	0.75051	0.337	0.71391	0.387	0.67909	0.437	0.64597	0.487	0.61447
0.288	0.74976	0.338	0.71320	0.388	0.67841	0.438	0.64533	0.488	0.61385
0.289	0.74901	0.339	0.71248	0.389	0.67773	0.439	0.64468	0.489	0.61324
0.290	0.74826	0.340	0.71177	0.390	0.67706	0.440	0.64404	0.490	0.61263
0.291	0.74752	0.341	0.71106	0.391	0.67638	0.441	0.64339	0.491	0.61201
0.292	0.74677	0.342	0.71035	0.392	0.67570	0.442	0.64275	0.492	0.61140
0.293	0.74602	0.343	0.70964	0.393	0.67503	0.443	0.64211	0.493	0.61079
0.294	0.74528	0.344	0.70893	0.394	0.67435	0.444	0.64147	0.494	0.61018
0.295	0.74453	0.345	0.70822	0.395	0.67368	0.445	0.64082	0.495	0.60957
0.296	0.74379	0.346	0.70751	0.396	0.67301	0.446	0.64018	0.496	0.60896
0.297	0.74304	0.347	0.70681	0.397	0.67233	0.447	0.63954	0.497	0.60835
0.298	0.74230	0.348	0.70610	0.398	0.67166	0.448	0.63890	0.498	0.60774
0.299	0.74156	0.349	0.70539	0.399	0.67099	0.449	0.63827	0.499	0.60714

$a = \lambda t = 0.500$ $a = \lambda t = 0.749$

λt	$e^{-\lambda t}$	λt	$e^{-\lambda t}$	λt	$e^{-\lambda t}$	λt	$e^{-\lambda t}$	λt	$e^{-\lambda t}$
0.500	0.60653	0.550	0.57695	0.600	0.54881	0.650	0.52205	0.700	0.49659
0.501	0.60592	0.551	0.57637	0.601	0.54826	0.651	0.52152	0.701	0.49609
0.502	0.60532	0.552	0.57580	0.602	0.54772	0.652	0.52100	0.702	0 49559
0.503	0.60471	0.553	0.57522	0.603	0.54717	0.653	0.52048	0.703	0.49510
0.504	0.60411	0.554	0.57465	0.604	0.54662	0.654	0.51996	0.704	0.49460
0.505	0.60351	0.555	0.57407	0.605	0.54607	0.655	0.51944	0.705	0.49411
0.506	0.60290	0.556	0.57340	0.606	0.54553	0.656	0.51892	0.706	0.49361
0.507	0.60230	0.557	0.57293	0.607	0.54498	0.657	0.51840	0.707	0.49312
0.508	0.60170	0.558	0.57235	0.608	0.54444	0.658	0.51789	0.708	0.49263
0.509	0.60110	0.559	0.57178	0.609	0.54389	0.659	0.51737	0.709	0.49214
0.510	0.60050	0.560	0.57121	0.610	0.54335	0.660	0.51685	0.710	0.49164
0.511	0.59990	0.561	0.57064	0.611	0.54281	0.661	0.51663	0.711	0.49115
0.512	0.59930	0.562	0.57007	0.612	0.54227	0.662	0.51582	0.712	0.49066
0.513	0.59870	0.563	0.56950	0.613	0.54172	0.663	0.51530	0.713	0.49017
0.514	0.59810	0.564	0.56893	0.614	0.54118	0.664	0.51479	0.714	0.48968
0.515	0.59750	0.565	0.56836	0.615	0.54064	0.665	0.51427	0.715	0.48919
0.516	0.59690	0.566	0.56779	0.616	0.54010	0.666	0.51376	0.716	0.48870
0.517	0.59631	0.567	0.56722	0.617	0.53956	0.667	0.51325	0.717	0.48821
0.518	0.59571	0.568	0.56666	0.618	0.53902	0.668	0.51273	0.718	0.48773
0.519	0.59512	0.569	0.56609	0.619	0.53848	0.669	0.51222	0.719	0.48724
0.520	0.59452	0.570	0.56553	0.620	0.53794	0.670	0.51171	0.720	0.48675
0.521	0.59393	0.571	0.56496	0.621	0.53741	0.671	0.51120	0.721	0.48627
0.522	0.59333	0.572	0.56440	0.622	0.53687	0.672	0.51069	0.722	0.48578
0.523	0.59274	0.573	0.56383	0.623	0.53633	0.673	0.51018	0.723	0.48529
0.524	0.59215	0.574	0.56327	0.624	0.53580	0.674	0.50967	0.724	0.48421
0.525	0.59156	0.575	0.56270	0.625	0.53526	0.675	0.50916	0.725	0.48432
0.526	0.59096	0.576	0.56214	0.626	0.53473	0.676	0.50865	0.726	0.48384
0.527	0.59037	0.577	0.56158	0.627	0.53419	0.677	0.50814	0.727	0.48336
0.528	0.58978	0.578	0.56102	0.628	0.53366	0.678	0.50763	0.728	0.48287
0.529	0.58919	0.579	0.56046	0.629	0.53312	0.679	0.50712	0.729	0.48239
0.530	0.58860	0.580	0.55990	0.630	0.53259	0.680	0.50662	0.730	0.48191
0.531	0.58802	0.581	0.55934	0.631	0.53206	0.681	0.50611	0.731	0.48143
0.532	0.58743	0.582	0.55878	0.632	0.53153	0.682	0.50560	0.732	0.48095
0.533	0.58684	0.583	0.55822	0.633	0.53100	0.683	0.50510	0.733	0.48047
0.534	0.58626	0.584	0.55766	0.634	0.53047	0.684	0.50459	0.734	0.47999
0.535	0.58567	0.585	0.55711	0.635	0.52994	0.685	0.50409	0.735	0.47951
0.536	0.58508	0.586	0.55655	0.636	0.52941	0.686	0.50359	0.736	0.47903
0.537	0.58450	0.587	0.55599	0.637	0.52888	0.687	0.50308	0.737	0.47855
0.538	0.58391	0.588	0.55544	0.638	0.52835	0.688	0.50258	0.738	0.47807
0.539	0.58333	0.589	0.55488	0.639	0.52782	0.689	0.50208	0.739	0.47759
0.540	0.58275	0.590	0.55433	0.640	0.52729	0.690	0.50158	0.740	0.47711
0.541	0.58217	0.591	0.55377	0.641	0.52677	0.691	0.50107	0.741	0.47664
0.542	0.58158	0.592	0.55322	0.642	0.52624	0.692	0.50057	0.742	0.47616
0.543	0.58100	0.593	0.55267	0.643	0.52571	0.693	0.50007	0.743	0.47568
0.544	0.58042	0.594	0.55211	0.644	0.52519	0.694	0.49957	0.744	0.47521
0.545	0.57984	0.595	0.55156	0.645	0.52466	0.695	0.49907	0.745	0.47473
0.546	0.57926	0.596	0.55101	0.646	0.52414	0.696	0.49858	0.746	0.47426
0.547	0.57868	0.597	0.55046	0.647	0.52361	0.697	0.49808	0.747	0.47379
0.548	0.57810	0.598	0.54991	0.648	0.52309	0.698	0.49758	0.748	0.47331
0.549	0.57753	0.599	0.54936	0.649	0.52257	0.699	0.49708	0.749	0.47284

Table A10 (Continued)

$a = \lambda t = 0.750$ $a = \lambda t = 0.999$

λt	$e^{-\lambda t}$	λt	$e^{-\lambda t}$	λt	$e^{-\lambda t}$	λt	$e^{-\lambda t}$	λt	$e^{-\lambda t}$
0.750	0.47237	0.800	0.44933	0.850	0.42741	0.900	0.40657	0.950	0.38674
0.751	0.47189	0.801	0.44888	0.851	0.42699	0.901	0.40616	0.951	0.38635
0.752	0.47142	0.802	0.44843	0.852	0.42656	0.902	0.40576	0.952	0.38596
0.753	0.47095	0.803	0.44798	0.853	0.42613	0.903	0.40535	0.953	0.38558
0.754	0.47048	0.804	0.44754	0.854	0.42571	0.904	0.40495	0.954	0.38520
0.755	0.47001	0.805	0.44709	0.855	0.42528	0.905	0.40454	0.955	0.38481
0.756	0.46954	0.806	0.44664	0.856	0.42486	0.906	0.40414	0.956	0.38443
0.757	0.46907	0.807	0.44619	0.857	0.42443	0.907	0.40373	0.957	0.38404
0.758	0.46860	0.808	0.44575	0.858	0.42401	0.908	0.40333	0.958	0.38366
0.759	0.46813	0.809	0.44530	0.859	0.42359	0.909	0.40293	0.959	0.38328
0.760	0.46767	0.810	0.44486	0.860	0.42316	0.910	0.40252	0.960	0.38289
0.761	0.46720	0.811	0.44441	0.861	0.42274	0.911	0.40212	0.961	0.38251
0.762	0.46673	0.812	0.44397	0.862	0.42232	0.912	0.40172	0.962	0.38213
0.763	0.46627	0.813	0.44353	0.863	0.42189	0.913	0.40132	0.963	0.38175
0.764	0.46580	0.814	0.44308	0.864	0.42147	0.914	0.40092	0.964	0.38136
0.765	0.46533	0.815	0.44264	0.865	0.42105	0.915	0.40052	0.965	0.38098
0.766	0.46487	0.816	0.44220	0.866	0.42063	0.916	0.40012	0.966	0.38060
0.767	0.46440	0.817	0.44175	0.867	0.42021	0.917	0.39972	0.967	0.38022
0.768	0.46394	0.818	0.44131	0.868	0.41979	0.918	0.39932	0.968	0.37984
0.769	0.46348	0.819	0.44087	0.869	0.41937	0.919	0.39892	0.969	0.37946
0.770	0.46301	0.820	0.44043	0.870	0.41895	0.920	0.39852	0.970	0.37908
0.771	0.46255	0.821	0 43999	0.871	0.41853	0.921	0.39812	0.971	0.37870
0.772	0.46209	0.822	0.43955	0.872	0.41811	0.922	0.39772	0.972	0.37833
0.773	0.46163	0.823	0.43911	0.873	0.41770	0.923	0.39733	0.973	0.37795
0.774	0.46116	0.824	0.43867	0.874	0.41728	0.924	0.39693	0.974	0.37757
0.775	0.46070	0.825	0.43823	0.875	0.41686	0.925	0.39653	0.975	0.37719
0.776	0.46024	0.826	0.43780	0.876	0.41645	0.926	0.39614	0.976	0.37682
0.777	0.45978	0 827	0.43736	0.877	0.41603	0.927	0.39574	0.977	0.37644
0.778	0.45932	0.828	0.43692	0.878	0.41561	0.928	0.39534	0.978	0.37606
0.779	0.45886	0.829	0.43649	0.879	0.41520	0.929	0.39495	0.979	0.37569
0.780	0.45841	0.830	0.43605	0.880	0.41478	0.930	0.39455	0.980	0.37531
0.781	0.45795	0.831	0.43561	0.881	0.41437	0.931	0.39416	0.981	0.37494
0.782	0.45749	0.832	0.43517	0.882	0.41395	0.932	0.39377	0.982	0.37456
0.783	0.45703	0.833	0.43474	0.883	0.41354	0.933	0.39337	0.983	0.37419
0.784	0.45658	0.834	0.43431	0.884	0.41313	0.934	0.39298	0.984	0.37381
0.785	0.45612	0.835	0.43387	0.885	0.41271	0.935	0.39259	0.985	0.37344
0.786	0.45566	0.836	0.43344	0.886	0.41230	0.936	0.39219	0.986	0.37309
0.787	0.45520	0.837	0.43301	0.887	0.41189	0.937	0.39180	0.987	0.37269
0.788	0.45478	0.838	0.43257	0.888	0.41148	0.938	0.39141	0.988	0.37323
0.789	0.45430	0.839	0.43214	0.889	0.41107	0.939	0.39102	0.989	0.37195
0.790	0.45384	0.840	0.43171	0.890	0.41066	0.940	0.39063	0.990	0.37158
0.791	0.45339	0.841	0.43128	0.891	0.41025	0.941	0.39024	0.991	0.37121
0.792	0.45294	0.842	0.43085	0.892	0.40984	0.942	0.38985	0.992	0.37083
0.793	0.45249	0.843	0.43042	0.893	0.40943	0.943	0.38946	0.993	0.37046
0.794	0.45203	0.844	0.42999	0.894	0.40902	0.944	0.38907	0.994	0.37009
0.795	0.45158	0.845	0.42956	0.895	0.40861	0.945	0.38868	0.995	0.36972
0.796	0.45113	0.846	0.42913	0.896	0.40820	0.946	0.38829	0.996	0.36935
0.797	0.45068	0.847	0.42870	0.897	0.40779	0.947	0.38790	0.997	0.36898
0.798	0.45023	0.848	0.42827	0.898	0.40738	0.948	0.38752	0.998	0.36861
0.799	0.44978	0.849	0.42784	0.899	0.40698	0.949	0.38713	0.999	0.36825

Table A11 *Summation of Terms of Poisson's Exponential Binomial Limit.**

						c				
a	0	1	2	3	4	5	6	7	8	9
0.02	980	1,000								
0.04	961	999	1,000							
0.06	942	998	1,000							
0.08	923	997	1,000							
0.10	905	995	1,000							
0.15	861	990	999	1,000						
0.20	819	982	999	1,000						
0.25	779	974	998	1,000						
0.30	741	963	996	1,000						
0.35	705	951	994	1,000						
0.40	670	938	992	999	1,000					
0.45	638	925	989	999	1,000					
0.50	607	910	986	998	1,000					
0.55	577	894	982	998	1,000					
0.60	549	878	977	997	1,000					
0.65	522	861	972	996	999	1,000				
0.70	497	844	966	994	999	1,000				
0.75	472	827	959	993	999	1,000				
0.80	449	809	953	991	999	1,000				
0.85	427	791	945	989	998	1,000				
0.90	407	772	937	987	998	1,000				
0.95	387	754	929	984	997	1,000				
1.00	368	736	920	981	996	999	1,000			
1.1	333	699	900	974	995	999	1,000			
1.2	301	663	879	966	992	998	1,000			
1.3	273	627	857	957	989	998	1,000			
1.4	247	592	833	946	986	997	999	1,000		
1.5	223	558	809	934	981	996	999	1,000		
1.6	202	525	783	921	976	994	999	1,000		
1.7	183	493	757	907	970	992	998	1,000		
1.8	165	463	731	891	964	990	997	999	1,000	
1.9	150	434	704	875	956	987	997	999	1,000	
2.0	135	406	677	857	947	983	995	999	1,000	
2.2	111	355	623	819	928	975	993	998	1,000	
2.4	091	308	570	779	904	964	988	997	999	1,000
2.6	074	267	518	736	877	951	983	995	999	1,000
2.8	061	231	469	692	848	935	976	992	998	999
3.0	050	199	423	647	815	916	966	988	996	999

* From Statistical Quality Control by E. L. Grant. Copyright 1952 by McGraw-Hill, Inc. Adapted with permission of McGraw-Hill Book Company.

Table A11 (Continued) ***Summation of Terms of Poisson's Exponential Binomial Limit.***

	c									
a	0	1	2	3	4	5	6	7	8	9
3.2	041	171	380	603	781	895	955	983	994	998
3.4	033	147	340	558	744	871	942	977	992	997
3.6	027	126	303	515	706	844	927	969	988	996
3.8	022	107	269	473	668	816	909	960	984	994
4.0	018	092	238	433	629	785	889	949	979	992
4.2	015	078	210	395	590	753	867	936	972	989
4.4	012	066	185	359	551	720	844	921	964	985
4.6	010	056	163	326	513	686	818	905	955	980
4.8	008	048	143	294	476	651	791	887	944	975
5.0	007	040	125	265	440	616	762	867	932	968
5.2	006	034	109	238	406	581	732	845	918	960
5.4	005	029	095	213	373	546	702	822	903	951
5.6	004	024	082	191	342	512	670	797	886	941
5.8	003	021	072	170	313	478	638	771	867	929
6.0	002	017	062	151	285	446	606	744	847	916
6.2	002	015	054	134	259	414	574	716	826	902
6.4	002	012	046	119	235	384	542	687	803	886
6.6	001	010	040	105	213	355	511	658	780	869
6.8	001	009	034	093	192	327	480	628	755	850
7.0	001	007	030	082	173	301	450	599	729	830
7.2	001	006	025	072	156	276	420	569	703	810
7.4	001	005	022	063	140	253	392	539	676	788
7.6	001	004	019	055	125	231	365	510	648	765
7.8	000	004	016	048	112	210	338	481	620	741
8.0	000	003	014	042	100	191	313	453	593	717
8.5	000	002	009	030	074	150	256	386	523	653
9.0	000	001	006	021	055	116	207	324	456	587
9.5	000	001	004	015	040	089	165	269	392	522
10.0	000	000	003	010	029	067	130	220	333	458
10.5	000	000	000	007	021	050	102	179	279	397
11.0	000	000	001	005	015	038	079	143	232	341
11.5	000	000	001	003	011	028	060	114	191	289
12.0	000	000	001	002	008	020	046	090	155	242
12.5	000	000	000	002	005	015	035	070	125	201
13.0	000	000	000	001	004	011	026	054	100	166
13.5	000	000	000	001	003	008	019	041	079	135
14.0	000	000	000	000	002	006	014	032	062	109
14.5	000	000	000	000	001	004	010	024	048	088
15.0	000	000	000	000	001	003	008	018	037	070

Table A12 *Table of Areas Under the Normal Curve from 0 to Z.**

z	0.00	0.01	0.02	0.03	0.04	0.05	0.06	0.07	0.08	0.09
0.0	0.0000	0.0040	0.0080	0.0120	0.0159	0.0199	0.0239	0.0279	0.0319	0.0359
0.1	0.0398	0.0438	0.0478	0.0517	0.0557	0.0596	0.0636	0.0675	0.0714	0.0753
0.2	0.0793	0.0832	0.0871	0.0910	0.0948	0.0987	0.1026	0.1064	0.1103	0.1141
0.3	0.1179	0.1217	0.1255	0.1293	0.1331	0.1368	0.1406	0.1443	0.1480	0.1517
0.4	0.1554	0.1591	0.1628	0.1664	0.1700	0.1736	0.1772	0.1808	0.1844	0.1879
0.5	0.1915	0.1950	0.1985	0.2019	0.2054	0.2088	0.2123	0.2157	0.2190	0.2224
0.6	0.2257	0.2291	0.2324	0.2357	0.2389	0.2422	0.2454	0.2486	0.2518	0.2549
0.7	0.2580	0.2612	0.2642	0.2673	0.2704	0.2734	0.2764	0.2794	0.2823	0.2852
0.8	0.2881	0.2910	0.2939	0.2967	0.2995	0.3023	0.3051	0.3078	0.3106	0.3233
0.9	0.3159	0.3186	0.3212	0.3238	0.3264	0.3289	0.3315	0.3340	0.3365	0.3389
1.0	0.3413	0.3438	0.3461	0.3485	0.3508	0.3531	0.3554	0.3577	0.3599	0.3621
1.1	0.3643	0.3665	0.3686	0.3718	0.3729	0.3749	0.3770	0.3790	0.3810	0.3830
1.2	0.3849	0.3869	0.3888	0.3907	0.3925	0.3944	0.3962	0.3980	0.3997	0.4015
1.3	0.4032	0.4049	0.4066	0.4083	0.4099	0.4115	0.4131	0.4147	0.4162	0.4177
1.4	0.4192	0.4207	0.4222	0.4236	0.4251	0.4265	0.4279	0.4292	0.4306	0.4319
1.5	0.4332	0.4345	0.4357	0.4370	0.4382	0.4394	0.4406	0.4418	0.4430	0.4441
1.6	0.4452	0.4463	0.4474	0.4485	0.4495	0.4505	0.4515	0.4525	0.4535	0.4545
1.7	0.4554	0.4564	0.4573	0.4582	0.4591	0.4599	0.4608	0.4616	0.4625	0.4633
1.8	0.4641	0.4649	0.4656	0.4664	0.4671	0.4678	0.4686	0.4693	0.4699	0.4706
1.9	0.4713	0.4719	0.4726	0.4732	0.4738	0.4744	0.4750	0.4758	0.4762	0.4767
2.0	0.47725	0.4778	0.4783	0.4788	0.4793	0.4798	0.4803	0.4808	0.4812	0.4817
2.1	0.4821	0.4826	0.4830	0.4834	0.4838	0.4842	0.4846	0.4850	0.4854	0.4857
2.2	0.4861	0.4865	0.4868	0.4871	0.4875	0.4878	0.4881	0.4884	0.4887	0.4890
2.3	0.4893	0.4896	0.4898	0.4901	0.4904	0.4906	0.4909	0.4911	0.4913	0.4916
2.4	0.4918	0.4920	0.4922	0.4925	0.4927	0.4929	0.4931	0.4932	0.4934	0.4936
2.5	0.4938	0.4940	0.4941	0.4943	0.4945	0.4946	0.4948	0.4949	0.4951	0.4952
2.6	0.4953	0.4955	0.4956	0.4957	0.4959	0.4960	0.4961	0.4962	0.4963	0.4964
2.7	0.4965	0.4966	0.4967	0.4968	0.4969	0.4970	0.4971	0.4972	0.4973	0.4974
2.8	0.4974	0.4975	0.4976	0.4977	0.4977	0.4978	0.4979	0.4980	0.4980	0.4981
2.9	0.4981	0.4982	0.4983	0.4984	0.4984	0.4984	0.4985	0.4985	0.4986	0.4986
3.0	0.49865	0.4987	0.4987	0.4988	0.4988	0.4988	0.4989	0.4989	0.4989	0.4990

* This table was adapted from F. C. Kent. Elements of Statistics, McGraw-Hill Book Company, New York, 1924. Used by Permission.

Table A13 *Table of χ^2 Values.**

Degrees of Freedom, df	PROBABILITY OF A DEVIATION GREATER THAN χ^2						
	0.01	0.02	0.05	0.10	0.20	0.30	0.50
1	6.635	5.412	3.841	2.706	1.642	1.074	0.455
2	9.210	7.824	5.991	4.605	3.219	2.408	1.386
3	11.341	9.837	7.815	4.642	3.665	3.665	2.366
4	13.277	11.668	9.488	7.779	5.989	4.878	3.357
5	15.086	13.388	11.070	9.236	7.289	6.064	4.351
6	16.812	15.033	12.592	10.645	8.558	7.231	5.348
7	18.475	16.622	14.067	12.017	9.803	8.383	6.346
8	20.090	18.168	15.507	13.362	11.030	9.524	7.344
9	21.666	19.679	16.919	14.684	12.242	10.656	8.343
10	23.209	21.161	18.307	15.987	13.442	11.781	9.342
11	24.725	22.618	19.675	17.275	14.631	12.899	10.341
12	26.217	24.054	21.026	18.549	15.812	14.011	11.340
13	27.688	25.472	22.362	19.812	16.985	15.119	12.340
14	29.141	26.873	23.685	21.064	18.151	16.222	13.339
15	30.578	28.259	24.996	22.307	19.311	17.322	14.339
16	32.000	29.633	26.296	23.542	20.465	18.418	15.338
17	33.409	30.995	27.587	24.769	21.615	19.511	16.338
18	34.805	32.346	28.869	25.989	22.760	20.601	17.338
19	36.191	33.687	30.144	27.204	23.900	21.689	18.338
20	37.566	35.020	31.410	28.412	25.038	22.775	19.337
21	38.932	36.343	32.671	29.615	26.171	23.858	20.337
22	40.289	37.659	33.924	30.813	27.301	24.939	21.337
23	41.638	38.968	35.172	32.007	28.429	26.018	22.337
24	42.980	40.270	36.415	33.196	29.553	27.096	23.337
25	44.314	41.566	37.652	34.382	30.675	28.172	24.337
26	45.642	42.856	38.885	35.563	31.795	29.246	25.336
27	46.963	44.140	40.113	36.741	32.912	30.319	26.336
28	48.278	45.419	41.337	37.916	34.027	31.391	27.336
29	49.588	46.693	42.557	39.087	35.139	32.461	28.336
30	50.892	47.962	43.773	40.256	36.250	33.530	29.336

* This table is adapted from Table IV of Fisher and Yates, *Statistical Tables for Biological, Agricultural and Medical Research*, published by Longman Group Ltd., London (previously published by Oliver & Boyd, Edinburgh), and by permission of the authors and publishers.

Table A13 (Continued)

Degree of Freedom, df	PROBABILITY OF A DEVIATION GREATER THAN χ^2					
	0.70	0.80	0.90	0.95	0.98	0.99
1	0.148	0.0642	0.0158	0.00393	0.000628	0.000157
2	0.713	0.446	0.211	0.103	0.0404	0.0201
3	1.424	1.005	0.584	0.352	0.185	0.115
4	2.195	1.649	1.064	0.711	0.429	0.297
5	3.000	2.343	1.610	1.145	0.752	0.554
6	3.828	3.070	2.204	1.635	1.134	0.872
7	4.671	3.822	2.833	2.167	1.564	1.239
8	5.527	4.594	3.490	2.733	2.032	1.646
9	6.393	5.380	4.168	3.325	2.532	2.088
10	7.267	6.179	4.865	3.940	3.059	2.558
11	8.148	6.989	5.578	4.575	3.609	3.053
12	9.034	7.807	6.304	5.226	4.178	3.571
13	9.926	8.634	7.042	5.892	4.765	4.107
14	10.821	9.467	7.790	6.571	5.368	4.660
15	11.721	10.307	8.547	7.261	5.985	5.229
16	12.624	11.152	9.312	7.962	6.614	5.812
17	13.531	12.002	10.085	8.672	7.255	6.408
18	14.440	12.857	10.865	9.390	7.906	7.015
19	15.352	13.716	11.651	10.117	8.567	7.633
20	16.266	14.578	12.443	10.851	9.237	8.260
21	17.182	15.445	13.240	11.591	9.915	8.897
22	18.101	16.314	14.041	12.338	10.600	9.542
23	19.021	17.187	14.848	13.091	11.293	10.196
24	19.943	18.062	15.659	13.848	11.992	10.856
25	20.867	18.940	16.473	14.611	12.697	11.524
26	21.792	19.820	17.292	15.379	13.409	12.198
27	22.719	20.703	18.114	16.151	14.125	12.879
28	23.647	21.588	18.939	16.928	14.847	13.565
29	24.577	22.475	19.768	17.708	15.574	14.256
30	25.508	23.364	20.599	18.493	16.306	14.953

Failure Analysis Report		No. _____
System _____ Name _____ S/N ___ Operating Time _____ Hrs.	Subsystem_____ Name _____ S/N ___ Operating Time _____ Hrs.	Unit _____ Name _____ S/N ___ Operating Time _____ Hrs.
Reported By:	Date of Malfunction	Test Location

Description of Observed Malfunction

Analysis Details

Active Maintenance Time

Diagnosis _____ Access _____ Total Active Time _____
Repair_____ Checkout _____ Total Man Hours _____

Failure Severity	Failure Code
1 ☐ Critical 2 ☐ Major 3 ☐ Minor	☐☐☐

Recommended Corrective Action

Figure A1 *Typical failure analysis report form (Courtesy RCA Corp.).*

Failure Rate Data Form

Equipment _____ DWG No _____
Group _____ DWG No _____
Unit _____ DWG No _____
Assembly _____ DWG No _____
Subassembly _____ DWG No _____

Environmental Service _____
Condition _____
Temperature _____
Rev _____ Date _____

Item No.	Component/Part Number	Component/Part Name	Stress Ratio	Failure Rate	Qty.	$F/10^6$ Hrs × Qty	Notes

Total _____

Sheet No. _____ of _____

Figure A2 *Commercial failure rate data form (Courtesy RCA Corp.)*

Table A14 *Table of Values of* $\ln\left(\frac{\theta_0}{\theta_1}\right)$

$\frac{\theta_0}{\theta_1}$	$\ln\left(\frac{\theta_0}{\theta_1}\right)$
1.0	0.000
1.2	0.182
1.4	0.336
1.5	0.405
1.6	0.470
1.8	0.588
2.0	0.693
2.2	0.789
2.4	0.875
2.6	0.956
2.8	1.030
3.0	1.099
3.5	1.253
4.0	1.386

θ_0 = specified MTBF
θ_1 = minimum acceptable MTBF

Table A15 *Table of Common Values for* $\ln\left(\frac{\beta}{1-\alpha}\right)$.

	$\ln\left(\frac{\beta}{1-\alpha}\right)$
$\alpha = 0.05$ $\beta = 0.05$	-2.944
$\alpha = 0.05$ $\beta = 0.10$	-2.251
$\alpha = 0.10$ $\beta = 0.05$	-2.890
$\alpha = 0.10$ $\beta = 0.10$	-2.197

α = producer's risk
β = consumer's risk

Table A16 *Table of Common Values for* $\ln\left(\frac{1-\beta}{\alpha}\right)$.

	$\ln\left(\frac{1-\beta}{\alpha}\right)$
$\alpha = 0.05$ $\beta = 0.05$	2.944
$\alpha = 0.05$ $\beta = 0.10$	2.890
$\alpha = 0.10$ $\beta = 0.05$	2.251
$\alpha = 0.10$ $\beta = 0.10$	2.197

α = producer's risk
β = consumer's risk

APPENDIX B

MIL-I-45208A Inspection System Requirements and MIL-Q-9858A Quality Program Requirements

MILITARY SPECIFICATION

INSPECTION SYSTEM REQUIREMENTS

This specification has been approved by the Department of Defense and is mandatory for use by the Departments of the Army, the Navy, the Air Force and the Defense Supply Agency.

1. SCOPE

1.1 Scope. This specification establishes requirements for contractors' inspection systems. These requirements pertain to the inspections and tests necessary to substantiate product conformance to drawings, specifications and contract requirements and to all inspections and tests required by the contract. These requirements are in add:.ion to those inspections and tests set forth in applicable specifications and other contractual documents.

1.2 Applicability.

1.2.1 *Applicability.* This specification shall apply to all suppliers or services when referenced in the item specification, contract or order.

1.2.2 *Relation to Other Contract Requirements.* The inspection system requirements set forth in this specification shall be satisfied in addition to all detail requirements contained in the statement of work or in other parts of the contract. The contractor is responsible for compliance with all provisions of the contract and for furnishing specified articles which meet all requirements of the contract. To the extent of any inconsistency between the contract schedule or its general provisions and this specification the contract schedule and the general provisions shall control.

1.2.3 *Options.* This specification contains fewer requirements than specification MIL–Q–9858, Quality Program Requirements. The contractor may use, at his option, the requirements of MIL–Q–9858, in whole or in part, whenever this specification is specified, provided no increase in price or fee is involved. This option permits one uniform system in the event the contractor is already complying with MIL–Q–9858.

2. APPLICABLE DOCUMENTS

2.1 The following documents of the issue in effect on date of invitations for bids form a part of this specification to the extent specified herein.

SPECIFICATIONS
MILITARY

MIL–Q–9858	Quality Program Requirements
MIL–C–45662	Calibration System Requirements

2.2 Amendments and Revisions. Whenever this specification is amended or revised subsequent to its contractually effective date, the contractor may follow or authorize his subcontractors to follow the amended or revised document provided no increase in price or fee is required. The contractor shall not be required to follow the amended or revised document except as a change in contract. If the contractor elects to follow the amended or revised document, he shall notify the Contracting Officer in writing of this election. When the contractor elects to follow the provisions of an amendment or revision, he must follow them in full.

2.3 Ordering Government Documents. Copies of specifications, standards and drawings required by contractors in connection with specific procurements may be obtained from the procuring agency or as otherwise directed by the Contracting Officer.

3. REQUIREMENTS

3.1 Contractor Responsibilities. The contractor shall provide and maintain an inspection system which will assure that all supplies and services submitted to the Government for acceptance conform to contract requirements whether manufactured or processed by the contractor, or procured from subcontractors or vendors. The contractor shall perform or have performed the inspections and tests required to substantiate product conformance to drawing, specifications and contract requirements and shall also perform or have performed all inspections and tests otherwise required by the contract. The contractor's inspection system shall be documented and shall be available for review by the Government Representative prior to the initiation of production and throughout the life of the contract. The Government at its option may furnish written notice of the acceptability or non-acceptability of the inspection system. The contractor shall notify the Government Representative in writing of any change to his inspection system. The inspection system shall be subject to disapproval if changes thereto would result in nonconforming product.

3.2 Documentation, Records and Corrective Action.

3.2.1 *Inspection and Testing Documentation.* Inspection and testing shall be prescribed by clear, complete and current instructions. The instructions shall assure inspection and test of materials, work in process and completed articles as required by the item specification and the contract. In addition, criteria for approval and rejection of product shall be included.

3.2.2 *Records.* The contractor shall maintain adequate records of all inspections and tests. The records shall indicate the nature and number of observations made, the number and type of deficiencies found, the quantities approved and rejected and the nature of corrective action taken as appropriate.

3.2.3 *Corrective Action.* The contractor shall take prompt action to correct assignable conditions which have resulted or could result in the submission to the Government of supplies and services which do not conform to (1) the quality assurance provisions of the item specification, (2) inspections and tests required by the contract, and (3) other inspections and tests required to substantiate product conformance.

3.2.4 *Drawings and Changes.* The contractor's inspection system shall provide for procedures which will assure that the latest applicable drawings, specifications and instructions required by the contract, as well as authorized changes thereto, are used for fabrication, inspection and testing.

3.3 Measuring and Test Equipment. The contractor shall provide and maintain gages and other measuring and testing devices necessary to assure that supplies conform to the technical requirements. In order to assure continued accuracy, these devices shall be calibrated at established intervals against certified standards which have known valid relationships to national standards. If production tooling, such as jigs, fixtures, templates, and patterns is used as a media of inspection, such devices shall also be proved for accuracy at established intervals. Calibration of inspection equipment shall be in accordance with MIL–C–45662. When required, the contractor's measuring and testing equipment shall be made available for use by the Government Representative to determine conformance of product with contract requirements. In addition, if conditions warrant, contractor's personnel shall be made available for operation of such devices and for verification of their accuracy and condition.

3.4 Process Controls. Process control procedures shall be an integral part of the inspection system when such inspections are a part of the specification or the contract.

3.5 Indication of Inspection Status. The contractor shall maintain a positive system for identifying the inspection status of supplies. Identification may be accomplished by means of stamps, tags, routing cards, move tickets, tote box cards or other control devices. Such controls shall be of a design distinctly different from Government inspection identification.

3.6 Government-furnished Material. When material is furnished by the Government, the contractor's procedures shall include as a minimum the following:

    (a) Examination upon receipt, consistent with practicability, to detect damage in transit;

    (b) Inspection for completeness and proper type;

(c) Periodic inspection and precautions to assure adequate storage conditions and to guard against damage from handling and deterioration during storage;

(d) Functional testing, either prior to or after installation, or both, as required by contract to determine satisfactory operation;

(e) Identification and protection from improper use or disposition; and

(f) Verification of quantity.

3.6.1 *Damaged Government-furnished Material.* The contractor shall report to the Government Representative any Government-furnished material found damaged, malfunctioning or otherwise unsuitable for use. In the event of damage or malfunction during or after installation, the contractor shall determine and record probable cause and necessity for withholding material from use.

3.7 Nonconforming Material. The contractor shall establish and maintain an effective and positive system for controlling nonconforming material, including procedures for the identification, segregation, presentation and disposition of reworked or repaired supplies. Repair of nonconforming supplies shall be in accordance with documented procedures acceptable to the Government. The acceptance of nonconforming supplies is the prerogative of and shall be as prescribed by the Government. All nonconforming supplies shall be positively identified to prevent use, shipment and intermingling with conforming supplies. Holding areas, mutually agreeable to the contractor and the Government Representative, shall be provided by the contractor.

3.8 Qualified Products. The inclusion of a product on the Qualified Products List only signifies that at one time the manufacturer made a product which met specification requirements. It does not relieve the contractor of his responsibility for furnishing supplies that meet all specification requirements or for performing specified inspections and tests for such material.

3.9 Sampling Inspection. Sampling inspection procedures used by the contractor to determine quality conformance of supplies shall be as stated in the contract or shall be subject to approval by the Government.

3.10 Inspection Provisions. Alternative inspection procedures and inspection equipment may be used by the contractor when such procedures and equipment provide, as a minimum, the quality assurance required in the contractual documents. Prior to applying such alternative inspection procdures and inspection equipment, the contractor shall describe them in a written proposal and shall demonstrate for the approval of the Government Representative that their effectiveness is equal to or better than the contractual quality assurance procedure. In cases of dispute as to whether certain procedures of the contractor's inspection system provide equal assurance, the procedures of this specification, the item specification and other contractual documents shall apply.

3.11 Government Inspection at Subcontractor or Vendor Facilities. The Government reserves the right to inspect at source supplies or services not manufactured or performed within the contractor's facility. Government inspection shall not constitute acceptance; nor shall it in any way replace contractor inspection or otherwise relieve the contractor of his responsibility to furnish an acceptable end item. When inspection at subcontractors' plants is performed by the Government, such inspection shall not be used by contractors as evidence of effective inspection by such subcontractors. The purpose of this inspection is to assist the Government Representative at the contractor's facility to determine the conformance of supplies or services with contract requirements. Such inspection can only be requested by or under authorization of the Government Representative.

3.11.1 *Government Inspection Requirements.* When Government inspection is required, the contractor shall add to his purchasing document the following statement:

> "Government inspection is required prior to shipment from your plant. Upon receipt of this order, promptly notify the Government Representative who normally services your plant so that appropriate planning for Government inspection can be accomplished."

3.11.2 *Purchasing Documents.* When, under authorization of the Government Representative, copies of the purchasing document are to be furnished directly by the subcontractor or vendor to the Government Representative at his facility rather than through Government channels, the contractor shall add to his purchasing document a statement substantially as follows:

"On receipt of this order, promptly furnish a copy to the Government Representative who normally services your plant or, if none, to the nearest Army, Navy, Air Force, or Defense Supply Agency inspection office. In the event the representative or office cannot be located, our purchasing agent should be notified immediately."

3.11.3 *Referenced Data.* All documents and referenced data for purchases applying to a Government contract shall be available for review by the Government Representative to determine compliance with the requirements for the control of such purchases. Copies of purchasing documents required for Government inspection purposes shall be furnished in accordance with the instructions of the Government Representative.

3.12 Receiving Inspection. Subcontracted or purchased supplies shall be subjected to inspection after receipt, as necessary, to assure conformance to contract requirements. The contractor shall report to the Government Representative any nonconformance found on Government source-inspected supplies and shall require his supplier to coordinate with his Government Representative on corrective action.

3.13 Government Evaluation. The contractor's inspection system and supplies generated by the system shall be subject to evaluation and verification inspection by the Government Representative to determine its effectiveness in supporting the quality requirements established in the detail specification, drawings and contract and as prescribed herein.

4. QUALITY ASSURANCE PROVISIONS

This section is not applicable to this specification.

5. PREPARATION FOR DELIVERY

This section is not applicable to this specification.

6. NOTES

6.1 Intended Use. This specification will apply to the procurement of supplies and services specified by the military procurement agencies.

6.2 Order Data. Procurement documents should specify the title, number and date of this specification.

Custodians:
  Army—Munitions Command
  Navy—Office of Naval Material
  Air Force—Hq USAF
  DSA—Hq DSA

Preparing activity:
  Army—Munitions Command

MILITARY SPECIFICATION

QUALITY PROGRAM REQUIREMENTS

This specification has been approved by the Department of Defense and is mandatory for use by the Departments of the Army, the Navy, the Air Force and the Defense Supply Agency.

1. SCOPE

1.1 Applicability. This specification shall apply to all supplies (including equipments, sub-systems and systems) or services when referenced in the item specification, contract or order.

1.2 Contractual Intent. This specification requires the establishment of a quality program by the contractor to assure compliance with the requirements of the contract. The program and procedures used to implement this specification shall be developed by the contractor. The quality program, including procedures, processes and product shall be documented and shall be subject to review by the Government Representative. The quality program is subject to the disapproval of the Government Representative whenever the contractor's procedures do not accomplish their objectives. The Government, at its option, may furnish written notice of the acceptability of the contractor's quality program.

1.3 Summary. An effective and economical quality program, planned and developed in consonance with the contractor's other administrative and technical programs, is required by this specification. Design of the program shall be based upon consideration of the technical and manufacturing aspects of production and related engineering design and materials. The program shall assure adequate quality throughout all areas of contract performance; for example, design, development, fabrication, processing, assembly, inspection, test, maintenance, packaging, shipping, storage and site installation.

All supplies and services under the contract, whether manufactured or performed within the contractor's plant or at any other source, shall be controlled at all points necessary to assure conformance to contractual requirements. The program shall provide for the prevention and ready detection of discrepancies and for timely and positive corrective action. The contractor shall make objective evidence of quality conformance readily available to the Government Representative. Instructions and records for quality must be controlled.

The authority and responsibility of those in charge of the design, production, testing, and inspection of quality shall be clearly stated. The program shall facilitate determinations of the effects of quality deficiencies and quality costs on price. Facilities and standards such as drawings, engineering changes, measuring equipment and the like which are necessary for the creation of the required quality shall be effectively managed. The program shall include an effective control of purchased materials and subcontracted work. Manufacturing, fabrication and assembly work conducted within the contractor's plant shall be controlled completely. The quality program shall also include effective execution of responsibilities shared jointly with the Government or related to Government functions, such as control of Government property and Government source inspection.

1.4 Relation to Other Contract Requirements. This specification and any procedure or document executed in implementation thereof, shall be in addition to and not in derogation of other contract requirements. The quality program requirements set forth in this specification shall be satisfied in addition to all detail requirements contained in the statement of work or in other parts of the contract. The contractor is responsible for compliance with all provisions of the contract and for furnishing specified supplies and services which meet all the requirements of the contract. If any inconsistency exists between the contract schedule or its general provisions and this specification, the contract schedule and the general provisions shall control. The contractor's quality program shall be planned and used in a manner to support reliability effectively.

1.5 Relation to MIL–I–45208. This specification contains requirements in excess of those in specification MIL–I–45208, Inspection System Requirements, inasmuch as total conformance to contract requirements is obtained best by controlling work operations, manufacturing processes as well as inspections and tests.

2. SUPERSEDING, SUPPLEMENTATION AND ORDERING

2.1 Applicable Documents. The following documents of the issue in effect on date of the solicitation form a part of this specification to the extent specified herein.

SPECIFICATIONS
MILITARY

MIL–I–45208 —Inspection System Requirements

MIL–C–45662 —Calibration System Requirements

2.2 Amendments and Revisions. Whenever this specification is amended or revised subsequent to its contractually effective date, the contractor may follow or authorize his subcontractors to follow the amended or revised document provided no increase in price or fee is required. The contractor shall not be required to follow the amended or revised document except as a change in contract. If the contractor elects to follow the amended or revised document, he shall notify the Contracting Officer in writing of this election. When the contractor elects to follow the provisions of an amendment or revision, he must follow them in full.

2.3 Ordering Government Documents. Copies of specifications, standards and drawings required by contractors in connection with specific procurements may be obtained from the procuring agency, or as otherwise directed by the Contracting Officer.

3. QUALITY PROGRAM MANAGEMENT

3.1 Organization. Effective management for quality shall be clearly prescribed by the contractor. Personnel performing quality functions shall have sufficient, well-defined responsibility, authority and the organizational freedom to identify and evaluate quality problems and to initiate, recommend or provide solutions. Management regularly shall review the status and adequacy of the quality program. The term "quality program requirements" as used herein identifies the collective requirements of this specification. It does not mean that the fulfillment of the requirements of this specification is the responsibility of any single contractor's organization, function or person.

3.2 Initial Quality Planning. The contractor, during the earliest practical phase of contract performance, shall conduct a complete review of the requirements of the contract to identify and make timely provision for the special controls, processes, test equipments, fixtures, tooling and skills required for assuring product quality. This initial planning will recognize the need and provide for research, when necessary, to update inspection and testing techniques, instrumentation and correlation of inspection and test results with manufacturing methods and processes. This planning will also provide appropriate review and action to assure compatibility of manufacturing, inspection, testing and documentation.

3.3 Work Instructions. The quality program shall assure that all work affecting quality (including such things as purchasing, handling, machining, assembling, fabricating, processing, inspection, testing, modification, installation, and any other treatment of product, facilities, standards or equipment from the ordering of materials to dispatch of shipments) shall be prescribed in clear and complete documented instructions of a type appropriate to the circumstances. Such instructions shall provide the criteria for performing the work functions and they shall be compatible with acceptance criteria for workmanship. The instructions are intended also to serve for supervising, inspecting and managing work. The preparation and maintenance of and compliance with work instructions shall be monitored as a function of the quality program.

3.4 Records. The contractor shall maintain and use any records or data essential to the economical and effective operation of his quality program. These records shall be available for review by the Government Representative and copies of individual records shall be furnished him upon request. Records are considered one of the principal forms of objective evidence of quality. The quality program shall assure that records are complete and reliable. Inspection and testing records shall, as a minimum, indicate the nature of the observations together with the number of observations made and the number and type of deficiencies found. Also, records for monitoring work performance and for inspection and testing shall indicate the acceptability of work or products and the action taken in connection with deficiencies. The quality program shall provide for the analysis and use of records as a basis for management action.

3.5 Corrective Action. The quality program shall detect promptly and correct assignable conditions adverse to quality. Design, purchasing, manufacturing, testing or other operations which could result in or have resulted in defective supplies, services, facilities, technical data, standards or other elements of contract performance which could create excessive losses or costs must be identified and changed as a result of the quality program. **Corrective action will extend to the performance of all suppliers and vendors and will be responsive to data and product** forwarded **from users. Corrective action** shall include **as a minimum:**

(a) Analysis of data and examination of product scrapped or reworked to determine extent and causes;

(b) Analysis of trends in processes or performance of work to prevent nonconforming product; and

(c) Introduction of required improvements and corrections, an initial review of the adequacy of such measures and monitoring of the effectiveness of corrective action taken.

3.6 Costs Related to Quality. The contractor shall maintain and use quality cost data as a management element of the quality program. These data shall serve the purpose of identifying the cost of both the prevention and correction of nonconforming supplies (e. g., labor and material involved in material spoilage caused by defective work, correction of defective work and for quality control exercised by the contractor at subcontractor's or vendor's facilities). The specific quality cost data to be maintained and used will be determined by the contractor. These data shall, on request, be identified and made available for "on site" review by the Government Representative.

4. FACILITIES AND STANDARDS

4.1 Drawings, Documentation and Changes. A procedure shall be maintained that concerns itself with the adequacy, the completeness and the currentness of drawings and with the control of changes in design. With respect to the currentness of drawings and changes, the contractor shall assure that requirements for the effectivity point of changes are met and that obsolete drawings and change requirements are removed from all points of issue and use. Some means of recording the effective points shall be employed and be available to the Government.

With respect to design drawings and design specifications, a procedure shall be maintained that shall provide for the evaluation of their engineering adequacy and an evaluation of the adequacy of proposed changes. The evaluation shall encompass both the adequacy in relation to standard engineering and design practices and the adequacy with respect to the design and purpose of the product to which the drawing relates.

With respect to supplemental specifications, process instructions, production engineering instructions, industrial engineering instructions and work instructions relating to a particular design, the contractor shall be responsible for a review of their adequacy, currentness and completeness. The quality program must provide complete coverage of all information necessary to produce an article in complete conformity with requirements of the design.

The quality program shall assure that there is complete compliance with contract requirements for proposing, approving, and effecting of engineering changes. The quality program shall provide for monitoring effectively compliance with contractual engineering changes requiring approval by Government design authority. The quality program shall provide for monitoring effectively the drawing changes of lesser importance not requiring approval by Government design authorities.

Delivery of correct drawings and change information to the Government in connection with data acquisition shall be an integral part of the quality program. This includes full compliance with contract requirements concerning rights and data both proprietary and other. The quality program's responsibility for drawings and changes extend to the drawings and changes provided by the subcontractors and vendors for the contract.

4.2 Measuring and Testing Equipment. The contractor shall provide and maintain gages and other measuring and testing devices necessary to assure that supplies conform to technical requirements. These devices shall be calibrated against certified measurement standards which have known valid relationships to national standards at established periods to assure continued accuracy. The objective is to assure that inspection and test equipment is adjusted, replaced or repaired before it becomes inaccurate. The calibration of measuring and testing equipment shall be in conformity with military specification MIL–C–45662. In addition, the contractor shall insure the use of only such subcontractor and vendor sources that depend upon calibration systems which effectively control the accuracy of measuring and testing equipment.

4.3 Production Tooling Used as Media of Inspection. When production jigs, fixtures, tooling masters, templates, patterns and such other devices are used as media of inspection, they shall be proved for accuracy prior to release for use. These devices shall be proved again for accuracy at intervals formally established in a manner to cause their timely adjustment, replacement or repair prior to becoming inaccurate.

4.4 Use of Contractor's Inspection Equipment. The contractor's gages, measuring and testing devices shall be made available for use by the Government when required to determine conformance with contract requirements. If conditions warrant, contractor's personnel shall be made available for operation of such devices and for verification of their accuracy and condition.

4.5 Advanced Metrology Requirements. The quality program shall include timely identification and report to the Contracting Officer of any precision measurement need exceeding the known state of the art.

5. CONTROL OF PURCHASES

5.1 Responsibility. The contractor is responsible for assuring that all supplies and services procured from his suppliers (subcontractors and vendors) conform to the contract requirements. The selection of sources and the nature and extent of control exercised by the contractor shall be dependent upon the type of supplies, his supplier's demonstrated capability to perform, and the quality evidence made available. To assure an adequate and economical control of such material, the contractor shall utilize to the fullest extent objectives evidence of quality furnished by his suppliers. When the Government elects to perform inspection at a supplier's plant, such inspection shall not be used by contractors as evidence of effective control of quality by such suppliers. The inclusion of a product on the Qualified Products List only signifies that at one time the manufacturer made a product which met specification requirements. It does not relieve the contractor of his responsibility for furnishing supplies that meet all specification requirements or for the performance of specified inspections and tests for such material. The effectiveness and integrity of the control of quality by his suppliers shall be assessed and reviewed by the contractor at intervals consistent with the complexity and quantity of product. Inspection of products upon delivery to the contractor shall be used for assessment and review to the extent necessary for adequate assurance of quality. Test reports, inspection records, certificates and other suitable evidence relating to the supplier's control of quality should be used in the contractor's assessment and review. The contractor's responsibility for the control of

purchases includes the establishment of a procedure for (1) the selection of qualified suppliers, (2) the transmission of applicable design and quality requirements in the Government contracts and associated technical requirements, (3) the evaluation of the adequacy of procured items, and (4) effective provisions for early information feedback and correction of nonconformances.

5.2 Purchasing Data. The contractor's quality program shall not be acceptable to the Government unless the contractor requires of his subcontractors a quality effort achieving control of the quality of the services and supplies which they provide. The contractor shall assure that all applicable requirements are properly included or referenced in all purchase orders for products ultimately to apply on a Government contract. The purchase order shall contain a complete description of the supplies ordered including, by statement or reference, all applicable requirements for manufacturing, inspecting, testing, packaging, and any requirements for Government or contractor inspections, qualification or approvals. Technical requirements of the following nature must be included by statement or reference as a part of the required clear description: all pertinent drawings, engineering change orders, specifications (including inspection system or quality program requirements), reliability, safety, weight, or other special requirements, unusual test or inspection procedures or equipment and any special revision or model identification. The description of products ordered shall include a requirement for contractor inspection at the subcontractor or vendor source when such action is necessary to assure that the contractor's quality program effectively implements the contractor's responsibility for complete assurance of product quality. Requirements shall be included for chemical and physical testing and recording in connection with the purchase of raw materials by his suppliers. The purchase orders must also contain a requirement for such suppliers to notify and obtain approval from the contractor of changes in design of the products. Necessary instructions should be provided when provision is made for direct shipment from the subcontractor to Government activities.

6. MANUFACTURING CONTROL

6.1 Materials and Materials Control.

Supplier's materials and products shall be subjected to inspection upon receipt to the extent necessary to assure conformance to technical requirements. Receiving inspection may be adjusted upon the basis of the quality assurance program exercised by suppliers. Evidence of the suppliers' satisfactory control of quality may be used to adjust the amount and kind of receiving inspection.

The quality program shall assure that raw materials to be used in fabrication or processing of products conform to the applicable physical, chemical, and other technical requirements. Laboratory testing shall be employed as necessary. Suppliers shall be required by the contractor's quality program to exercise equivalent control of the raw materials utilized in the production of the parts and items which they supply to the contractor. Raw material awaiting testing must be separately identified or segregated from already tested and approved material but can be released for initial production, providing that identification and control is maintained. Material tested and approved must be kept identified until such time as its identity is necessarily obliterated by processing. Controls will be established to prevent the inadvertent use of material failing to pass tests.

6.2 Production Processing and Fabrication.

The contractor's quality program must assure that all machining, wiring, batching, shaping and all basic production operations of any type together with all processing and fabricating of any type is accomplished under controlled conditions. Controlled conditions include documented work instructions, adequate production equipment, and any special working environment. Documented work instructions are considered to be the criteria for much of the production, processing and fabrication work. These instructions are the criteria for acceptable or unacceptable "workmanship". The quality program will effectively monitor the issuance of and compliance with all of these work instructions.

Physical examination, measurement or tests of the material or products processed is necessary for each work operation and must also be conducted under controlled conditions. If physical inspection of processed material is impossible or disadvantageous, indirect control by monitoring processing methods, equipment and personnel shall be provided. Both physical inspection and process monitoring shall be provided when control is inadequate without both, or when contract or specification requires both.

Inspection and monitoring of processed material or products shall be accomplished in any suitable systematic manner selected by the contractor. Methods of inspection and monitoring shall be corrected any time their unsuitability with reasonable evidence is demonstrated. Adherence to selected methods for inspection and monitoring shall be complete and continuous. Corrective measures shall be taken when noncompliance occurs.

Inspection by machine operators, automated inspection gages, moving line or lot sampling, setup or first piece approval, production line inspection station, inspection or test department, roving inspectors — any other type of inspection — shall be employed in any combination desired by the contractor which will adequately and efficiently protect product quality and the integrity of processing.

Criteria for approval and rejection shall be provided for all inspection of product and monitoring of methods, equipment, and personnel. Means for identifying approved and rejected product shall be provided.

Certain chemical, metallurgical, biological, sonic, electronic, and radiological processes are of so complex and specialized a nature that much more than the ordinary detailing of work documentation is required. In effect, such processing may require an entire work specification as contrasted with the normal work operation instructions established in normal plant-wide standard production control issuances such as job operation routing books and the like. For these special processes, the contractors' quality program shall assure that the process control procedures or specifications are adequate and that processing environments and the certifying, inspection, authorization and monitoring of such processes to the special degree necessary for these ultraprecise and super-complex work functions are provided.

6.3 Completed Item Inspection and Testing.

The quality program shall assure that there is a system for final inspection and test of completed products. Such testing shall provide a measure of the overall quality of the completed product and shall be performed so that it simulates, to a sufficient degree, product end use and functioning. Such simulation frequently involves appropriate life and endurance tests and qualification testing. Final inspection and testing shall provide for reporting to designers any unusual difficulties, deficiencies or questionable conditions. When modifications, repairs or replacements are required after final inspection or testing, there shall be reinspection and retesting of any characteristics affected.

6.4 Handling, Storage and Delivery. The quality program shall provide for adequate work and inspection instructions for handling, storage, preservation, packaging, and shipping to protect the quality of products and prevent damage, loss, deterioration, degradation, or substitution of products. With respect to handling, the quality program shall require and monitor the use of procedures to prevent handling damage to articles. Handling procedures of this type include the use of special crates, boxes, containers, transportation vehicles and any other facilities for materials handling. Means shall be provided for any necessary protection against deterioration or damage to products in storage. Periodic inspection for the prevention and results of such deterioration or damage shall be provided. Products subject to deterioration or corrosion during fabrication or interim storage shall be cleaned and preserved by methods which will protect against such deterioration or corrosion. When necessary, packaging designing and packaging shall include means for accommodating and maintaining critical environments within packages, e.g., moisture content levels, gas pressures. The quality program shall assure that when such packaging environments must be maintained, packages are labeled to indicate this condition. The quality program shall monitor shipping work to assure that products shipped are accompanied with required shipping and technical documents and that compliance with Interstate Commerce Commission rules and other applicable shipping regulations is effected to assure safe arrival and identification at destination. In compliance with contractual requirements, the quality program shall include monitoring provisions for protection of the quality of products during transit.

6.5 Nonconforming Material. The contractor shall establish and maintain an effective and positive system for controlling nonconforming material, including procedures for its identification, segregation, and disposition. Repair or rework of nonconforming material shall be in accordance with documented procedures acceptable to the Government. The acceptance of nonconforming supplies is a prerogative of and shall be as prescribed by the Government and may involve a monetary adjustment. All nonconforming supplies shall be positively identified to prevent unauthorized use, shipment and intermingling with conforming supplies. Holding areas or procedures mutually agreeable to the contractor and the Government Representative shall be provided by the contractor. The contractor shall make known to the Government upon request the data associated with the costs and losses in connection with scrap and with rework necessary to reprocess nonconforming material to make it conform completely.

6.6 Statistical Quality Control and Analysis. In addition to statistical methods required by the contract, statistical planning, analysis, tests and quality control procedures may be utilized whenever such procedures are suitable to maintain the required control of quality. Sampling plans may be used when tests are destructive, or when the records, inherent characteristics of the product or the noncritical application of the product, indicate that a reduction in inspection or testing can be achieved without jeopardizing quality. The contractor may employ sampling inspection in accordance with applicable military standards and sampling plans (e.g., from MIL–STD–105, MIL–STD–414, or Handbooks H 106, 107 and 108). If the contractor uses other sampling plans, they shall be subject to review by the cognizant Government Representative. Any sampling plan used shall provide valid confidence and quality levels.

6.7 Indication of Inspection Status. The contractor shall maintain a positive system for identifying the inspection status of products. Identification may be accomplished by means of stamps, tags, routing cards, move tickets, tote box cards or other normal control devices. Such controls shall be of a design distinctly different from Government inspection identification.

7. COORDINATED GOVERNMENT/ CONTRACTOR ACTIONS

7.1 Government Inspection at Subcontractor or Vendor Facilities. The Government reserves the right to inspect at source supplies or services not manufactured or performed with the contractor's facility. Government inspection shall not constitute acceptance; nor shall it in any way replace contractor inspection or otherwise relieve the contractor of his responsibility to furnish an acceptable end item. The purpose of this inspection is to assist the Government Representative at the contractor's facility to determine the conformance of supplies or services with contract requirements. Such inspection can only be requested by or under authorization of the Government Representative. When Government inspection is required, the contractor shall add to his purchasing document the following statement:

"Government inspection is required prior to shipment from your plant. Upon receipt of this order, promptly notify the Government Representative who normally services your plant so that appropriate planning for Government inspection can be accomplished."

When, under authorization of the Government Representative, copies of the purchasing document are to be furnished directly by the subcontractor or vendor to the Government Representative at his facility rather than through Government channels, the contractor shall add to his purchasing document a statement substantially as follows:

"On receipt of this order, promptly furnish a copy to the Government Representative who normally services your plant, or, if none, to the nearest Army, Navy, Air Force, or Defense Supply Agency inspection office. In the event the representative or office cannot be located, our purchasing agent should be notified immediately."

All documents and referenced data for purchases applying to a Government contract shall be available for review by the Government Representative to determine compliance with the requirements for the control of such purchases. Copies of purchasing documents required for Government purposes shall be furnished in accordance with the instructions of the Government Representative. The contractor shall make available to the Government Representative reports of any nonconformance found on Government source inspected supplies and shall (when requested) require the supplier to coordinate with his Government Representative on corrective action.

7.2 Government Property.

7.2.1 *Government-furnished Material.* When material is furnished by the Government, the contractor's procedures shall include at least the following:

(a) Examination upon receipt, consistent with practicability to detect damage in transit;

(b) Inspection for completeness and proper type;

(c) Periodic inspection and precautions to assure adequate storage conditions and to guard against damage from handling and deterioration during storage;

(d) Functional testing, either prior to or after installation, or both, as required by contract to determine satisfactory operation;

(e) Identification and protection from improper use or disposition; and

(f) Verification of quantity.

7.2.2 *Damaged Government-furnished Material.* The contractor shall report to the Government Representative any Government-furnished material found damaged, malfunctioning, or otherwise unsuitable for use. In the event of damage or malfunctioning during or after installation, the contractor shall determine and record probable cause and necessity for withholding material from use.

7.2.3 *Bailed Property.* The contractor shall, as required by the terms of the Bailment Agreement, establish procedures for the adequate storage, maintenance and inspection of bailed Government property. Records of all inspections and maintenance performed on bailed property shall be maintained. These procedures and records shall be subject to review by the Government Representative.

8. NOTES

(The following information is provided solely for guidance in using this specification. It has no contractual significance.)

8.1 Intended Use. This specification will apply to complex supplies, components, equipments and systems for which the requirements of MIL–I–45208 are inadequate to provide needed quality assurance. In such cases, total conformance to contract requirements cannot be obtained effectively and economically solely by controlling inspection and testing. Therefore, it is essential to control work operations and manufacturing processes as well as inspections and tests. The purpose of this control is not only to assure that particular units of hardware conform to contractual requirements, but also to assure interface compatibility among these units of hardware when they collectively comprise major equipments, sub-systems and systems.

8.2 Exemptions. This specification will not be applicable to types of supplies for which MIL–I–45208 applies. The following do not normally require the application of this specification:

(a) Personal services, and

(b) Research and development studies of a theoretical nature which do not require fabrication of articles.

8.3 Order Data. Procurement documents should specify the title, number and date of this specification.

Custodians:
   Army—Munitions Command
   Navy—Office of Naval Material
   Air Force—Hq USAF
   DSA—Hq DSA

Preparing Activity:
   Air Force—Hq USAF

☆ U. S. GOVERNMENT PRINTING OFFICE: 1972 -714-539/2555

BIBLIOGRAPHY

1. BAZOVSKY, IGOR, *Reliability Theory and Practice*. Prentice-Hall, Inc., Englewood Cliffs, N.J., 1961.

2. CALABRO, S. R., *Reliability Principles and Practices*, McGraw-Hill Book Company, New York, 1962.

3. ENGINEERING AND STATISTICAL STAFF OF ARINC RESEARCH CORP., edited by W. H. Von Alven, *Reliability Engineering*. Prentice-Hall, Inc., Englewood Cliffs, N.J., 1964.

4. SANDLER, G. H., *System Reliability Engineering*. Prentice-Hall, Inc., Englewood Cliffs, N.J., 1963.

5. MYERS, R. H., WONG, K. L., AND GORDY, H. M., *Reliability Engineering for Electronic Systems*. John Wiley & Sons, Inc., New York, 1964.

6. AMSTADTER, B. L., *Reliability Mathematics: Fundamentals, Practices, Procedures*. McGraw-Hill Book Company, New York, 1971.

7. BLANCHARD, B. S., JR., AND LOWERY, E. E., *Maintainability, Principles and Practices*. McGraw-Hill Book Company, New York, 1969.

8. FREUND, J. E., *Mathematical Statistics*. Prentice-Hall, Inc., Englewood Cliffs, N.J., 1962.

9. EHRENFELD, SYLVAIN, AND LITTANER, S. B., *Introduction to Statistical Method*. McGraw-Hill Book Company, New York, 1964.

10. MOLINA, E. C., *Poisson's Exponential Binomial Limit*. D. Van Nostrand Company, Inc., Princeton, N.J., 1942.

11. DODGE, H. F., AND ROMIG, H. G., *Sampling Inspection Tables, Single and Double Sampling*. John Wiley & Sons, Inc., New York, 1959.

12. DUNCAN, A. J., *Quality Control and Industrial Statistics*. Richard D. Irwin, Inc., Homewood, Ill., 1959.

13. GRANT, E. L., *Statistical Quality Control*. McGraw-Hill Book Company, New York, 1952.

14. FEIGENBAUM, A. V., *Total Quality Control.* McGraw-Hill Book Company, New York, 1961.
15. JURAN, J. M., AND GRYNA, F. M., JR., *Quality Planning and Analysis.* McGraw-Hill Book Company, New York, 1970.
16. LANDERS, R. R., *Reliability and Product Assurance: A Manual for Engineering and Management.* Prentice-Hall, Englewood Cliffs, N. J., 1963.
17. SAATY, T. L., *Elements of Queueing Theory with Applications.* McGraw-Hill Book Company, New York, 1961.
18. GUEDENKO, B. V., BELYAYER, Y. K., AND SOLOVYEV, A. D., *Mathematical Methods of Reliability Theory.* Academic Press, New York, 1969.
19. PIERUSCHKA, E., *Principles of Reliability.* Prentice-Hall, Englewood Cliffs, N.J., 1963.
20. RIGGS, J. L., *Economic Decision Models for Engineers and Managers.* McGraw-Hill Book Company, New York, 1968.
21. CHURCHMAN, C. W., ACKOFF, R. L., ARNOFF, E. L., *Introduction to Operations Research.* John Wiley & Sons, Inc., New York, 1957.
22. HANSEN, B. H., *Quality Control; Theory and Applications.* Prentice-Hall, Englewood Cliffs, N.J., 1963.
23. MIL-HDBK-217A, Military Standardization Handbook, Reliability Stress and Failure Rate Data for Electronic Equipment, December 1, 1965.
24. MIL-HDBK-217B, Military Standardization Handbook, Reliability Prediction of Electronic Equipment, September 20, 1974.
25. MIL-M-24365, Military Specification, Maintenance Engineering Analysis, General Specification, Department of Defense, May 26, 1969.
26. MIL-STD-414, Military Standard, Sampling Procedures and Tables for Inspection by Variables for Percent Defective, June 11, 1957.
27. MIL-STD-781B, Military Standard, Reliability Tests Exponential Distribution, Department of Defense, Washington, D.C., November 15, 1967.
28. MIL-STD-105D, Military Standard, Sampling Procedures and Tables for Inspection by Attributes, April 29, 1963.
29. MIL-STD-202E, Military Standard, Test Methods for Electronic and Electrical Component Parts, April 16, 1973.
30. MIL-HDBK-472, Military Standardization Handbook, Maintainability Prediction, Department of Defense, May 24, 1966.
31. MIL-I-45208A, Military Specification, Inspection System Requirements, Department of Defense, December 16, 1963.
32. MIL-Q-9858A, Military Specification, Quality Program Requirements, Department of Defense, December 16, 1963.
33. NAVSHIP 0967-312-8010, Maintainability Design Criteria Handbook for Designers of Shipboard Electronic Equipment, Department of the Navy, July 1972.
34. OR-30, Integrated Logistic Support Program Requirements, Naval Ordnance Systems Command, Department of the Navy, March 3, 1969.
35. H108, Quality Control and Reliability Handbook, Sampling Procedures and Tables for Life and Reliability Testing; Office of the Assistant Secretary of Defense, Washington, D.C., April 19, 1960.
36. MIL-C-11015D, Capacitors, Fixed, Ceramic Dielectric (General Purpose), General Specification for, December 6, 1967.

37. MIL-R-55182E, Resistors, Fixed, Film, Established Reliability, General Specification for, May 11, 1972.

38. MIL-R-10509F, Resistors, Fixed Film (High Stability), General Specification for, June 15, 1965.

39. MIL-T-39013, Transformers and Inductors (Audio and Power), Established Reliability, General Specification for, March 2, 1970.

40. MIL-C-39014, Capacitors, Fixed, Ceramic Dielectric (General Purpose), Established Reliability, General Specification for, August 21, 1970.

41. MIL-STD-471A, Military Standard, Maintainability Verification/Demonstration/Evaluation, Department of Defense, March 27, 1973.

42. MIL-M-26512C, Maintainability Requirements for Weapon Systems and Subsystems, December 13, 1963.

43. NAVSHIPS 0967-312-8010 Maintainability Design Criteria Handbook for Designers of Shipboard Electronic Equipment, April 1962.

44. NHB 5300.4 (1B) Quality Program Provisions for Aeronautical and Space Systems Contractors, National Aeronautics and Space Administration, April 1969.

45. Report CR 114391 Parts, Materials and Processes Experience Summary, National Aeronautics and Space Administration, February 24, 1972.

46. Pamphlet 705-1 Research and Development of Material, Maintainability Engineering Headquarters, Department of the Army, June 1966.

47. Report: Reliability of Military Electronic Equipment, by the Advisory Group on Reliability of Electronic Equipment, Office of the Assistant Secretary of Defense, June 4, 1957.

INDEX

INDEX